SINA

Sixty-Two Years of Uncertainty

Historical, Philosophical, and Physical Inquiries into the Foundations of Quantum Mechanics

NATO ASI Series

Advanced Science Institutes Series

A series presenting the results of activities sponsored by the NATO Science Committee, which aims at the dissemination of advanced scientific and technological knowledge, with a view to strengthening links between scientific communities.

The series is published by an international board of publishers in conjunction with the NATO Scientific Affairs Division

A	**Life Sciences**	Plenum Publishing Corporation
B	**Physics**	New York and London
C	**Mathematical**	Kluwer Academic Publishers
	and Physical Sciences	Dordrecht, Boston, and London
D	**Behavioral and Social Sciences**	
E	**Applied Sciences**	
F	**Computer and Systems Sciences**	Springer-Verlag
G	**Ecological Sciences**	Berlin, Heidelberg, New York, London,
H	**Cell Biology**	Paris, and Tokyo

Recent Volumes in this Series

Volume 221—Guidelines for Mastering the Properties of Molecular Sieves:
Relationship between the Physicochemical Properties of Zeolitic
Systems and Their Low Dimensionality
edited by Denise Barthomeuf, Eric G. Derouane,
and Wolfgang Hoelderich

Volume 222—Relaxation in Complex Systems and Related Topics
edited by Ian A. Campbell and Carlo Giovannella

Volume 223—Particle Physics: *Cargèse 1989*
edited by Maurice Lévy, Jean-Louis Basdevant, Maurice Jacob,
David Speiser, Jacques Weyers, and Raymond Gastmans

Volume 224—Probabilistic Methods in Quantum Field Theory and Quantum Gravity
edited by P. H. Damgaard, H. Hüffel, and A. Rosenblum

Volume 225—Nonlinear Evolution of Spatio-Temporal Structures in
Dissipative Continuous Systems
edited by F. H. Busse and L. Kramer

Volume 226—Sixty-Two Years of Uncertainty: Historical, Philosophical, and
Physical Inquiries into the Foundations of Quantum Mechanics
edited by Arthur I. Miller

Volume 227—Dynamics of Polyatomic Van Der Waals Complexes
edited by Nadine Halberstadt and Kenneth C. Janda

Volume 228—Hadrons and Hadronic Matter
edited by Dominique Vautherin, F. Lenz, and J. W. Negele

Series B: Physics

Sixty-Two Years of Uncertainty

Historical, Philosophical,
and Physical Inquiries
into the Foundations
of Quantum Mechanics

Edited by

Arthur I. Miller

Cambridge University
Cambridge, England

Plenum Press
New York and London
Published in cooperation with NATO Scientific Affairs Division

Proceedings of a NATO Advanced Study Institute on
Sixty-Two Years of Uncertainty: Historical, Philosophical,
and Physical Inquiries into the Foundations of Quantum Mechanics,
held August 5–15, 1989,
in Erice, Sicily, Italy

Library of Congress Cataloging-in-Publication Data

NATO Advanced Study Institute on Sixty-Two Years of Uncertainty (1989
  : Erice, Sicily)
   Sixty-two years of uncertainty : historical, philosophical, and
physical inquiries into the foundations of quantum mechanics /
edited by Arthur I. Miller.
       p.   cm. -- (NATO ASI series. B, Physics ; v. 226)
   "Proceedings of a NATO Advanced Study Institute on Sixty-Two Years
of Uncertainty: historical, philosophical, and physical inquiries
into the foundations of quantum mechanics, held August 5-15th, 1989,
in Erice, Sicily, Italy"--T.p. verso.
   "Published in cooperation with NATO Scientific Affairs Division."
   Includes bibliographical references and index.
   ISBN 0-306-43608-6
   1. Quantum theory--Congresses. 2. Heisenberg uncertainty
principle--Congresses.   I. Miller, Arthur I. II. North Atlantic
Treaty Organization. Scientific Affairs Division. III. Title.
IV. Series.
QC173.96.N385   1989
530.1'2--dc20   90-7603
                                                          CIP

© 1990 Plenum Press, New York
A Division of Plenum Publishing Corporation
233 Spring Street, New York, N.Y. 10013

Printed in the United States of America

PREFACE

   This volume contains proceedings from the International School of History
of Science, **Sixty-Two Years of Uncertainty: Historical Philosophical
and Physical Inquiries into the Foundations of Quantum Mechanics**,
convened at the Ettore Majorana Centre for Scientific Culture, Erice, Sicily, 5-15
August 1989. In response to the high state of enthusiasm from the sixty-one
participants there were six to eight lectures each day, beginning at 9:00 AM and
often ending at 7:00 PM. Vigorous discussions took place at every opportunity,
even including the delightful excursions.

   The papers presented here are by the twelve invited lecturers (in some
cases with coauthors) with a contribution from Philip Pearle.

   All of us attending the conference express our appreciation to the
exemplary staff of the Ettore Majorana Centre, and particularly to the Centre's
Director, Professor Antonino Zichichi, for superb hospitality which made this
conference a memorable intellectual and cultural experience. It is a pleasure to
acknowledge financial support from the North Atlantic Treaty Organization
(NATO) Scientific Affairs Division.

Arthur I. Miller, Director
International School of History of Science
Ettore Majorana Centre for Scientific
          Culture
Erice, Sicily

CONTENTS

OPENING REMARKS

Arthur I. Miller

Department of Physics
Harvard University
Cambridge, MA 02138
USA

and

Department of Philosophy
University of Lowell
Lowell, MA 01854
USA

In these brief opening remarks I should like to convey my principal reason for convening this meeting. We read often that in 1927 the formalism and interpretation of nonrelativistic quantum mechanics was firmly settled and etched in stone under the rubric Copenhagen Interpretation. Most physicists relegate the ensuing Bohr-Einstein debates to heroic tales about the distant past of our culture. Yet sixty-two years after publication of Werner Heisenberg's uncertainty principle paper there is a not so small group of scholars who rightly consider that certain fundamental issues of quantum mechanics remain unsettled.

So, about two years ago I thought that it would be useful to convene a different sort of meeting, one which brings together scholars who would consider taking an interdisciplinary approach to fundamental issues in quantum mechanics. The issues that remain unsettled are important because they bear directly on our scientific and, so too, philosophical understanding of the world in which we live. And these issues bear on the problem of the construction of knowledge itself. For is it not the case that more than any other theory in the history of recent science, quantum mechanics has radically changed our notion of what constitutes physical reality? For example, atomic entities can be simultaneously wave and particle which defies our modes of mental representation of physical objects; the time-dependent Schrödinger equation describes the evolution in space and time of a probability function and not the space coordinates as in Newtonian physics; the measurement operation links inextricably object and measurement apparatus, forever altering the object; and there are long-range correlations between particles emitted from a common source. These results, among others, are radically counterintuitive to those associated with any other physical theory and to the way in which we interpret visually and linguistically the world in which we live. How did these nonclassical and therefore nonintuitive

Sixty-Two Years of Uncertainty
Edited by A. I. Miller
Plenum Press, New York, 1990

notions enter physical theory? What were the early reactions to them? Are these early reactions connected with present interpretational problems? Proper treatment of these questions seem just the proper grist for the combined mill of historians, philosophers and physicists. As Abner Shimony has noted (1989) owing to developments particularly in quantum mechanics, the "twentieth century is one of the golden ages of metaphysics."

During the next few days we will hear much about important technical developments toward better understanding the foundations of quantum mechanics. But, as we all know, we ought not lose sight of conceptual analysis. For has it not been the case throughout the history of science that great advances have been made along this route? As Werner Heisenberg wrote in his first epistemological study of the quantum mechanics in September (1926), "Let us turn...from the mathematical elaboration of the theory to the physical significance of this formalism, that is, to a discussion of statements that can be made concerning the reality and the laws of particles."

So, for example, ought we to accept as axiomatic that there are long-range correlations and there are things such as collapse of the wave function with apparently superluminal speed, and that is that? As John Bell put it as long as we do not understand the "wave packet reduction...we do not have an exact and unambiguous formulation of our most fundamental physical theory"(1987).

Perhaps we shall have to be content with axiomatics. But curious creatures such as we try to "understand" these phenomena better by in some way exploring how we can extend our modes of intuition into the subatomic realm. Surely we will have to widen our circle of inquiry to include disciplines other than physics such as history and philosophy both broadly defined to include cognitive psychology as well.

So, with some sense of alternatives or lack thereof for the present quantum mechanics, I look forward to our exploring together over the next few days interdisciplinary approaches to the interpretation of our most fundamental physical theory.

REFERENCES

Bell, J., 1987, Collected Papers on Quantum Philosophy: Speakable and Unspeakable in Quantum Mechanics, Cambridge University Press, Cambridge.
Heisenberg, W., 1926, Die Quantenmechanik, Die Naturwissenschaften, 14, 899.
Shimony, A., 1989, Search for a World View which can Accommodate our Knowledge of Microphysics, in Philosophical Consequences of Quantum Theory: Reflections on Bell's Theorem J.T. Cushing and E. McMullin eds., Notre Dame Press, South Bend IN.

IMAGERY, PROBABILITY AND THE ROOTS OF WERNER HEISENBERG'S

UNCERTAINTY PRINCIPLE PAPER

Arthur I. Miller

Department of Physics
Harvard University
Cambridge, MA 02138
USA

and

Department of Philosophy
University of Lowell
Lowell, MA 01854
USA

I should like to set the stage for this meeting by exploring how the concept of probability was transformed by developments in atomic physics during 1913-1927. In outline I will proceed as follows: I will begin with a survey of the visual imagery, causality and probability of classical physics; then I turn to Niels Bohr's 1913 theory of the atom and the emergence in 1924 of nonclassical notions of probability; the contrast between the quantum and wave mechanics in 1926; Max Born's theory of scattering; Werner Heisenberg on fluctuations, discontinuity and probability; transformation theory and word meanings; the uncertainty principle paper; and then conclude with Niels Bohr's concept of complementarity.[*]

We will find a connection between physicists' changing conceptions of mental imagery of phenomena and concepts of probability that at first were <u>imposed</u> on atomic phenomena and then <u>emerged</u> from the new atomic physics. These conceptual changes affected the roots and contents of Heisenberg's uncertainty principle paper (Heisenberg, 1927). In conclusion we inquire whether we can better understand this connection with concepts from theories of mental representation of knowledge.

1. VISUAL IMAGERY, CLASSICAL CAUSALITY AND PROBABILITY

Prior to and into the first decade of Bohr's 1913 atomic theory, physicists dealt with physical systems in which the usual space and time pictures of classical physics were assumed trustworthy, for example, electrons that are supposed to move like billiard balls and light that behaves like water waves. In the German scientific milieu this visual imagery was accorded a reality status higher than viewing merely

---

[*] The secondary literature on the history of atomic physics is enormous. Here I will take the liberty to refer the reader to the bibliographies in Hendry (1984), Jammer (1966, 1974) and Miller (1986, 1988).

Sixty-Two Years of Uncertainty
Edited by A. I. Miller
Plenum Press, New York, 1990

with the senses and was referred to as "customary intuition [gewöhnliche Anschauung]." Customary intuition is the visual imagery that is abstracted from phenomena that we have actually witnessed in the world of sense perceptions. The concept of customary intuition was much debated during 1923-1927 by physicists like Bohr, Heisenberg, Pauli and Erwin Schrödinger, all of whom used this term with proper Kantian overtones, and all of whom lamented its loss in the new quantum mechanics.

Customary intuition is associated with the strong causality of classical mechanics. According to the law of causality since initial conditions can be ascertained with in-principle perfect accuracy, then a system's continuous development in space and time can be traced with in-principle perfect accuracy. Any limitations to the accuracy of meaurements are assumed not to be intrinsic to the phenomena, that is, they are assumed to be systematic measurement errors that can be made to vanish. Thus, in classical physics we have the connection pictures-causality-conservation laws.

In the first decade of the 20th century the consensus among physicists was that a method would be found to extend our intuition from classical physics into the domain of the atom. They believed that laws governing the behavior of individual atoms would not be statistical. For example, Ernest Rutherford's law for how many of a large number of atoms undergo radioactive decay in a certain time period is a statistical law in the sense of classical physics, where statistics and probability were interpreted as reflecting our ignorance of the underlying dynamics of individual processes. Rather, some complex form of the causal Newtonian mechanics would in time be formulated for Rutherford's model of the atom as a nucleus surrounded by electrons.

How important imagery was (and still is) to physicists is clear from Bohr's seminal papers of 1913 (Bohr, 1913). Despite his theory's violation of classical mechanics, Bohr emphasized that the mathematical symbols from classical mechanics permitted visualization of the atom as a miniscule Copernican sytem. Although suitably quantized laws of classical mechanics are used to calculate the electron's allowed orbits, or stationary states, classical mechanics cannot depict or describe the electron in transit. So in transit the orbital electron behaves like the Cheshire cat, that is, the quantum jump or "essential discontinuity," is unvisualizable. In contrast, classical electrodynamics could not at all account for the characteristics of radiation emitted in the transition.

2. THE CORRESPONDENCE PRINCIPLE

In 1918 Bohr proposed a method to extend classical electrodynamics into the realm of the atom by means of what he would call in 1920 the "correspondence principle." This principle is based on Einstein's A and B coefficients from his 1916-1917 quantum theory of radiation. The A and B coefficients are the probabilities for an atom to make transitions that are spontaneous or induced by external radiation, respectively. Einstein assumed the statistical laws for these processes to be like that of "radioactivity." So the A and B coefficients reflected ignorance of the mechanism of individual atomic transitions. Consequently, for Bohr, Einstein's A and B coefficients were just the prescription for dealing with the unvisualizable "essential discontinuities."

Bohr's procedure goes as follows (Bohr, 1918): Orbits very far from the nucleus are very close together. Hence, in transitions between orbits whose principal quantum number $n \gg 1$ the quantum frequency of the emitted radiation v_q is nearly equal to the classical frequency v_c of the electron's revolution in either of these orbits, that is

$$v_q = v_c = \sum_k \tau_k v_k \tag{1}$$

where the classical frequency is the sum of higher harmonics and the τ_k are integers.

For the purpose of studying the response of atoms to radiation, classical electrodynamics represents the atom as comprised of harmonically bound electrons. The atom's dipole moment is

$$P(t) = \sum_{\tau_i} C_{\tau 1 \ldots \tau s} \exp\{2\pi i(w_1 + \ldots + w_s)\} \tag{2}$$

where $w_i = v_i t$ are angle variables and the coefficients of the fourier expansion $C_{\tau 1 \ldots \tau s}$ are functions of action variables J_i. So, the spectrum of emitted radiation from an atom predicted by classical theory differs completely from the one measured and the one predicted by Bohr's atomic theory.

According to classical physics, from Eq.(2) the rate at which radiation is emitted is

$$\frac{dE_{\tau 1 \ldots \tau s}}{dt} = \frac{(2\pi v_c)^4}{3c^3} \mid C_{\tau 1 \ldots \tau s} \mid^2 \tag{3}$$

Bohr drew upon the correspondence principle (Eq. (1)) to rewrite Eq. (2) as

$$P(t) = \sum_q U_q \exp\{2\pi i v_q t\} \tag{4}$$

where the summation is over all quantum jumps, and then to express the rate at which radiation is emitted for a spontaneous transition between stationary states i and k as

$$\frac{dE_{ik}}{dt} = h v_{ik} A_{ik} = \frac{(2\pi v_q)^4}{3c^3} e^2 \mid C_q \mid^2 \tag{5}$$

In this way the magnitude squared of the suitably quantized amplitude for the atom's dipole moment became proportional to Einstein's A coefficient with its classical meaning.

3. THE HARMONIC OSCILLATOR REPRESENTATION, DISPERSION AND NON-CLASSICAL PROBABILITY

By 1923 the picture of a planetary atom was beginning to whither away. Besides its lack of success in dealing with atoms more complex than hydrogen, the problem of dispersion altered dramatically Bohr's theory of the atom because the response of atomic electrons to incident light could not always be correlated with their simple motion in Keplerian orbits. In 1923 Bohr proposed that "fundamental difficulties" facing his theory all had as their common denominator the problem of the interaction of light with atoms (Bohr, 1923).

The key point was to reconcile essential discontinuities of atomic physics with the inherent continuity of classical electrodynamics. One approach that Bohr suggested involved the light quantum and maintaining energy and momentum conservation in individual processes. But this was an unsatisfactory solution because the "picture [Bild] of light quanta precludes explaining interference." This had been the principal criticism against the light quantum ever since its invention by Einstein in 1905. Yet the undeniable usefulness of the light quantum for explaining certain phenomena reinforced Bohr's belief that a contradiction-free description of atomic processses could not be arrived at by "use of conceptions

5

borrowed from classical electrodynamics." Since in classical physics the conservation laws are linked with a continuous space-time description then, continued Bohr, these laws may "not possess unlimited validity." Presently, however, he was not prepared to take this step.

Bohr's guide in the atomic domain would be the correspondence principle, upon which he based the "coupling mechanism." According to Bohr's coupling mechanism atoms respond to incident light like an ensemble of harmonic oscillators each of which emits continuous radiation with the frequency of a possible atomic transition. Consequently, according to the coupling mechanism atoms and radiation are in stationary states. This method permitted Bohr to renounce the "so-called hypothesis of light quanta." Yet to Bohr the coupling mechanism was only a first approximation for treating radiation because it ran counter to the accepted dualistic picture of light and matter in which there are source particles and spreading spherical waves of radiation.

In 1924 the coupling mechanism provided Bohr, Hendrik Kramers and John C. Slater the means to avoid interpreting the Compton effect in terms of light quanta (Bohr, Kramers and Slater, 1924). To Bohr the tension between the two conceptions of light would have to be resolved on the basis of the wave theory. For although there are essential discontinuities in atomic physics, our "customary intuition [Anschauung]" requires that light be a wave phenomenon. In order to exclude light quanta Bohr, Kramers and Slater resorted to combining the most exreme consequence of the first method of 1923 (renouncing energy conservation) with the oscillators in the coupling principle, to which they referred as "virtual oscillators." Besides emitting real radiation in spherical waves in response to incident radiation, the virtual oscillators were assumed to emit a field carrying only the probability for inducing atomic transitions. The virtual radiation field of one atom could induce an upward atomic transition in another atom without the source atom undergoing the corresponding downward transition, thereby violating energy conservation and causality in individual processes. In this way they were able to reconcile discontinuous atomic transitions with the continuous radiation field. Bohr considered such a radical version of his theory necessary in order to avoid the paradoxical circumstance of an entity being both wave and particle simultaneously.

Bohr, Kramers and Slater interpreted the Compton effect as follows: Each illuminated electron in the target crystal emits coherent secondary wavelets that can be understood as the usual sort of light scattered from a virtual oscillator. But as a consequence of the virtual radiation field the scattered electron has a probability of having momenta in any direction. In this way the Compton effect can be understand as a continuous process.

Bohr, Kramers and Slater did not use the term "picture" of the atom to mean visualization. The reason is that it is impossible to visualize an electron in a stationary state as represented by as many oscillators as there are transitions to and from this state. Rather, they meant the term "picture" to refer to the interpretation of the mathematical framework. The picture of the Copernican atom had been imposed on the 1913-1923 Bohr theory owing to Bohr's use of the language (semantics) of "ordinary mechanics" (Bohr, 1913). The 1924 Bohr, Kramers and Slater version of Bohr's theory started the movement toward defining the image of atomic theory to be synonymous or given by its mathematical scheme. Another advance of Bohr, Kramers and Slater is that they raised the concept of probability from a strictly mathematical entity to one that actually produced physical phenomena such as atomic transitions.

Heisenberg, among others, was much impressed by the "intermediate kinds of reality" (AHQP, 13 February 1963)[*] offered by the virtual oscillator representation

---

[*] Quotations from AHQP (Archive for History of Quantum Physics) are taken from interviews of Werner Heisenberg by Thomas S. Kuhn.

for example, it freed atomic electrons from their planetary orbits and transformed the concept of probability into a causative agent. This situation augured to Heisenberg that "cheap solutions would not be found" (AHQP, 13 February 1963). While subsequent work on dispersion by Born, Heisenberg and Kramers used virtual oscillators, neither violations of energy nor momentum conservation were well received.

By interpreting Eq. (4) as the virtual oscillator representation for a bound electron, Kramers set out a program in which Bohr's theory contained only measurable quantities, that is, no reference to the bound electron's orbit. The intensity of a spectral line is given by the magnitude squared of the amplitude in Eq. (4) and the line's measured frequency is v_q. The Kramers-Heisenberg paper (Kramers, 1925), completed in December 1924 with their famous dispersion relation, turned out to be the high water mark of the Bohr theory. No further progress was made.

In mid-1925 fundamental conceptual problems focused on lack of visualization of atomic phenomena: owing to the virtual oscillator representation bound electrons had lost their localization and visualizability; owing to Bose-Einstein statistics (as it was interpreted in 1926) free electrons had lost their distinguishability and individuality too; and then there was the wave-particle duality of light and matter. Lack of visualizability entailed linguistic problems as well. For example, the defining equation for a light quantum is $E = hv$. Although the quantity E connotes localization, v is a "radiation frequency defined by experiments on interference phenomena" (Bohr, Kramers and Slater, 1924).

4. QUANTUM MECHANICS AND WAVE MECHANICS IN 1926

Faced in 1925 with experimental refutation of the Bohr-Kramers-Slater version of his atomic theory, and the possibility that the light quantum might be real, Bohr reluctantly renounced "intuitive pictures" of atomic processes, while accepting the conservation laws for individual atomic processes (Bohr, 1925).

Suffice it to say that the virtual oscillator representation was central to Heisenberg's invention of the new quantum mechanics or matrix mechanics in June 1925, based "exclusively on relations between quantities which in principle are [empirically] observable" (Heisenberg, 1925).

Although renunciation of the picture of a bound electron had been a necessary prerequisite to Heisenberg's invention of the new quantum mechanics, the lack of an "intuitive" [anschauliche] interpretation was of great concern to Bohr, Born and Heisenberg. This concern emerges from their scientific papers of the period 1925-1927 (see Miller, 1986).

With publication in early 1926 of Erwin Schrödinger's wave mechanics the quest for some sort of visualization of atomic processes intensified and took a subjective turn in the published scientific literature. Schrödinger (1926) wrote that he formulated the wave mechanics because he "felt discouraged not to say repelled...by lack of visualizability [Anschaulichkeit]" of the quantum mechanics. He offered a visual representation based on the customary intuition of atomic processes occurring without discontinuities as wave phenomena.

To summarize: In mid-1926 there were two seemingly dissimilar atomic theories. Heisenberg's quantum mechanics was corpuscular based and yet renounced any visualization of the bound corpuscle itself. Its mathematical apparatus was unfamiliar to most physicists. Wave mechanics was a continuum theory based on matter as waves. Its familiar mathematical apparatus led to a calculational breakthrough and its claim to restore customary intuition was welcomed by many physicists including Einstein.

Heisenberg thought otherwise. On 8 June 1926 he wrote to Wolfgang Pauli (1979): "The more I reflect on the physical portion of Schrödinger's theory the more disgusting I find it....What Schrödinger writes on the visualizability of his theory...I consider trash. The great accomplishment of Schrödinger's theory is the calculation of matrix elements."

5. BORN'S THEORY OF SCATTERING

The tension between the quantum and wave mechanics increased with the appearance of Born's quantum theory of scattering in mid-1926. To Born (1926b) neither scattering problems nor transitions in atoms can be understood using quantum mechanics which denies "exact representation of processes in space and time," or wave mechanics which denies visualization in phenomena with more than one particle. Problems concerning scattering and transitions require the "construction of new concepts," and for his vehicle Born chose to use wave mechanics which allows for at least the possibility of visualization.

One new concept Born proposed is from unpublished speculations of Einstein, namely, that light quanta are guided by a wave field (ghost field) that carries only probability, providing the means to account for interference using light quanta. Born boldly assumed the "complete analogy" between a light quantum and an electron in order to postulate the interpretation that the "de Broglie-Schrödinger waves," that is, the wave function in three dimensional space, is the "guiding field" for the electron. He attributed physical reality to the magnitude squared of Schrödinger's wave function as had Einstein for the intensity of the ghost field. Born (1926a) went on to propose that $|\psi|^2$ is the probability for a scattered electron to be found within a differential element $d\Omega$ of solid angle.

6. TOWARD AN INTERPRETATION OF P AND Q

Pauli (1979) supplied some key observations on Born's results in a letter of 19 October 1926 to Heisenberg who was in Copenhagen:

--- Born's probability interpretation for scattering should be generalized to the statement that the quantity $|\psi(q_1...q_f)|^2 dq_1...dq_f$ is the probability that the coordinates q_k of a particle will be between q_k and $q_k + dq_k$. So, "we must look at this probability as in principle observable."

--- Next there is a "dim point": "The p's must be taken as controlled, the q's as uncontrolled." He arrived at this conclusion from noting that in Born's scattering theory off-diagonal matrix elements are calculated using wave functions related by fourier transforms. Consequently, wrote Pauli, "One can see the world with p-eyes and one can see it with q-eyes, but if one opens both eyes together one can go astray."

Pauli went on to express the need for systematic means to relate matrix elements and wave functions in their various representations.

Pauli's letter was studied in Copenhagen by Bohr, Heisenberg and P.A.M. Dirac. The problem at hand was to relate Born's scattering theory to quantum mechanics and then to demonstrate that Schrödinger's theory possesses discontinuities just like quantum mechanics.

On 28 October 1926 Heisenberg wrote to Pauli that concerning the so-called "dim point I should like to believe that your p-waves have just as great a physical reality as the q-waves. The equation $pq-qp = h/2\pi i$ thus corresponds always in the wave representation to the fact that it is impossible to speak of a monochromatic wave at a fixed point in time (or in a very short time interval)....Analogously, it is impossible to talk of the position of a particle of fixed velocity."

In summary thus far, by 28 October 1926 Heisenberg and Pauli recognized the impossibility to measure q and p exactly in the same experiment. They connected this result to the commutation relations (see, too, discussion of these letters in Hendry, 1984).

On 4 November 1926 Heisenberg wrote to Pauli (1979) that "in general every scheme that satisfied pq - qp = h/2πi is correct and physically useful, so one has a completely free choice as to how to fulfill this equation, with, matrices, operators, or anything else." Heisenberg concluded that the "problem of canonical transformations in the wave representation is as good as solved." But this would have to await Dirac's transformation theory.

7. HEISENBERG ON FLUCTUATIONS, DISCONTINUITY AND PROBABILITY

Meanwhile, since June 1926 Heisenberg had been enraged over the successes of Schrödinger's wave mechanics and Born's assessment of the quantum mechanics as an incomplete theory that required a new hypothesis introduced, no less, with wave mechanics. In response Heisenberg wrote, "Fluctuation Phenomena and Quantum Mechanics," (completed 6 November 1926), which he recalled as having received little attention but "for myself it was a very important paper" (AHQP, 22 February 1963). It is a paper written by an angry man in which Born's theory of scattering is not cited and Schrödinger is sharply criticized.

Heisenberg (1926a) set out show that a probability interpretation for the canonical transformation matrix emerges naturally from quantum mechanics; and that atomic phenomena cannot be understood without discontinuities. For help in demonstrating these points Heisenberg turned to Einstein's (1909) paper on fluctuation phenomena in which Einstein argued that fluctuations implied discontinuities which meant corpuscular concepts. Heisenberg reversed Einstein's line of reasoning thus: Since corpuscular atomic systems exhibit discontinuities, that is the existence of stationary states, and since quantum mechanics provided "quantitative description" of such systems, then one should be able to deduce fluctuations from quantum mechanics. Heisenberg studied two identical atoms, one in a state n, the other in a state m, coupled by a symmetrical interaction. In earlier papers he had demonstrated that this system behaves like two coupled oscillators. As Pauli had suggested, Heisenberg assumed that the time mean or average value $\overline{f}$ of the operator f in the state α is given by its diagonal matrix elements, and so

$$\overline{f(E_\alpha)} = \sum_\delta |S_{\alpha\delta}|^2 \ f(E_\delta) = \frac{1}{2} f(E_n) + \frac{1}{2} f(E_m) \tag{6}$$

and so either atom is one-half of the time in state n or m, where S is a unitary transformation matrix. Since

$$\sum |S_{\alpha\delta}|^2 = 1, \tag{7}$$

then $|S|^2$ is the probability for occurrence of $f(E_m)$ or $f(E_n)$. Consequently, whereas in classical mechanics two coupled oscillators exchange energy continuously, such is not the case in quantum mechanics where only two energy states E_m and E_n have a "physical meaning." Therefore, concluded Heisenberg, a probability interpretation emerges naturally from quantum mechanics and can be understood only if there are quantum jumps or discontinuous energy changes.

8. TRANSFORMATION THEORY AND WORD MEANINGS

Despite the success of the new quantum mechanics (e.g., calculations of the anomalous Zeeman effect and helium atom spectrum), the physical meaning was unclear of the intermediate manipulations that produced results to be compared

9

with experiment. That is, the mathematical symbols of the quantum mechanics (syntax) did not yet possess unambiguous meanings (semantics).

During the latter part of 1926 into the spring of 1927 at Copenhagen, Bohr and Heisenberg struggled to find a physical interpretation of the quantum mechanics. Heisenberg's review paper of September 1926 "Die Quantenmechanik" enables us to glimpse their struggles. He stressed that our "customary intuition" cannot be extrapolated into the atomic realm because the "electron and the atom possess not any degree of physical reality as the objects of daily experience. Investigation of the type of physical reality which is proper to electrons and atoms is precisely the subject of quantum mechanics" (1926b). In Heisenberg's view fundamental problems in quantum mechanics had moved into the realm of philosophy. After repeated warnings throughout this paper against intuitive interpretations for quantum mechanics, Heisenberg concluded that "there has been missing in our picture [Bild] of the structure of matter any substantial progress toward a contradiction-free intuitive [anschaulich] interpretation of experiments." What could he have meant by a "contradiction-free intuitive interpretation"? Any reply would have to await Dirac's transformation theory.

On 23 November 1926 Heisenberg reported to Pauli (1979) that Dirac "has managed an extremely broad generalization of my fluctuation paper." This is Dirac's transformation theory paper (1926) that provided the mathematical framework missing from Heisenberg's and Pauli's attempts to relate measurements of canonically conjugate variables. Central to Dirac's paper is that Born's probability amplitude is the transformation function between different representations, for example, position and energy. Actually Heisenberg had discovered this property of the transformation matrix for the discrete case in his fluctuation paper (1926a). Heisenberg's thoughts toward a contradiction-free interpretive framework for quantum mechanics began to crystallize. Throughout he remained focused on the mathematical formalism of the quantum mechanics with its essential discontinuities and nonvisualizability. For both Bohr and Heisenberg linguistic (semantic) difficulties persisted of the same sort as in mid-1925, as Heisenberg described in this letter to Pauli: "That the world is continuous I consider more than ever as totally unacceptable. But as soon as it is discontinuous, all our words that we apply to the description of facts are so many c numbers. What the words 'wave' or 'corpuscle' mean we know not any more" (see Miller, 1989a).

9. THE UNCERTAINTY PRINCIPLE PAPER

By the end of February 1927 Heisenberg found the connection between measurement of kinematical quantities and Bohr's insistence since 1913 on how unclear the terminology from classical physics becomes when used in a theory of phenomena for a realm beyond sense perceptions. Heisenberg described these results in his paper (completed March 1927) "On the Intuitive [anschauliche] Contents of the Quantum-Theoretical Kinematics and Mechanics" (1927). How important was the concept of intuition to Heisenberg is indicated by its inclusion into the title of this classic paper in the history of ideas. This is the paper where he found the new "intuitive interpretation of the various phenomena," for which he had searched (Heisenberg, 1926b) by redefining the concept of "intuition."

Heisenberg's line of argumentation is: "The present paper sets up exact definitions of the words position, velocity, energy, etc. (of an electron)." How can we accomplish this? From our experience with the general theory of relativity we know that the means to extend "intuitively based" concepts (in the classical meaning of this term) into large space-time regions is "derivable neither from our laws of thought nor from experiment." Presently, attempts to obtain an intuitive interpretation of quantum mechanics are full of contradictions because of the "struggle of opinions concerning discontinuum and continuum theory, particles and waves," which implies that "it is not possible to interpret quantum mechanics in the customary kinematical terms." The "necessity of revising kinematical and mechanical concepts appears to follow directly from the basic equations of quantum mechanics; particularly"

$$pq - qp = h/2\pi i \qquad (8)$$

from which we "have good reason to be suspicious about uncritical application of the words 'position' and 'momentum'."

Consequently, may we not say that Heisenberg has redefined the concept of intuition [Anschauung] with the equations of quantum mechanics? After all, the Kantian notion of intuition entails a visualization that had led physicists astray. Heisenberg separated intuition from visualization by basing all deliberations on unvisualizable particles and essential discontinuities. The means to extend the concept of intuition into small regions of space-time is the mathematics of quantum mechanics because it gives restrictions on perception-laden terms such as position and momentum. What are these restrictions? They are the uncertainty relations which Heisenberg goes on to develop with thought experiments. These experiments illustrate his view that concepts such as position of an electron and stationary state of an atom derive meaning from experimental measurement.

Among the well known thought experiments is the Γ-ray microscope experiment which provides a rough derivation of the uncertainty principle for position and momentum

$$p_1 q_1 \sim h, \qquad (9)$$

where p_1 and q_1 are errors in determination of momentum and position. Heisenberg offers as a substantiation of Eq. (9) that "$[p_1 q_1 \sim h]$ is the precise expression for the facts which one previously tried to describe by dividing phase space into cells of size h." In fact, in a letter to Pauli (5 November 1926), written prior to Dirac's transformation theory, Heisenberg had speculated on a relation such as Eq. (9) by analogy with statistical mechanics. In this letter Heisenberg also related to Pauli discussions with Bohr on the possibility that the essential discreteness of quantum mechanics is a glimpse of the discreteness of the space-time metric, which means the impossibility of measuring jointly or separately momentum and position to any arbitrary degree of accuracy.

Heisenberg provides a more rigorous derivation of Eq. (9) with Dirac's transformation theory, to which he gives the following intuitive interpretation in terms of principal axis transformations from classical mechanics. A matrix associated with an operator is diagonal in a reference system that is along a principal axis. The type of experiment performed on a physical system specifies a certain direction that may or may not be along a principal axis. If not, then there is a certain probable error or inaccuracy denoted by the transformation formulae to principal axes. For example, measuring the energy of a system throws the system into a state where the position q has a probability distribution given by the transformation matrix which can be interpreted as the cosine of the angle of inclination between two principal axes. Consequently, experiments divide physical quantities into "known and unknown (alternatively: more or less precisely known variables)." The relationship of results from two experiments that effect different divisions into known and unknown can only be a statistical one. This was evidently a key point in discussions between Bohr and Heisenberg because it concerns puzzles that swirl about the definition of a stationary state.

To pursue this point further Heisenberg analyzes a Stern-Gerlach experiment in which a beam of atoms collimated with a single slit passes through two successive inhomogeneous magnetic field regions F_1 and F_2. Before entering F_1 the atomic beam is prepared in the stationary state n with energy E_n. The probability for a transition into a state f after passing through F_2 depends on whether an experiment was actually performed between F_1 and F_2 to determine the stationary state of atoms in this region. If such an experiment was actually performed then the probability for a transition from state n to state f is

$$\sum_m |c_{nm}|^2 |d_{mf}|^2 \tag{10}$$

where c_{nm} (d_{mf}) is the probability for transition between stationary states n and m (m and f). If no actual experiment was performed between F_1 and F_2, then the transition probability from n to f is

$$|\sum_m c_{nm} \, d_{mf}|^2 \tag{11}$$

which is different from Eq. (10) owing to interference terms.

With emphasis on particles and discontinuities, Heisenberg preferred not to interpret Eq.(11) as an "interference of probabilities," but due to the difference between experimental setups. He attributes a collapse of the wave function interpretation to Eq. (10): "'state m' we select from the abundance of various possibilities (c_{nm}) a single one," thereby limiting the possibilities for all subsequent experiments. The stationary state measurement destroys the phase relationships of the c_{nm} as must be the case because the phase and energy of a stationary state are canonically conjugate quantities.

Reverting from Eq.(11) to Eq.(10) means assuming that the stationary state measurement between F_1 and F_2 has actually been done, introducing unknown phases into each term in Eq.(11). Phase averaging reduces Eq. (11) to Eq.(10). Consequently, phase averaging relates the two experiments statistically through the quantity $|d_{mf}|^2$ and not $|\sum_m c_{nm} \, d_{mf}|^2$, in agreement with Heisenberg's intuitive interpretation of Dirac's transformation theory. Actually in the end result the interference terms would vanish anyway because a third Stern-Gerlach setup is required with pole faces along the beam's direction of motion in order to measure the stationary states after F_2. Heisenberg goes on to give an example of such a measurement which, however, needed corrections by Bohr: Heisenberg neglected to include the wave-particle duality of matter (see Bohr, 1928 and Heisenberg, 1930).

What conclusions does Heisenberg draw from these deliberations?

Since the uncertainty relations placed limits on the accuracy to which initial conditions could be determined then invalid is the causal law from classical mechanics which required both visualization and the continuous development of physical systems.

The wave function collapse interpretation of the measurement process reveals the "deep meaning of the linearity of the Schrödinger equations." Any attempts at replacing them with nonlinear equations are "hopeless." Unfortunately, Heisenberg did not elaborate on this point here or in any extant correspondence.

From where does the quantum theoretic statistics emerge? Heisenberg preferred an interpretation that he attributes to Dirac, namely, that the "statistic is induced by our experiments." However, Heisenberg cautions, we should not conclude that quantum mechanics is "an essentially statistical theory in the sense that only statistical conclusions can be drawn from specified data." For example, exact conclusions can be drawn from the conservation laws of energy and momentum. But owing to the uncertainty relations, speculations that there "is a 'real' world hidden behind the perceived statistical world [are] fruitless and sterile. Physics should describe formally only the connection of perceptions."

12

During the month of February 1927 when Heisenberg wrote the uncertainty principle paper Bohr was away from Copenhagen on vacation. Upon return Bohr was critical over Heisenberg's neglect of the wave-particle duality of matter and light which led Heisenberg to conclude that observational uncertainties were rooted exclusively on the presence of discontinuities. In this way, for example, Heisenberg had reached erroneous conclusions for the Γ-ray microscope experiment, among other Gedanken experiments. In a Note Added in Proof Heisenberg (1927) acknowledged comments of this sort, although he made no move to correct them in the uncertainty principle paper itself.

10. COMPLEMENTARITY

On 16 September 1927 at the International Congress of Physics at Como, Italy, Bohr presented his complementarity view, honed in heated discussions with Heisenberg (see, too, Holton, 1973; Jammer, 1966). Since our customary intuition cannot be extended into the atomic domain, then the "classical mode of description must be generalized" (Bohr, 1928). Our usual "causal space-time description" depends on the smallness of Planck's constant. But in the atomic domain Planck's constant links the measuring apparatus to the system under investigation in a way that "is completely foreign to the classical theories." This is how intrinsic statistics enter quantum theory. In the atomic domain the notion of an undisturbed system developing in space and time is an abstraction and "there can be no question of causality in the ordinary sense of the word," that is, strong causality. Instead of renouncing the causal law like Heisenberg, Bohr linked causality to the predictive powers of the conservation laws of energy and momentum and not to space-time pictures which are relegated to the role of restricted metaphors.

Bohr went on to reason that just as the large value of the velocity of light had prevented our realizing the relativity of time, the minuteness of Planck's constant rendered paradoxical the wave-particle duality of matter and light. Since Planck's constant places restrictions on the use of our language in the atomic domain, then so too on our customary intuition or visual imagery, which enables us to describe only things that are either continuous or discontinuous but not both. Rather, stressed Bohr, the wave and particle modes of light and matter are neither contradictory nor paradoxical, but complementary in the extreme, that is, mutually exclusive. Yet both modes or sides are required for a complete description of the atomic entity. Heisenberg's uncertainty relations turned out to be a particular case of complementarity because, for example, the quantities p and x are not mutually exclusive.

Although Heisenberg agreed with the complementarity principle's restrictions on metaphors from the world of perceptions, he remained wary of them owing to their previous disservices. In a letter of 16 May 1927 to Pauli, Heisenberg wrote that there are "presently between Bohr and myself differences of opinion on the word "intuitive [anschauliche]." This divergence of opinion widened through Heisenberg's subsequent scientific work (see Miller, 1985, 1988).

CONCLUSION

Can we not interpret the results from this historical case study to be indicative of a switch of Heisenberg's mental representation of knowledge? This switch went beyond merely inverting the Kantian notion of perception in which Anschauung is accorded a higher status than Anschaulichkeit. Let me summarize this case study using terminology from concepts of mental representation. Until 1924 Bohr and Heisenberg focused on the content of a mental representation -- that is, what is being represented, which in this case is the Anschauung or the visualization from classical physics that was imposed on atomic theory. Starting in 1924, owing to the Bohr-Kramers-Slater version of Bohr's atomic theory, Bohr and Heisenberg began to shift toward emphasis on the format of a representation by permitting the mathematics of the theory to give a purely descriptive representation of the atomic domain. (The format of a mental representation is its encoding.) Yet even after Heisenberg's invention of the new quantum mechanics physicists lamented over loss of visual imagery. In 1927 Heisenberg redefined the concept of intuition by separating it from visualization -- that is, "intuition" had no visual content. Rather, visualizability or Anschaulichkeit displaced Anschauung. Whereas Anschauung is a product of our cognitive apparatus, Anschaulichkeit pertains to intrinsic properties of subatomic entities that are, to use Einstein's terminology, "out there" regardless of whether we set up experimental apparatus. Bohr continued to advocate the usefulness of restricted Anschauungen.

Suffice it to say that in the course of his scientific research in nuclear physics, in 1932 Heisenberg found a clue to the depictive mode of visualizability, a mode that would in time enable us to imagine things we have not seen, needless to say, within the restrictive framework of our sense perceptions (see Miller, 1985, 1986, 1989a, 1989b). As Heisenberg recalled of his own research: "The picture changes over and over again, its so nice to see how such pictures change" (AHQP, 11 February 1963).

In conclusion, could not the circle of inquiry in fundamental problems of quantum mechanics be widened to include an analysis of "intuition"? Such an analysis could elucidate apparently "unintuitive" content of the theory such as long range correlations.

ACKNOWLEDGEMENT

This research is supported in part by a grant from the National Science Foundation.

REFERENCES

Bohr, N., 1913, On the Constitution of Atoms and Molecules, Phil. Mag., 26: 1, 476, 857.
---   1918, On the Quantum Theory of Line Spectra, Kgl. Danske Vid. Selsk Skr.
         nat.-mat. Afd, IV: 1.
---   1923, Über die Anwendung der Quantentheorie auf den Atombau. I. Die
         Grundpostulate der Quantentheorie, ZsP, 13: 117; and 1924 Proc. Camb. Phil.
         Soc. (Supplement).
---   1925, Atomic Theory and Mechanics, Nature (Supplement).
--- and Kramers, H and Slater, J.C., 1924, The Quantum Theory of Radiation, Phil.
         Mag., 47: 785.
---   1928, The Quantum Postulate and the Recent Development of Atomic Theory,
         Nature (Supplement): 580.
Born, M., 1926a, Zur Quantenmechanik der Stossvorgänge, ZsP, 37: 863.
---   1926b, Zur Quantenmechanik der Stossvorgänge, ZsP, 38: 803.
Dirac, P.A.M., 1926, The Physical Interpretation of the Quantum Mechanics, Proc.
         Roy. Soc. (A), 113: 621.
Einstein, A., 1909, Entwickelung unserer Anschauungen über das Wesen und die
         Konstitution der Strahlung, Phys. Z., 10: 817.
Heisenberg, W., 1925, Über quantentheoretische Umdeutung kinematischer und
         mechanischer Beziehungen, ZsP, 33: 879.
---   1926a, Schwankungserscheinungen und Quantenmechanik, ZsP, 40:
         501.
---   1926b, Die Quantenmechanik, Die Naturwissenschaften, 14, 899.
---   1927, Über den anschaulichen Inhalt der quantentheoretischen
         Kinematik und Mechanik, ZsP, 43: 172.
---   1930, Die physikalischen Prinzipien der Quantentheorie, Herzl,
         Leipzig; 1930, translated by C. Eckart and F.C. Hoyt as The Physical Principles
         of the Quantum Theory, Dover, New York
Hendry, J., 1984, The Creation of Quantum Mechanics and the Bohr-Pauli Dialogue,
         Reidel, Boston.

Holton, G., 1973, The Roots of Complementarity, in: G. Holton, Thematic Origins of Scientific Thought: Kepler to Einstein, Harvard University Press, Cambridge, MA.

Jammer, M. 1966, The Conceptual Development of Quantum Mechanics, McGraw-Hill, New York.

--- 1974, The Philosophy of Quantum Mechanics, Wiley, New York

Kramers, H., 1924, The Law of Dispersion and Bohr's Theory of Spectra, Nature, 113: 673.

--- and W. Heisenberg, 1925, Über die Streuung von Strahlung durch Atome, ZsP, 31: 681.

Miller, A.I., 1985, Werner Heisenberg and the Beginning of Nuclear Physics, Physics Today, 38: 60.

--- 1986, Imagery in Scientific Thought: Creating 20th-Century Physics, MIT Press, Cambridge, MA.

--- 1988, On the Origins of the Copenhagen Interpretation, in: Niels Bohr: Physics and the World, H. Feshbach, T. Matsui, and A. Oleson eds., Harwood, New York.

--- 1989a, Discussion of Thomas S. Kuhn's Paper: 'Possible Worlds in History of Science,'in: Possible Worlds in Humanities, Arts and Sciences: Proceedings of Nobel Symposium 65, S. Allén ed., W. de Gruyter, New York.

--- 1989b, Imagery Metaphor and Physical Reality, in: Psychology of Science: Contributions to Metascience, B. Gholson, W.R. Shadish, R.A. Neimeyer, and A.C. Houts, eds., Cambridge University Press, Cambridge, UK.

--- 1991, Source Book in Quantum Field Theory, Harvard University Press, Cambridge, MA.

Pauli, W., 1979, Wissenschaftlicher Briefwechsel mit Bohr, Einstein, Heisenberg, U.A., Band I: 1919-1929, Springer-Verlag, Berlin, A. Hermann, K.v. Meyenn, and V.F. Weisskopf, eds.

Schrödinger, E., 1926, Über das Verhältnis der Heisenberg-Born-Jordanschen Quantenmechanik zu der meinen, Ann. Phys., 70: 734.

AGAINST "MEASUREMENT"

J.S. Bell

CERN, Geneva

1.INTRODUCTION

Surely, after 62 years, we should have an exact formulation of some serious part of quantum mechanics? By "exact" I do not of course mean "exactly true". I mean only that the theory should be fully formulated in mathematical terms, with nothing left to the discretion of the theoretical physicist... until workable approximations are needed in applications. By "serious" I mean that some substantial fragment of physics should be covered. Nonrelativistic "particle" quantum mechanics, perhaps with the inclusion of the electromagnetic field and a cut-off interaction, is serious enough. For it covers "a large part of physics and the whole of chemistry" [1]. I mean too, by "serious", that "apparatus" should not be separated off from the rest of the world into black boxes, as if it were not made of atoms and not ruled by quantum mechanics.

The question, "....should we not have an exact formulation....?", is often answered by one or both of two others. I will try to reply to them:

Why bother?

Why not look it up in a good book?

2.WHY BOTHER?

Perhaps the most distinguished of "why bother?"'ers has been Dirac [2]. He divided the difficulties of quantum mechanics into two classes, those of the first class and those of the second. The second class difficulties were essentially the infinities of relativistic quantum field theory. Dirac was very disturbed by these, and was not impressed by the "renormalization" procedures by which they are circumvented. Dirac tried hard to eliminate these second class difficulties, and urged others to do likewise. The first class difficulties concerned the role of the "observer", "measurement", and so on. Dirac thought that these problems were not ripe for solution, and should be left for later. He expected developments in the theory which would make these problems look quite different. It would be a waste of effort to worry over much about them now, especially since we get along very well in practice without solving them.

Dirac gives at least this much comfort to those who are troubled by these questions: he sees that they exist and are difficult. Many other distinguished physicists do not. It seems to me that it is among the most sure-footed of quantum physicists, those who have it *in their bones*, that one finds the greatest impatience with the idea that the "foundations of quantum mechanics" might need some attention. Knowing what is right by instinct, they can become

Sixty-Two Years of Uncertainty
Edited by A. I. Miller
Plenum Press, New York, 1990

a little impatient with nitpicking distinctions between theorems and assumptions. When they do admit some ambiguity in the usual formulations, they are likely to insist that ordinary quantum mechanics is just fine "for all practical purposes". I agree with them about that:

ORDINARY QUANTUM MECHANICS

(as far as I know)

IS JUST FINE

FOR ALL PRACTICAL PURPOSES.

Even when I begin by insisting on this myself, and in capital letters, it is likely to be insisted on repeatedly in the course of the discussion. So it is convenient to have an abreviation for the last phrase:

FOR ALL PRACTICAL PURPOSES = FAPP

I can imagine a practical geometer, say an architect, being impatient with Euclid's fifth postulate, or Playfair's axiom: ....of *course* in a plane, through a given point, you can draw only one straight line parallel to a given straight line...at least FAPP. The reasoning of such a natural geometer might not aim at pedantic precision, and new assertions, known in the bones to be right, even if neither among the originally stated assumptions nor derived from them as theorems, might come in at any stage. Perhaps these particular lines in the argument should, in a systematic presentation, be distinguished by this label

....*FAPP*

and the conclusions likewise:

....*QED FAPP*

I expect that mathematicians have classified such fuzzy logics. Certainly they have been much used by physicists.

But is there not something to be said for the approach of Euclid? Even now that we know that Euclidean geometry is (in some sense) not quite true? Is it not good to know what follows from what, even if it is not really necessary FAPP? Suppose for example that quantum mechanics were found to *resist* precise formulation. Suppose that when formulation beyond FAPP is attempted, we find an unmovable finger obstinately pointing outside the subject....to the Mind of the Observer, to God, or even only Gravitation? Would not that be *very very interesting?*

But I must say at once that it is not mathematical precision, but physical, with which I will be concerned here. I am not squeamish about delta functions. From the present point of view, the approach of von Neumann's book is not preferable to that of Dirac's.

3.WHY NOT LOOK IT UP IN A GOOD BOOK?

But *which* good book? In fact it is seldom that a "no problem" person is, on reflection, willing to endorse a treatment already in the literature. Usually the good unproblematic formulation is still in the head of the person in question, who has been too busy with practical things to put it on paper. I think that this reserve, as regards the formulations already in the good books, is well founded. For the good books known to me are not much concerned with physical precision. This is clear already from their vocabulary.

Here are some words which, however legitimate and necessary in application, have no place in a *formulation* with any pretension to physical precision:

system

apparatus

environment

microscopic, macroscopic

reversible, irreversible

observable

information

measurement

The concepts "system", "apparatus", "environment", immediately imply an artificial division of the world, and an intention to neglect, or take only schematic account of, the interaction across the split. The notions of "microscopic" and "macroscopic" defy precise definition. So also do the notions of "reversible" and "irreversibile". Einstein said that it is theory which decides what is "observable". I think he was right.... "observation" is a complicated and theory-laden business. Then that notion should not appear in the *formulation* of fundamental theory. *Information? Whose* information? Information about *what* ?

On this list of bad words from good books, the worst of all is "measurement". It must have a section to itself.

4.AGAINST "MEASUREMENT"

When I say that the word "measurement" is even worse than the others, I do not have in mind the use of the word in phrases like "measure the mass and width of the Z boson". I do have in mind its use in the fundamental interpretive rules of quantum mechanics. For example, here they are as given by Dirac [3]:

"... any result of a measurement of a real dynamical variable is one of its eigenvalues..."

".... if the measurement of the observable.... is made a large number of times the average of all the results obtained will be.... "

".... a measurement always causes the system to jump into an eigenstate of the dynamical variable that is being measured.... "

It would seem that the theory is exclusively concerned about "results of measurement", and has nothing to say about anything else. What exactly qualifies some physical systems to play the role of "measurer"? Was the wavefunction of the world waiting to jump for thousands of millions of years until a single-celled living creature appeared? Or did it have to wait a a little longer, for some better qualified system... with a Ph.D.? If the theory is to apply to anything but highly idealized laboratory operations, are we not obliged to admit that more or less "measurement-like" processes are going on more or less all the time, more or less everywhere? Do we not have jumping then all the time?

The first charge against "measurement", in the fundamental axioms of quantum mechanics, is that it anchors there the shifty split of the world into "system" and "apparatus". A second charge is that the word comes loaded with meaning from everday life, meaning which is entirely inappropriate in the quantum context. When it is said that something is "measured" it is difficult not to think of the result as referring to some *preexisting property* of the object in question. This is to disregard Bohr's insistence that in quantum phenomena the apparatus as well as the system is essentially involved. If it were not so, how could we understand, for example, that "measurement" of a component of "angular momentum".... *in an arbitrarily chosen direction*.... yields one of a discrete set of values? When one forgets the role of the apparatus, as the word "measurement" makes all too likely, one despairs of ordinary logic.... hence "quantum logic". When one remembers the role of the apparatus, ordinary logic is just fine.

In other contexts, physicists have been able to take words from everyday language and use them as technical terms with no great harm done. Take for example the "strangeness", "charm", and "beauty" of elementary particle physics. No one is taken in by this "kiddy talk".... as Bruno Touschek called it. Would that it were so with "measurement". But in fact the word has had such a damaging effect on the discussion, that I think it should now be banned altogether in quantum mechanics.

5. THE ROLE OF EXPERIMENT

Even in a lowbrow practical account, I think it would be good to replace the word "measurement", in the formulation, by the word "experiment". For the latter word is altogether less misleading. However the idea that quantum mechanics, our most fundamental physical theory, is exclusively even about the results of experiments would remain disapponting.

In the beginning natural philosophers tried to understand the world around them. Trying to do that they hit upon the great idea of contriving artificially simple situations in which the number of factors involved is reduced to a minimum. Divide and conquer. Experimental science was born. But experiment is a tool. The aim remains: to understand the world. To restrict quantum mechanics to be exclusively about piddling laboratory operations is to betray the great entreprise. A serious formulation will not exclude the big world outside the laboratory.

6. THE QUANTUM MECHANICS OF LANDAU AND LIFSHITZ

Let us have a look at the good book "Quantum Mechanics",by L.D.Landau and E.M. Lifshitz [4]. I can offer three reasons for this choice:

1) It is indeed a good book.

2) It has a very good pedigree. Landau sat at the feet of Bohr. Bohr himself never wrote a systematic account of the theory. Perhaps that of Landau and Lifshitz is the nearest to Bohr that we have.

3) It is the only book on the subject in which I have read every word.

This last came about because my friend John Sykes enlisted me as technical assistant when he did the English translation. My recommendation of this book has nothing to do with the fact that one percent of what you pay for it comes to me.

LL emphasize, following Bohr, that quantum mechanics requires for its formulation "classical concepts".... a classical world which intervenes on the quantum system, and in which experimental results occur:

"....It is in principle impossible.... to formulate the basic concepts of quantum mechanics without using classical mechanics." (LL2)

"....The possibility of a quantitative description of the motion of an electron requires the presence also of physical objects which obey classical mechanics to a sufficient degree of accuracy." (LL2)

"....the "classical object" is usually called *apparatus* and its interaction with the electron is spoken of as *measurement*. However it must be emphasized that we are here not discussing a process....in which the physicist-observer takes part. By *measurement*, in quantum mechanics, we understand any process of interaction beteween classical and quantum objects, ocurring apart from and independently of any observer. The importance of the concept of measurement in quantum mechanics was elucidated by N.Bohr." (LL2)

And with Bohr they insist again on the inhumanity of it all:

".... Once again we emphasize that, in speaking of "performing a measurement", we refer to the interaction of an electron with a classical "apparatus", which in no way presupposes the presence of an external observer." (LL3)

"....Thus quantum mechanics occupies a very unusual place among physical theories: it contains classical mechanics as a limiting case, yet at the same time it requires this limiting case for its own formulation...." (LL3)

"....consider a system consisting of two parts: a classical apparatus and an electron.... The states of the apparatus are described by quasiclassical wavefunctions $\Phi_n(\xi)$, where the suffix n corresponds to the "reading" g_n of the apparatus, and ξ denotes the set of its coordinates. The classical nature of the apparatus appears in the fact that, at any given instant, we can say with certainty that it is in one of the known states Φ_n with some definite value of the quantity g; for a quantum system such an assertion would of course be unjustified." (LL21)

"....Let $\Phi_0(\xi)$ be the wavefunction of the initial state of the apparatus....and$\Psi(q)$ of the electron....the initial wave function of the whole system is the product

$$\Psi(q)\Phi_0(\xi)$$

....After the measuring process....we obtain a sum of the form

$$\sum_n A_n(q)\Phi_n(\xi)$$

where the $A_n(q)$ are some functions of q." (LL22)

"The classical nature of the apparatus, and the double role of classical mechanics as both the limiting case and the foundation of quantum mechanics, now make their appearance. As has been said above, the classical nature of the apparatus means that, at any instant, the quantity g (the "reading of the apparatus") has some definite value. This enables us to say that the state of the system apparatus + electron after the measurement will in actual fact be described, not by the entire sum.... but by only the one term which corresponds to the "reading" g_n of the apparatus,

$$A_n(q)\Phi_n(\xi)$$

It follows from this that $A_n(q)$ is proportional to the wave function of the electron after the measurement.... " (LL22)

This last is (a generalization of) the Dirac jump, not an assumption here but a theorem. Note however that it has become a theorem only in virtue of *another* jump being assumed.... that of a "classical" apparatus into an eigenstate of its "reading". It will be convenient later to refer to this last, the *spontaneous* jump of a macroscopic system into a definite macroscopic configuration, as the LL jump. And the *forced* jump of a quantum system as a result of "measurement"..... *an external intervention* .... as the Dirac jump. I am not implying that these men are the inventors of these concepts. They have used them in references that I can give.

According to LL (LL24), measurement (I think they mean the LL jump)

".... brings about a new state.... Thus the very nature of the process of measurement involves a far-reaching principle of irreversibility....causes the two directions of time to be physically non-equivalent, i.e. creates a difference between the future and the past."

The LL formulation, with vaguely defined wave function collapse, when used with good taste and discretion, is adequate FAPP. It remains that the theory is ambiguous in principle, about exactly when and exactly how the collapse occurs, about what is microscopic and what is macroscopic, what quantum and what classical. We are allowed to ask: is such ambiguity dictated by experimental facts? Or could theoretical physicists do better if they tried harder?

7. THE QUANTUM MECHANICS OF K.GOTTFRIED

The second good book that I will look at here is that of Kurt Gottfried [5]. Again I can give three reasons for this choice:

1) It is indeed a good book. The CERN library had four copies. Two have been stolen... already a good sign. The two that remain are falling apart from much use.

2) It has a very good pedigree. Kurt Gottfried was inspired by the treatments of Dirac and Pauli. His personal teachers were J.D.Jackson, J.Schwinger, V.F.Weisskopf, and J.Goldstone. As consultants he had P.Martin, C.Schwartz, W.Furry, and D.Yennie.

3) I have read some of it more than once.

This last came about as follows. I have often had the pleasure of discussing these things with Viki Weisskopf. Always he would end up with "you should read Kurt Gottfried". Always I would say "I *have* read Kurt Gottfried". But Viki would always say again next time "you should read Kurt Gottfried". So finally I read again some parts of K.G., and again, and again, and again.

At the beginning of the book there is a declaration of priorities (KG1):

"....The creation of quantum mechanics in the period 1924-28 restored logical consistency to its rightful place in theoretical physics. Of even greater importance, it provided us with a theory that appears to be in complete accord with our empirical knowledge of all nonrelativistic phenomena...."

The first of these two propositions, admittedly the less important, is actually given rather little attention in the book. One can regret this a bit, in the rather narrow context of the particular present enquiry....into the possibility of precision. More generally, KG's priorities are those of all right-thinking people.

The book itself is above all pedagogical. The student is taken gently by the hand, and soon finds himself or herself *doing* quantum mechanics, without pain,... and almost without thought. The essential division of KG's world into system and apparatus, quantum and classical, a notion that might disturb the student, is gently implicit rather than brutally explicit. No explicit guidance is then given as to how in practice this shifty division is to be made. The student is simply left to pick up good habits by being exposed to good examples.

KG declares that the task of the theory is (KG16)

"....to predict the results of measurements on the system..."

The basic structure of KG's world is then

$$W = S + R$$

where S is the quantum system, and R is the rest of the world... from which measurements on S are made. When your *only* interpretative axioms are about measurement results (or findings(KG11)) you absolutely *need* such a base R from which measurements can be made. There can be no question then of identifying the quantum system S with the whole world W. There can be no question... without changing the axioms... of getting rid of the shifty split. Sometimes some authors of "quantum measurement" theories seem to be trying to do just that. It is like a snake trying to swallow itself by the tale. It can be done... up to a point. But it becomes uncomfortable for the spectators even before it becomes painful for the snake.

But there is something which can and must be done... to analyse theoretically not *removing* the split, which can not be done with the usual axioms, but *shifting* it. This is taken up in KG's chapter IV: "The Measurement Process... ". Surely "apparatus" can be seen as made of atoms? And it often happens that we do not know, or not well enough, either a priori or by experience, the functioning of some system that we would regard as "apparatus". The theory can help us with this only if we take this "apparatus" A out of the rest of the world R and treat it together with S as part of an enlarged quantum system S':

$$R = A + R'$$

$$S + A = S'$$

$$W = S' + R'$$

The original axioms about "measurement" (whatever they were exactly) are then applied not at the S/A interface, but at the A/R' interface... where for some reason it is regarded as more safe to do so. In real life it would not be possible to find *any* such point of division which would be *exactly* safe. For example, strictly speaking it would not be *exactly* safe to take it between the counters, say, and the computer... slicing neatly through some of the atoms of the wires. But with some idealization, which might "...be highly stylized and not do justice to the enormous complexity of an actual laboratory experiment..." (KG165), it might be possible to find more than one not too implausible way of dividing the world up. Clearly it is necessary to check that different choices give consistent results (FAPP). A disclaimer towards the end of KG's chapter IV suggests that that, and only that, is the modest aim of that chapter (KG189):

"....we emphasize that our discussion has merely consisted of several demonstrations of internal consistency....".

But reading reveals other ambitions.

Neglecting the interaction of A with R', the joint system $S' = S + A$ is found to end, in virtue of the Schrödinger equation, after the "measurement" on S by A, in a state

$$\Psi = \sum_n c_n \Psi_n$$

where the states Ψ_n are supposed each to have a definite apparatus pointer reading g_n. The corresponding density matrix is

$$\rho = \sum_n \sum_m c_n c_m^* \Psi_n \Psi_m^*$$

At this point KG insists very much on the fact that A, and so S', is a macroscopic system. For macroscopic systems, he says, (KG186)

"....$tr A\hat{\rho} = tr A\rho$ for all observables A known to occur in nature...."

where

$$\hat{\rho} = \sum_n |c_n|^2 \Psi_n \Psi_n^*$$

i.e. $\hat{\rho}$ is obtained from ρ by dropping interference terms involving pairs of macroscopically different states. Then (KG188)

"....we are free to replace ρ by $\hat{\rho}$ after the measurement, safe in the knowledge that the error will never be found...."

Now while quite uncomfortable with the concept "all known observables", I am fully convinced [6] of the practical elusiveness, even the absence FAPP, of interference between macroscopically different states. So let us go along with KG on this and see where it leads:

"...If we take advantage of the indistinguishability of ρ and $\hat{\rho}$ to say that $\hat{\rho}$ is the state of the system subsequent to measurement, the intuitive interpretation of c_m as a probability amplitude emerges without further ado. This is because c_m enters $\hat{\rho}$ only via $|c_m|^2$, and the latter quantity appears in $\hat{\rho}$ in precisely the same manner as probabilities do in classical statistical physics..."

I am quite puzzled by this. If one were not actually on the lookout for probabilities, I think the obvious interpretation of even $\hat{\rho}$ would be that the system is in a state in which the various Ψ's somehow *coexist*:

$$\Psi_1 \Psi_1^* \ and \ \Psi_2 \Psi_2^* \ and.......$$

This is not at all a *probability* interpretation, in which the different terms are seen not as *coexisting*, but as *alternatives*:

$$\Psi_1 \Psi_1^* \; or \; \Psi_2 \Psi_2^* \; or \ldots \ldots$$

The idea that elimination of coherence, in one way or another, implies the replacement of "and" by "or", is a very common one among solvers of the "measurement problem". It has always puzzled me.

It would be difficult to exagerate the importance attached by KG to the replacement of ρ by $\hat{\rho}$:

"....To the extent that nonclassical interference terms (such as $c_m c_{m'}^*$) are present in the mathematical expression for ρ .... the numbers c_m are intuitively uninterpretable, and the theory is an empty mathematical formalism...." (KG187).

But this suggests that the original theory, "an empty mathematical formalism", is not just being approximated.... but discarded and replaced. And yet elsewhere KG seem clear that it is in the business of approximation that he is engaged, approximation of the sort that introduces irreversibility in the passage from classical mechanics to thermodynamics:

"....In this connection one should note that in approximating ρ by $\hat{\rho}$ one introduces irreversibility, because the time reversed Schrödinger equation cannot retreive ρ from $\hat{\rho}$." (KG188)

New light is thrown on KG's ideas by a recent recapitulation[7], referred to in the following as KGR. This is dedicated to the proposition that (KGR1)

".... the laws of quantum mechanics yield the results of measurements..."

These laws are taken to be (KGR1):

"1) a pure state is described by some vector in Hilbert space from which expectation values of observables are computed in the standard way; and

2) the time evolution is a unitary transformation on that vector." (KGR1)

Not included in the laws is (KGR1) von Neumann's

".... infamous postulate: the measurement act "collapses" the state into one in which there are no interference terms between different states of the measurement apparatus...."

Indeed, (KGR1)

"the reduction postulate is an ugly scar on what would be a beautiful theory if it could be removed...."

Perhaps it is useful to recall here just how the infamous postulate is formulated by von Neumann [8]. If we look back we find that what vN actually *postulates* (vN347,418) is that "measurement".... an external intervention by R on S... causes the state

$$\phi = \sum_n c_n \phi_n$$

to jump, with various probabilities, into

$$\phi_1 \ or \ \phi_2 \ or.....$$

From the "or" here, replacing the "and", as a result of external intervention, vN *infers* that the resulting density matrix, averaged over the several possibilities, has no interference terms between states of the *system* which correspond to different measurement results (vN347). I would emphasize several points here:

1) Von Neumann presents the disappearance of coherence in the density matrix, not as a postulate, but as a *consequence* of a postulate. The *postulate* is made at the wavefunction level, and is just that already made by Dirac for example.

2) I can not imagine von Neumann arguing in the opposite direction, that lack of interference in the density matrix implies, without further ado, "or" replacing "and" at the wavefunction level. A special postulate to that effect would be required.

3) Von Neumann is concerned here with what happens to the state of the system that has suffered the measurement....an *external intervention*. In application to the extended system $S'(= S + A)$ von Neumann's collapse would not occur before external intervention from R'. It would be surprising if this consequence of external intervention on S' could be inferred from the purely internal Schrödinger equation for S'. Now KG's collapse, although justified by reference to "all known observables" at the S'/R' interface, occurs after measurement by A on S, *but before interaction across S'/R'*. Thus the collapse which KG discusses is not that which von Neumann infamously postulates. It is the LL collapse rather than that of von Neumann and Dirac.

The explicit assumption that expectation values are to be calculated in the usual way throws light on the subsequent falling out of the usual probability interpretation "without further ado". For the rules for calculating expectation values, applied to projection operators for example, yield the Born probabilities for eigenvalues. The mystery is then: what has the author actually *derived* rather than assumed. And why does he insist that probabilities appear only after the butchering of ρ into $\hat{\rho}$, the theory remaining an "empty mathematical formalism" so long as ρ is retained? Dirac, von Neumann, and the others, nonchalantly assumed the usual rules for expectation values, and so probabilities, in the context of the unbutchered theory. Reference to the usual rules for expectation values also makes clear what KG's probabilities are probabilities of. *They are probabilities of "measurement" results*, of external results of external interventions, from R' on S' in the application. We must not *drift* into thinking of them as probabilities of intrinsic properties of S'. independent of, or before, "measurement". Concepts like that have no place in the orthodox theory.

Having tried hard to understand what KG has written, I will finally permit myself some guesses about what he may may have in mind. I think that from the beginning KG tacitly *assumes* the Dirac rules at S'/R'. The Dirac von Neumann jump is included here. It is required to get the correlations between results of successive measurements. Then, for "all known observables", he sees that the "measurement" results at S'/R' are

AS IF

the LL jump had ocurred in S'. This is important, for it shows how, FAPP, we can get away with attributing definite classical properties to "apparatus" while believing it to be governed by quantum mechanics. But a jump assumption remains. LL derived the Dirac jump from the assumed LL jump. KG derives, FAPP, the LL jump from assumptions at the shifted split R'/S' which include the Dirac jump *there* .

It seems to me that there is then some conceptual drift in the argument. The qualification "as if (FAPP)" is dropped, and it is supposed that the LL jump really takes place. The drift is away from the "measurement"(... external intervention...) orientation of orthodox quantum mechanics towards the idea that systems, such as S' above, have *intrinsic properties* .... independently of and before observation. In particular the readings of experimental apparatus are supposed to be *really there* before they are read. This would explain KG's reluctance to interpret the unbutchered density matrix ρ, for the interference terms there could seem to imply the simultaneous existence of different readings. It would explain his need to collapse ρ into $\hat{\rho}$, in contrast with von Neumann and the others, *without* external intervention across the last split S'/R'. It would explain why he is anxious to obtain this reduction from the internal Schrödinger equation of S'. (It would not explain the curious reference to "all known observables"....at the S'/R' split. I have not been able to grasp all his ideas in a crystal clear way.) The resulting theory would be one in which some "macroscopic" "physical attributes" *have* values at all times, with a dynamics that is related somehow to the butchering of ρ into $\hat{\rho}$... which is seen as somehow not incompatible with the internal Schrödinger equation of the system. But the retention of the vague word "macroscopic" would reveal limited ambition as regards precision. To avoid the vague "microscopic" " macroscopic" distinction... another shifty split.... I think one would be lead to introduce variables which *have* values even on the smallest scale. If the exactness of the Schrödinger equation is maintained, I see this leading towards the picture of de Broglie and Bohm.

8. THE QUANTUM MECHANICS OF N.G.van KAMPEN

Let us look at one more good book, namely Physica A153(1988), and more specifically at the contribution: "Ten theorems about quantum mechanical measurements", by N.G.van Kampen [9]. This paper is distinguished especially by its robust common sense. The author has no patience with

".... such mind boggling fantasies as the many world interpretation...." (vK98)

He dismisses out of hand the notion of von Neumann, Pauli, Wigner,.... that "measurement" might be complete only in the mind of the observer:

"....I find it hard to understand that someone who arrives at such a conclusion does not seek the error in his argument." (vK101)

For vK

"....the mind of the observer is irrelevant.... *the quantum mechanical measurement is terminated when the outcome has been macroscopically recorded....*" (vK101)

Moreover, for vK, no special dynamics comes in to play at "measurement":

".... The measuring act is fully described by the Schrödinger equation for object system and apparatus together. The collapse of the wavefunction is a consequence rather than an additional postulate..." (vK97)

After the measurement the measuring instrument, according to the Schrödinger equation, will admittedly be in a superposition of different readings. For example Schrödinger's cat will be in a superposition

$$|cat> = a|life> + b|death>$$

And it might seem that we do have to deal with "and" rather than "or" here, because of interference:

"....for instance the temperature of the cat......the expectation value of such a quantity G ....is *not* a statistical average of the values G_{ll} and G_{dd} with probabilities $|a|^2$ and $|b|^2$, but contains cross terms between life and death...." (vK103)

But vK is not impressed:

"The answer to this paradox is again that *the cat is macroscopic*. Life and death are macrostates containing an enormous number of eigenstates $|l>$ and $|d>$.....

$$|cat> = \sum_l a_l |l> + \sum_d b_d |d>$$

....the cross terms in the expression for $<G>$.... as there is such a wealth of terms, all with different phases and magnitudes, they mutually cancel and their sum practically vanishes. This is the way in which the typical quantum mechanical interference becomes inoperative between macrostates.... " (vK103)

This argument for no interference is not, it seems to me, by itself immediatly convincing.... Surely it would be possible to find a sum of very many terms, with different amplitudes and phases, which is not zero? However I am convinced anyway that interference between macroscopically different states is very very elusive. Granting this, let me try to say what I think the argument to be, for the collapse as a "consequence" rather than an additional postulate.

The world is again divided into "system", "apparatus", and the rest:

$$W = S + A + R' = S' + R'$$

At first, the usual rules for quantum "measurements" are *assumed* at the S'/R' interface.... *including the collapse postulate*, which dictates correlations between results of "measurements" made at different times. But the "measurements" at S'/R' which can actually be done, FAPP, do not show interference between macroscopically different states of S'. It is as *if* the "and" in the superposition had already, *before* any such measurements, been replaced by "or". So the "and" *has* already been replaced by "or". It is as *if* it were so.... so it *is* so.

This may be good FAPP logic. If we are more pedantic, it seems to me that we do not have here the proof of a theorem, but a *change of the theory* .... at a strategically well chosen point. The change is from a theory which speaks *only* of the results of *external interventions* on the quantum system, S' in this discussion, to one in which that system is attributed *intrinsic properties* .... deadness or aliveness in the case of cats. The point is strategically well chosen in that the predictions for results of "measurements" across S'/R' will still be the same.... FAPP.

Whether by theorem or by assumption, we end up with a theory like that of LL, in which superpositions of macroscopically different states decay somehow into one of the members. We can ask as before just how and how often it happens. If we really had a theorem, the answers to these questions would be calculable. But the only possibility of calculation in schemes like those of KG and vK, involves shifting further the shifty split.... and the questions with it.

For most of the paper, vK's world seems to be the petty world of the laboratory, even one that is not treated very realistically:

".... in this connection the measurement is always taken to be instantaneous ...." (vK100)

But almost at the last moment a startling new vista opens up... an altogether more vast one:

"*Theorem IX: The total system is described throughout by the wave vector* Ψ *and has therefore zero entropy at all times....*

This ought to put an end to speculations about measurements being responsible for increasing the entropy of the universe. (It won't of course.)" (vK111)

So vK, unlike many other very practicle physicists, seems willing to consider the universe as a whole. His universe, or at any rate some "total system", has a wavefunction, and that wavefuntion satisfies a linear Schrödinger equation. It is clear however that this wavefunction cannot be the whole story of vK's totality. For it is clear that he expects the experiments in his laboratotries to give definite results, and his cats to be dead *or* alive. He believes then in variables X which identify the realities... in a way which the wavefunction.... without collapse.... can not. His complete kinematics is then of the de Broglie Bohm "hidden variable" dual type:

$$(\Psi(t, q), X(t))$$

For the dynamics, he has the Schrödinger equation for Ψ, but I do not know *exactly* what he has in mind for the X, which for him would be restricted to some "macroscopic" level. Perhaps indeed he would prefer to remain somewhat vague about this, for

"*Theorem IV: Whoever endows* ψ *with more meaning than is needed for computing observable phenomena is responsible for the consequences....*"(vK99)

9. TOWARDS A PRECISE QUANTUM MECHANICS

In the beginning, Schrödinger tried to interpret his wavefunction as giving somehow the density of the stuff of which the world is made. He tried to think of an electron as represented by a wavepacket......a wavefunction appreciably different from zero only over a small region in space. The extension of that region he thought of as the actual size of the electron.....his electron was a bit fuzzy. At first he thought that small wavepackets, evolving according to the Schrödinger equation, would remain small. But that was wrong. Wavepackets diffuse, and with the passage of time become indefinitely extended, according to the Schrödinger equation. But however far the wavefuncton has extended, the reaction of a detector to an electron remains spotty. So Schrödinger's "realistic" interpretation of his wavefunction did not survive.

Then came the Born interpretation. The wavefunction gives not the density of *stuff*, but gives rather (on squaring its modulus) the density of *probability*. Probability of what, exactly? Not of the electron *being* there, but of the electron being *found* there, if its position is "measured".

Why this aversion to "being" and insistence on "finding"? The founding fathers were unable to form a clear picture of things on the remote atomic scale. They became very aware of the intervening apparatus, and of the need for a "classical" base from which to intervene on the quantum system. And so the shifty split.

The kinematics of the world, in this orthodox picture, is given by a wavefunction (maybe more than one?) for the quantum part, and classical variables..... variables which *have* values.... for the classical part:

$$(\Psi(t, q\ldots\ldots), X(t)\ldots\ldots\ldots\ldots)$$

The $X's$ are somehow macroscopic. This is not spelled out very explicitly. The dynamics is not very precisely formulated either. It includes a Schrödinger equation for the quantum part, and some sort of classical mechanics for the classical part, and "collapse" recipes for their interaction.

It seems to me that the only hope of precision with this dual (Ψ, x) kinematics is to omit completely the shifty split, and let both Ψ and x refer to the world βas a whole. Then the $x's$ must not be confined to some vague macroscopic scale, but must extend to all scales. In the picture of de Broglie and Bohm, every particle is attributed a position $x(t)$. Then instrument pointers..... assemblies of particles, *have* positions, and experiments *have* results. The dynamics is given by the world Schrödinger equation *plus* precise "guiding"equations prescribing how the $x(t)'s$ move under the influence of Ψ. Particles are *not* attributed angular momenta, energies, etc., but *only* positions as functionsof time. Peculiar *"measurement"* results for angular momenta, energies, and so on, emerge as pointer positions in appropriate experimental setups. Considerations of the KG and vK type, on the absence (FAPP) of macroscopic interference, take their place here, and an important one, in showing how usually we do not have (FAPP) to pay attention to the whole world, but only to some subsystem, and can simplify the wavefunction....FAPP.

The Born-type kinematics (Ψ, X) has a duality that the original "density of stuff" picture of Schrödinger did not. The position of the particle there was just a feature of the wavepacket, not something in addition. The Landau Lifshitz approach can be seen as maintaining this simple nondual kinematics, but with the wavefunction compact on a macroscopic rather than microscopic scale. We know, they seem to say, that macroscopic pointers *have* definite positions. *And* we think there is nothing *but* the wavefunction. So the wavefunction *must* be narrow as regards macroscopic variables. The Schrödinger equation does not preserve such narrowness (as Schrödinger himself dramatized with his cat). So there must be some kind of "collapse" going on in addition, to restore macroscopic narrowness when momentarily it is violated In the same way, if we had modified Schrödinger's evolution somehow we might have prevented the spreading of his wavepacket-electrons. But actually the idea that an electron in a ground-state hydrogen atom is as big as the atom (which is then perfectly spherical) is perfectly tolerable.... and maybe even attractive. The idea that a macroscopic pointer can point simultaneously in different directions, or that a cat can have several of its nine lives at the same time, is harder to swallow. And if we have no extra variables X to express macroscopic definiteness, the wavefunction must be narrow in macroscopic directions in the configuration space. This the Landau-Lifshitz collapse brings about. It does so in a rather vague way, at rather vaguely specified times.

In the Ghirardi-Rimini-Weber scheme [10] this vagueness is replaced by mathematical precision. The Schrödinger wave function even for a single particle, is supposed to be unstable, with a prescribed mean life per particle, against spontaneous collapse of a prescribed form. The lifetime and collapsed extension are such that departures of the Schrödinger equation show up very rarely and very weakly in few-particle systems. But in macroscopic systems,as *a consequence of the prescribed equations* , pointers very rapidly point, and cats are killed or spared.

The orthodox approaches, whether the authors think they have made derivations or assumptions, are just fine FAPP........when used with the good taste and discretion picked up from exposure to good examples. At least two roads are open from there towards a precise theory, it seems to me. Both eliminate the shifty split. The deBroglie-Bohm type theories retain, exactly, the linear wave equation, and necessarily add complementary variables to express the non-waviness of the world on the macroscopic scale. The GRW type theories have nothing in their kinematics but the wavefunction. It gives the density (in a multidimensional configuration space!) of *stuff*. To account for the narrowness of that stuff in macroscopic dimensions, the linear Schrödinger equation has to be modified, in the GRW picture by a mathematically prescribed spontaneous collapse mechanism.

The big question, in my opinion, is which, if either, of these two precise pictures can be redeveloped in a Lorentz invariant way.

"....All historical experience confirms that men might not achieve the possible if they had not, time and time again, reached out for the impossible."

Max Weber

"....we do not know where we are stupid until we stick our necks out."

R.P.Feynman

REFRENCES

1. P.A.M.Dirac, Proc.Roy.Soc.A123(1929)714

2. P.A.M.Dirac, Scientific American 208(1963)No5(May)45

3. P.A.M.Dirac, "Quantum mechanics", third edition, Oxford University Press, Oxford 1948. (First edition 1930)

4. L.D.Landau and E.M.lifshitz, "Quantum mechanics", third edition, Pergamon Press, Oxford, 1977

5. K.Gottfried, "Quantum mechanics", Benjamin, New York, 1966

6. K.Gottfried, "Does quantum mechanics describe the collapse of the wavefunction?", presented at "62 years of uncertainty", Erice, August 5-15, 1989

7. J.S.Bell and M.Nauenberg, "The moral aspect of quantum mechanics", in: Preludes in theoretical physics (in honour of V.F.Weisskopf), North-Holland, Amsterdam, 1966, 278-286

8. J.von Neumann, "Mathematical fondations of quantum mechanics", Princeton University Press, Princeton, 1955. (German original 1932.)

9. N.G.van Kampen, "Ten theorems about quantum mechanical measurements", Physica A153(1988)97-113

10. See the contributions of Ghirardi, Rimini, Weber, Pearle, and Diosi, to these proceedings

AN EXPOSITION OF BELL'S THEOREM

Abner Shimony

Departments of Philosophy and Physics
Boston University
Boston, MA

The purpose of this lecture is to give a self-contained demonstration of a version of Bell's theorem and a discussion of the significance of the theorem and the experiments which it inspired. The lecture should be comprehensible to people who have had no previous acquaintance with the literature on Bell's theorem, but I hope that explicitness about premises and consequences will make it useful even to those who are familiar with the literature.

All versions of Bell's theorem are variations, and usually generalizations, of the pioneering paper of J.S. Bell of 1964, entitled "On the Einstein-Podolsky-Rosen Paradox." All of them consider an ensemble of pairs of particles prepared in a uniform manner, so that statistical correlations may be expected between outcomes of tests performed on the particles of each pair. If each pair in the ensemble is characterized by the same quantum state $\emptyset$, then the quantum mechanical predictions for correlations of the outcomes can in principle be calculated when the tests are specified. On the other hand, if it is assumed that the statistical behavior of the pairs is governed by a theory which satisfies certain independence conditions (always similar to the Parameter and Outcome Independence conditions stated below, though the exact details vary from version to version of Bell's theorem), then it is possible to derive a restriction upon the statistical correlations of the outcomes of tests upon the two particles. The restriction is stated in the form of an inequality, known by the collective name of "Bell's Inequality." Each version of Bell's theorem exhibits a choice of $\emptyset$ and of the tests upon the two particles such that the quantum mechanical predictions of correlations violates one of the Bell's Inequalities. The theorem therefore asserts that <u>no physical theory satisfying the specified independence conditions can agree in all circumstances with the predictions of quantum mechanics</u>. The theorem becomes physically significant when the experimental arrangement is such that relativistic locality <u>prima facie</u> requires that the independence conditions be satisfied. Because such arrangements are in principle possible (and, in fact, actually realizable, if certain reasonable assumptions are made), one can restate Bell's Theorem more dramatically as follows: <u>no local physical theory can agree in all circumstances with the predictions of quantum mechanics</u>. I shall now present a schematic arrangement which will allow the foregoing sketch to be filled out in detail.

Sixty-Two Years of Uncertainty
Edited by A. I. Miller
Plenum Press, New York, 1990

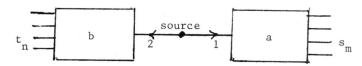

Fig. 1. An ensemble of particle pairs 1+2 is emitted
in a uniform manner from the source. Particle
1 enters an analyzer with a controllable parameter a, and the possible outcomes are s
(m = 1,2,...). Particle 2 enters an analyzer
with controllable parameter b, and the possible
outcomes are t_n (n = 1,2,...).

Figure 1 shows a source from which particle pairs, labeled 1 and 2, are
emitted in a uniform manner. The complete state of a pair 1+2 is denoted by
k, where k belongs to a space K of complete states. No assumption is made
about the structure of K, except that probability measures can be defined on
it. Because of the uniform experimental control of emission, it is reasonable to suppose that there is a definite probability measure w defined over
K which governs the ensemble of pairs; but the uniformity need not be such
that w is a delta-function, i.e., that every pair of the ensemble is in the
same complete state k. Particle 1 enters an analyzer with a controllable parameter a, which the experimenter can specify, for instance, by turning a
knob. Likewise, particle 2 enters an analyzer with a controllable parameter
b. The possible outcomes of the analysis of 1 are s_m(m = 1,2,...), and for
mathematical convenience all these values are assumed to lie in the interval
[-1, 1]. The possible values of the analysis of 2 are t_n(n = 1,2...), and
these values are assumed to lie in the same interval. It will be assumed that
when the parameters a and b and the complete state k are all specified, then
the probabilities of the various single and joint outcomes of analysis are
well-defined. Specifically,

$p^1(m/k,a,b)$ is the probability of the outcome s_m of the analysis of
particle 1, given the complete state k and the parame-
ters a and b;

$p^2(n/k,a,b)$ is the probability of the outcome t_n of the analysis
of particle 2, given the complete state k and the parameters a and b;

$p(m,n/k,a,b)$ is the probability of joint outcomes s_m and t_n, given the
complete state k and the parameters a and b;

$p^1(m/k,a,b,n)$ is the probability of the outcome s_m of the analysis of
particle 1, given the complete state k, the parameters
a and b, and the outcome t_n of the analysis of particle 2;

$p^2(n/k,a,b,m)$ is the probability of the outcome t_n of the analysis
of particle 2, given the complete state k, the parame-
ters a and b, and the outcome s_m of the analysis of
particle 1.

The general principles of probability theory, with no further assumptions, impose the following product rule:

$$p(m,n/k,a,b) = p^1(m/k,a,b)p^2(n/k,a,b,m) = p^2(n/k,a,b)p^1(m/k,a,b,n).$$

We now have sufficient notation to make explicit the independence conditions which were mentioned in the sketch above, and which were first made explicit by Jarrett (1984).

Parameter Independence:

$p^1(m/k,a,b)$ is independent of b, and hence may be written as $p^1(m/k,a)$,

$p^2(n/k,a,b)$ is independent of a, and hence may be written as $p^2(n/k,b)$.

Outcome Independence:

$p^1(m/k,a,b,n) = p^1(m/k,a,b)$,

$p^2(n/k,a,b,m) = p^2(n/k,a,b)$.

The conjunction of Parameter Independence and Outcome Independence implies the following factorization, which is crucial in the argument ahead:

$$p(m,n/k,a,b) = p^1(m/k,a)p^2(n/k,b). \tag{1}$$

Eq. (1) is often called "Bell's locality condition," but even though I have used this nomenclature myself, I now think that it is misleading, and a more neutral name is preferable.

Expectation values can be defined explicitly in terms of the outcomes s_m and t_n and appropriate probabilities:

$E^1(k,a) = \sum_m p^1(m/k,a)s_m$ is the expectation value of the outcome of analysis of particle 1, given complete state k and parameter a;

$E^2(k,b) = \sum_n p^2(n/k,b)t_n$ is the expectation value of the outcome of analysis of particle 2, given complete state k and parameter b;

$E(k,a,b) = \sum_{m,n} p(m,n/k,a,b)s_m t_n.$ is the expectation value of the product of the outcomes of analysis of the two particles, given k, a, and b.

These definitions, together with Eq. (1), immediately yield the following:

$$E(k,a,b) = E^1(k,a)E^2(k,b). \tag{2}$$

I shall now state and prove a simple mathematical lemma, which will bring us close to one of Bell's Inequalities.

Lemma: if x', y', x", and y" all belong to the interval [-1, 1], then S belongs to the interval [-2, 2], where S = x'y' + x'y" + x"y' - x"y".

The proof I shall now give will not be the inelegant one which I presented in Erice, but the elegant argument which N.David Mermin suggested after the lecture. The first step is to note that S is linear in each of its four variables and hence takes on its extreme values at corners of the domain, i.e. at (x',y',x",y") = (±1,±1,±1,±1). Clearly, at a corner the value of S must be an integer between -4 and 4. But S can also be written as

$$S = (x' + x")(y' + y") - 2x"y".$$

Since the two quantities in parentheses can only be 0 or ±2, and the last term is ±2, S cannot have values ±3 or ±4 at the corners.   Q.E.D.

The lemma is applied to our physical problem by identifying x' with $E^1(k,a')$, y' with $E^2(k,b')$, x" with $E^1(k,a")$, and y" with $E^2(k,b")$. Since each of the outcomes s_m and t_n lies in $[-1,1]$, so also do these four expectation values, so that the conditions of the lemma are satisfied. The conclusion of the lemma is then also satisfied, and when Eq. (2) is combined with the conclusion, the result is

$$-2 \leq E(k,a',b') + E(k,a',b") + E(k,a",b') - E(k,a",b") \leq 2. \qquad (3)$$

Now integrate Inequality (3) over the space K, using the probability distribution w throughout as a weighting, and we obtain

$$-2 \leq E_w(a',b') + E_w(a',b") + E_w(a",b') - E_w(a",b") \leq 2, \qquad (4)$$

where we have used the normalization condition

$$\int_K dw = 1, \qquad (5)$$

and we have defined the ensemble expectation value $E_w(a,b)$ as

$$E_w(a,b) = \int_K E(k,a,b)dw. \qquad (6)$$

Inequality (4) is Bell's Inequality, or, more accurately, it is the version of Bell's Inequalities which emerges in the present exposition.[1]

It is noteworthy that except for the assumption that probability measures can be defined on K there are no assumptions about the structure of the space K of complete states and no characterization of the complete states k. Also, no assumptions have been made about the probability measure w over K, except that the same w is used in integrating each of the terms in Inequality (4). Physically this one assumption would not be justified if the choice of the parameters a and b affected the emission of particle pairs by the source. That the w governing the ensemble of particle pairs emitted by a source is independent of the parameters of the analyzers is an independence condition distinct from Parameter Independence and Outcome Independence, which were used above, but somewhat similar to Parameter Independence.

In order to complete the proof of Bell's Theorem it is essential to find a realization of the schema of Figure 1 in which the quantum mechanical predictions are in conflict with Inequality (4). One realization which is easy to analyze takes particles 1 and 2 to be photons propagating respectively in z and -z directions and prepared by the source in the polarization state

$$\emptyset = 2^{-\frac{1}{2}}[u_x(1)u_x(2) + u_y(1)u_y(2)], \qquad (7)$$

and takes the analyzers to be linear polarization filters placed perpendicular to the z-axis in the paths of photons 1 and 2 respectively. The parameter a is the angle from the x-axis to the transmission axis of the first polarization filter, and b is similarly defined for the second filter. In Eq. (7) $u_x(1)$ is a normalized vector representing quantum state of linear polarization along the x-axis for photon 1; and $u_y(1)$, $u_x(2)$, and $u_y(2)$ have analogous meanings. $\emptyset$ is a superposition of a state in which both 1 and 2 are polarized along the x-axis and another state in which both 1 and 2 are polarized along the y-axis. Obviously, $\emptyset$ is a quantum state in which neither photon 1 nor photon 2 has a definite polarization with respect to the x-y axes, and yet the results of polarization measurements with respect to these axes is strictly correlated, for if photon 1 passes through a filter with transmission axis along x, so also will photon 2; and if photon 1 fails to pass through such a filter, photon 2 will likewise fail.

The two outcomes of analysis of photon 1 are passage and non-passage through the polarization filter, and these outcomes will conventionally be assigned the numerical values 1 and -1 respectively (these are the s_m of Fig. 1). Likewise, passage and non-passage of photon 2 through its filter will be assigned 1 and -1 respectively (the t_n of Fig. 1). In order to calculate the quantum mechanical expectation value of the product of the outcomes, which will be the counterpart of the expectation value of Eq. (4), it is essential to find an appropriate self-adjoint operator S_a corresponding to analyzing photon 1 with a filter having a transmission axis at the angle a, and an analogous self-adjoint operator T_b corresponding to analyzing photon 2. S_a is determined by the requirements that it be linear on the two-dimensional space of polarization states of photon 1 and have eigenvalues 1 and -1 respectively for states of linear polarization along the directions specified by a and $a+\pi/2$ respectively:

$$S_a u_a = u_a, \tag{8}$$

$$S_a u_{a+\pi/2} = -u_{a+\pi/2}. \tag{9}$$

The states u_a and $u_{a+\pi/2}$ are obtained by rotating $u_x(1)$ and $u_y(1)$ by the angle a:

$$u_a = \cos a\, u_x(1) + \sin a\, u_y(1), \tag{10}$$

$$u_{a+\pi/2} = -\sin a\, u_x + \cos a\, u_y(1). \tag{11}$$

It is then straightforward to compute the effect of S_a on $u_x(1)$ and $u_y(1)$:

$$S_a u_x(1) = \cos 2a\, u_x(1) + \sin 2a\, u_y(1), \tag{12}$$

$$S_a u_y(1) = \sin 2a\, u_x(1) - \cos 2a\, u_y(1). \tag{13}$$

The operator T_b is constructed in the same way, and

$$T_b u_x(2) = \cos 2b\, u_x(2) + \sin 2b\, u_y(2), \tag{14}$$

$$T_b u_y(2) = \sin 2b\, u_x(2) - \sin 2b\, u_y(2). \tag{15}$$

The quantum mechanical counterpart of Eq. (6) is obtained by taking the expectation value of the operator product $S_a T_b$ in the quantum mechanical state $\emptyset$ of Eq. (7):

$$E_\emptyset(a,b) = \langle\emptyset|S_a T_b|\emptyset\rangle = \tfrac{1}{2}\langle u_x(1)u_x(2) + u_y(1)u_y(2)|$$

$$[\cos 2a\, u_x(1) + \sin 2a\, u_y(1)]\,[\cos 2b\, u_x(2) + \sin 2b\, u_y(2)] +$$

$$[\sin 2a\, u_x(1) - \cos 2a\, u_y(1)]\,[\sin 2b\, u_x(2) - \cos 2b\, u_y(2)]\rangle =$$

$$\cos 2(b - a). \tag{16}$$

If we now choose a', b', a", b" to be respectively $\pi/4$, $\pi/8$, 0, and $3\pi/8$, then

$$E_\emptyset(a',b') = E_\emptyset(a',b'') = E_\emptyset(a'',b') = -E_\emptyset(a'',b'') = 0.707, \tag{17}$$

and therefore

$$E_\emptyset(a',b') + E_\emptyset(a',b'') + E_\emptyset(a'',b') - E_\emptyset(a'',b'') = 2.828,$$

in disaccord with Inequality (4) (Bell's Inequality).   Q.E.D.

More than ten experimental tests of Bell's Inequality have been performed by examining the correlation of linear polarizations of photon pairs, as outlined in the preceding paragraph, and several other tests have also been carried out.[2] In all these experiments the analyzers are separated by distances of the order of a meter or more, so that no obvious mechanism would exist whereby Parameter Independence or Outcome Independence would be violated. But it is also highly desirable to exclude the possibility of a mechanism which is not obvious, and this exclusion can be achieved only if the events of analysis have space-like separation and hence cannot be directly connected causally according to Relativity Theory. Only the experiment of Aspect, Dalibard, and Roger (1982) has realized this desideratum. In their experiment the choice between the values a' and a" of the analyzer of photon 1, and between the values b' and b" of the analyzer of photon 2, is effected by acousto-optical devices which switch from one value to the other in 10 nanoseconds; whereas the switch-analyzer assembly for photon 1 is separated from that for photon 2 by about 13 meters, which can be traversed by a relativistically permitted signal in no shorter time interval than 40 nanoseconds. One would therefore antecedently expect both Parameter and Outcome Independence to hold, and moreover the distribution w over the space K of complete states to be independent of the parameters. Aspect et al. found, however, that their measured expectation values E(a,b) violated In eq. (4) by 5 standard deviations, but were in good agreement with the predictions of quantum mechanics. If one disregards certain loopholes (which will be discussed below), then this experiment constitutes a spectacular confirmation of quantum mechanics at a point where it seems to be endangered, as well as a spectacular demonstration that there is some nonlocality in the physical world.

Since Bell's Inequality is violated by the results of Aspect et al., and the Inequality follows from Parameter and Outcome Independence together with the independence of w from the parameter values, one of these three premises must be false, and it is important to locate the false one. The natural way to obtain this information is to examine the implications of quantum mechanics, which after all was brilliantly confirmed by Aspect et al., as well as by most of the other experiments inspired by Bell's Theorem.

Outcome Independence is violated by the quantum mechanical predictions based upon $\emptyset$ of Eq. (7). Suppose that the angles a and b of the two polarization filters are both taken to be 0, i.e., their transmission axes are both along the x direction. The conditional probability of photon 2 passing through its filter if photon 1 passes through its filter is 1, but it is 0 if photon 1 fails to pass through its filter. Since these two conditional probabilities are different from each other, it is impossible for both of them to equal the unconditioned probability that photon 2 will pass through its filter ·(which, in fact, is obviously ½). Thus Outcome Independence, as defined above, is violated. It should be noted that a violation of Outcome Independence is predicted on the basis of any quantum state which is "entangled" (in Schrödinger's locution), that is, not expressible as a product of a quantum state of particle 1 and a quantum state of particle 2. For any entangled state of a two-particle system can be written in the form[3]

$$\psi = \sum_i c_i u_i(1) v_i(2), \tag{18}$$

where the $u_i(1)$ are orthonormal, the $v_i(2)$ are orthonormal, the sum of the absolute squares of the expansion coefficients c_i is unity, and the sum contains at least two terms with non-zero coefficients. By constructing self-adjoint operators S and T of which the u_i and the v_i are eigenstates with distinct eigenvalues, one obtains a violation of Outcome Independence.

The quantum mechanical predictions do not violate Parameter Independence if the Hamiltonian of the composite system can be written in the form

$$H_{tot} = H_1 + H_2, \tag{19}$$

where H_1 is the Hamiltonian of particle 1 alone (in the environment to which it is exposed), and H_2 is the Hamiltonian of particle 2 alone (in the environment to which it is exposed), with no interaction Hamiltonian, and with no influence of particle 1 upon the environment of particle and conversely. If the composite system 1+2 is prepared at the initial time 0 in the state $\emptyset(0)$, then the state $\emptyset(t)$ at a later time t is determined by the Hamiltonian of Eq. (19) and also $\emptyset(0)$. It is straightforward to prove[4] that the expectation value of any self-adjoint operator S on the space of states of particle 1 is independent of H_2, and the expectation value of any self-adjoint operator T on the space of states of particle 2 is independent of H_1. Now the choice of a parameter a of the analyzer of particle 1 is effectively the choice of the Hamiltonian of particle 1, and likewise concerning the choice of parameter b. Outcome Independence follows. This general argument may be made more intuitive by considering the special case of a pair of photons with $\emptyset$ of Eq. (7) as the state at time 0. At time t photon 1 impinges upon a polarization filter with one of two orientations of its transmission axis: (i) a = 0, or (ii) a = $\pi/4$. In either case, the filter upon which photon 2 will impinges will be taken to have its transmission axis along the x direction, i.e., parameter b is 0. We calculate the probability that photon 2 will pass through the filter in each of the two cases.

(i) Photon 1 has probability $\frac{1}{2}$ of passing through the filter with a = 0, in view of Eq. (7), and if it does so the term $u_x(1)u_x(2)$ is picked out of the superposition, so that the conditional probability that photon 2 will pass through its filter is 1. Photon 1 also has probability $\frac{1}{2}$ of not passing, in which case the term $u_y(1)u_y(2)$ is picked out, and the conditional probability that photon 2 will pass its filter is 0. The net probability of passage of photon 2 is $\frac{1}{2} \cdot 1 + \frac{1}{2} \cdot 0 = \frac{1}{2}$.

(ii) It is useful to rewrite Eq. (7) in the equivalent form

$$\emptyset = 2^{-\frac{1}{2}}[u_{x'}(1)u_{x'}(2) + u_{y'}(1)u_{y'}(2)], \tag{7A}$$

where x' is the direction in the x-y plane making an angle $\pi/4$ to both the x and y directions, and y' is perpendicular to x' in the x-y plane. (The equivalence of Eq. (7A) to Eq. (7) follows from Eqs. (10) and (11) and their counterparts for photon 2.) There is probability $\frac{1}{2}$ that photon 1 will pass through its filter, picking out the term $u_{x'}(1)u_{x'}(2)$, in which case the conditional probability that photon 2 will pass through its filter is $\cos^2 \pi/4$. There is also probability $\frac{1}{2}$ that photon 1 will not pass through its filter, in which case the term $u_{y'}(1)u_{y'}(2)$ is picked out, and the conditional probability that photon 2 will pass through its filter is $\cos^2 3\pi/4$. The net probability of passage of photon 2 is $\frac{1}{2} \cdot \frac{1}{2} + \frac{1}{2} \cdot \frac{1}{2} = \frac{1}{2}$. The equality of the net probabilities of passage in cases (i) and (ii) illustrates Parameter Independence.

Since the standard quantum mechanical treatment of polarization correlation assigns the same quantum state to all photon pairs of the ensemble of interest (either the $\emptyset$ of Eq. (7) or an appropriate variant of it), there is no question of a quantum mechanical violation of the third premiss utilized in deriving Bell's Inequality (i.e., that the distribution over the complete states is independent of the parameters a and b).

It is very interesting now to consider the relation between violations of Parameter Independence and Outcome Independence and relativistic locality.

Suppose that a violation of Parameter Independence occurred in the situation schematized by Fig. 1 because for some k and m

$$p^1(m/k,a,b') \neq p^1(m/k,a,b'').$$

(20)

Then one binary unit of information can be transmitted from the location of the second analyzer to the location of the first analyzer by making the choice between b' and b'' at the fromer location, in the following way. Aspect shows the choice between b' and b'' can be made extremely quickly. We can also suppose (as a thought experiment) that a large number of pairs of particles 1+2 are prepared in a time which is short compared to the time needed to choose between b' and b'', and also that the complete state of each of these pairs is k. Then the difference in probability in Inequality (20) will with near certainty, by the law of large numbers, produce a clear difference between the statistics of occurrence of the value s_m conditional upon the two choices of the parameter b. Hence with near certainty an observer of the outcomes s_m can infer whether the choice made at the other analyzer was b' or b''. By the hypothetical arrangement, this binary unit of information is transmitted at superluminal speed between the two analyzers. Hence, in principle a violation of the relativistic upper limit upon the speed of a signal can be obtained by exploiting failure of Parameter Independence in a situation where the analyses of the two particles are events with space-like separation. If there is a violation of Outcome Independence, a binary unit of information can also be transmitted, but it is easy to see that the transmission is slower than the speed of light. Suppose that for some k and m

$$p^1(m/k,a,b,n') \neq p^1(m/k,a,b,n'').$$

(21)

Again prepare a large number of pairs 1+2 in the state k, and for each pair analyze particle 2 with the same parameter setting b. While the analysis is being performed, particle 1 is to be placed "on hold," e.g., by being kept in a circular light guide. An antecedent agreement is made that particle 1 will be released only if the result of analysis of particle 2 is $t_{n'}$ or $t_{n''}$ but not both, and that a uniform decision will be made for all the pairs 1+2. An observer of the statistics of s_m can then infer with near certainty whether the choice has been made to release particles 1 of which the partners are analyzed with result $t_{n'}$ or with result $t_{n''}$, since the difference in probability in Inequality (21) will, by the law of large numbers, produce a difference in the statistics with near certainty. The transmission of a binary unit of information in this way, however, will be subluminal, because the analysis of particle 2 must be completed, then the result of the analysis must be transmitted to the ring where particle 1 is "on hold," then particle 1 must be released, then it must propagate towards its analyzer, and finally it must be analyzed. Clearly, this complex process takes longer than a straight radar signal between the two analyzers.

In an experimental arrangement like that of Aspect et al. a violation of either Parameter Independence or of Outcome Independence produces some tension with the relativity Theory. But the violation of Parameter Independence seems to be the more serious of the two, because it entails the possibility in principle of superluminal signalling. The fact that quantum mechanics does not violate Parameter Independence but does violate Outcome Independence is most remarkable on two counts: it does show that quantum mechanical entanglement can be responsible for a kind of causal relation between two events with space-like separation, but also that quantum mechanics can "co-exist peacefully" with relativity theory because of the impossibility of exploiting entanglement for the purpose of superluminal communication. By using the locution "peaceful coexistence" I do not wish to convey the impression that there is nothing problematic in the state of affairs which has been exhibited. A deeper analysis is certainly desirable.

It is possible that a deepened understanding of space-time structure will be required in order to clarify quantum mechanical nonlocality. Or it may be that the concept of "event," which has been borrowed from pre-quantum physics, will have to be radically modified. However, I have not yet seen a promising development of either of these suggestions.

Because the implications of Bell's Theorem and of the experiments which it inspired are philosophically momentous, it is important to pay attention to the loopholes in the experimental reasoning.

The first loophole is due to the periodicity of the switches which Aspect et al. employed to choose between a' and a" and between b' and b". The switches are not randomly turned off and on, but rather operate periodically with a total period of 20 nanoseconds. Even though relativity theory does not permit a direct causal connection between contemporaneous settings of the switches (where "contemporaneous" must of course be understood relative to some definite frame of reference, such as that of the laboratory), the periodicity may enable clever demons located in one analyzer to _infer_ the contemporaneous setting of the other switch and to regulate the outcome of analysis of the particle accordingly. The attribution of such a process of inductive reasoning to the demons would not violate relativity theory. In order to block this loophole it would be necessary to operate the switches stochastically. It has been suggested by Clauser, for example, that each should be controlled by the arrival of starlight gathered by a telescope pointed to a distant galaxy. Blocking the periodicity loophole seems to be experimentally feasible in principle, but it would greatly complicate an experiment that is already difficult and delicate. It remains to be seen whether any experimenter is sufficiently motivated to make the great effort that would be required. See also Zeilinger (1986).

The second loophole is due to the fact that actual particle detectors are not 100% efficient. In the foregoing discussion of Fig. 1 it was tacitly assumed that that if the outcome of analyzing particle 1 is s_m (e.g., the particle passes into channel m), then this fact can be known with certainty because the particle detectors are ideally efficient; and likewise concerning the analysis of particle 2. In the polarization correlation tests of Bell's Inequality the photodetectors were less than 20% efficient, and therefore fewer than 4% of the photon pairs that jointly pass through their respective filters are actually detected. It is not inconceivable that the passage rates satisfy Bell's Inequality, but that the counting rates agree with the predictions of quantum mechanics (in disaccord with the Inequality) because of peculiarities in the way that the complete states k determine the probability of detection. There are, in fact, several models[5] which preserve Parameter Independence and Outcome Independence and nevertheless yield counting rates in agreement with the predictions of quantum mechanics. Inefficiency of the particle detectors is crucial for these models.

The following argument shows that the detection loophole can be blocked in a polarization correlation experiment if technology improves and photodetectors of efficiency greater than 0.841 are constructed. The foregoing realization of the schema of Fig. 1 can be modified by taking the analyzers not to be polarization filters, which allow photons polarized along the transmission axis to pass but absorb those polarized in the perpendicular direction, but rather Wollaston prisms, which allow the first set of photons to emerge in one ray (the "ordinary ray") and the second to emerge in another (the "extraordinary ray"). Then three outcomes of analysis of photon 1 can be distinguished: detection in the ordinary ray, detection in the extraordinary ray, and non-detection; and the values of s_m assigned to these three outcomes can be conventionally taken to be 1, −1, and 0. The outcomes t_n of analysis of photon 2 will likewise have three values 1, −1, and 0, with analogous interpretations. If Parameter and Outcome

Independence are satisfied, and the distribution w is independent of the parameters, then Bell's Inequality (Inequality (4)) follows. The expression for the quantum mechanical expectation of the product of s_m and t_n is equal to the expectation value of the operator product $S_a T_b$, which was given by Eq. (16), multiplied by the probability of joint detection of a pair of photons emerging from the respective Wollaston prisms. If, for simplicity, we assume that the four photodetectors intercepting the ordinary and extraordinary rays from the two Wollaston prisms have the same efficiency η, then the probability of joint detection is η^2. Hence the expectation value of present interest is

$$E_\emptyset^{det}(a,b) = \eta^2 \cos 2(b-a).\tag{22}$$

For the choice of angles made before Eq. (17) we have

$$E_\emptyset^{det}(a',b') + E_\emptyset^{det}(a',b'') + E_\emptyset^{det}(a'',b') - E_\emptyset^{det}(a'',b'') = 2.828\,\eta^2.\tag{23}$$

Disaccord with Bell's Inequality results provided that

$$\eta > 0.841.\tag{24}$$

Mermin and Schwarz (1982) and Garg and Mermin (1987) have shown that the detection loophole can be blocked if a less stringent constraint is placed upon the efficiency of the photodetectors, namely,

$$\eta > 0.828,\tag{25}$$

but their arguments are more complex than the simple one just given and depend upon some additional (but empirically testable) symmetry assumptions.

Finally, I wish to point out that even though I have used polarization correlation to discuss Bell's Inequality, there is nothing about the Inequality that is intrinsically restricted to polarization experiments. That should be obvious, in fact, from the generality of Fig. 1 and of the proof given of Bell's Inequality. In my lecture on two-particle interferometry I shall show how an experiment performed on a pair of photons with entangled momentum states can test the Inequality.

FOOTNOTES

[1] This version of Bell's Inequality was first derived by Clauser, Horne, Shimony, and Holt (1969) in the special case where $p^1(m/k,a,b)$ and $p^2(n/k,a,b)$ are allowed to have only the values 1 and 0 (so called "deterministic" hidden variables theories). A derivation without this restriction was first given by Bell (1971) and in another way by Clauser and Horne (1974). The procedure in my lecture, making use of the simple mathematical lemma, was inspired by Clauser and Horne, although their lemma was different. Alain Aspect pointed out to me after the lecture that a proof exactly like mine, with the same lemma, is in the unpublished part of his doctoral thesis (1983).

[2] Summaries of experiments up to 1978 are given by Clauser and Shimony (1978) and later ones by Redhead (1987), pp. 107ff. Two important recent tests of Bell's Inequality, using photon pairs produced by parametric down-conversion are Ou and Mandel (1988) and Shih and Alley (1988).

[3] See, for example, von Neumann (1955), pp. 431-4.

[4] Eberhard (1977); Ghirardi, Rimini, and Weber (1980); and Page (1982).

[5]Clauser and Horne (1974), and Marshall, Santos, and Selleri (1983). The earlier model of Pearle (1970) achieves agreement with quantum mechanics only if the probability that a pair will be detected once it has passed through the pair of filters has a rather special form g(b−a), and the constant function (specifically with value η^2, as discussed above) is not of his required form; hence his model does not achieve all that those of Clauser and Horne and of Marshall et al. have established.

REFERENCES

Aspect, A., 1983, Thèse, Université de Paris-Sud (unpublished).
Aspect, A., Dalibard, J., and Roger, G., 1982, Physical Review Letters, 49: 1804.
Bell, J.S., 1964, Physics, 1: 195. Reprinted in Bell (1987).
Bell, J.S., 1971, in "Foundations of Quantum Mechanics," B. d'Espagnat, ed., Academic Press, New York, 171. Reprinted in Bell (1987).
Bell, J.S., 1987, "Speakable and Unspeakable in Quantum Mechanics," Cambridge University Press, Cambridge, England.
Clauser, J.F., Horne, M.A., Shimony, A., and Holt, R.A., 1969, Physical Review Letters, 26: 880.
Clauser, J.F. and Horne, M.A., 1974, Physical Review, D10: 526.
Clauser, J.F. and Shimony, A., 1978, Reports on Progress in Physics, 41: 1881.
Eberhard, P., 1977, Nuovo Cimento, 38B; 75.
Garg, A. and Mermin, N.D., 1987, Physical Review, D35: 3831.
Ghirardi, G.C., Rimini, A., And Weber,T., 1980, Lettere al Nuovo Cimento, 27: 293.
Jarrett, J., 1984, Nous, 18: 569.
Marshall, T.W., Santos, E., and Selleri, F., 1983, Physics Letters, 98A; 5.
Mermin, N.D. and Schwarz, G.M., 1982, Foundations of Physics, 12: 101.
Neumann, J. v., 1955, "Mathematical Foundations of Quantum Mechanics," Princeton University Press, Princeton, N.J..
Ou, Z.Y. and Mandel, L., 1988, Physical Review Letters, 61: 50.
Page, D., 1982, Physics Letters, 91A: 57.
Pearle, P., 1970, Physical Review, D2: 1418.
Redhead, M., 1987, "Incompleteness, Nonlocality, and Realism," Clarendon Press, Oxford, England.
Shih, Y.H. and Alley, C.O., 1988, Physical Review Letters, 61: 2921.
Zeilinger, A., 1986, Physics Letters A , 118: 1.

WAVE-PARTICLE DUALITY : A CASE STUDY

Alain Aspect

Collège de France et Laboratoire de Spectroscopie Hertzienne
de l'Ecole Normale Supérieure (associé au CNRS et à
l'Université Paris VI), 24 rue Lhomond, 75005 Paris

Philippe Grangier

Institut d'Optique, Bât. 503, Université Paris XI, 91405 Orsay

1. INTRODUCTION - WAVE PARTICLE DUALITY IN TEXT-BOOKS

1.1 Gedanken experiment

Many introductory courses in Quantum Mechanics -whether or not they choose an historical perspective- begin with an "experiment" exhibiting the wave-particle duality of the behaviour of matter [1]. This experiment is usually presented as in Fig. 1.a and Fig. 1.b.

The first setup (Fig. 1.a) shows that the rate of detection $N(x)$ is modulated according to a sine law, i.e. it exhibits an interference pattern. Such a phenomenon can be interpreted by invoking a wave that passes through both holes : it is well known that the resulting intensity then depends on the "path difference"

$$\Delta = [S \, T_1 \, D] - [S \, T_2 \, D] \tag{1}$$

and leads to a modulation depending on the interference order

$$p = \frac{\Delta}{\lambda} \tag{2}$$

where λ is the wavelength of the considered wave. The "particles" emitted by the source thus have a wave-like behaviour.

The second "experiment" is not always explicitly presented [2]. Its purpose is to prove that the source S really emits particles (if it was not the case, the discussion would be pointless). The particle-like behaviour is evidenced by the absence of coincidences, although the detectors D_1 and D_2 are fired at the same rate. The natural image for this behaviour is that of a particle that passes either through hole T_1 or through T_2, but not through both holes.

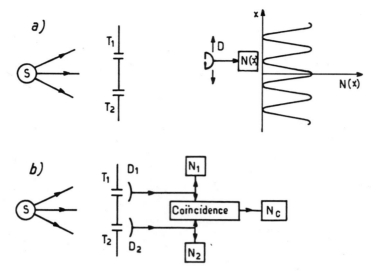

Fig. 1 . *Ideal wave — particle duality experiment. A source s emits indepent particles, one at a time. The particles fall on a screen with two holes T_1 and T_2. In the experiment (a), the movable detector D measures the detection rate as a function of its position and reveals the existence of an interference pattern. In experiment (b), the two detectors D_1 and D_2, just after the holes T_1 and T_2, feed singles an coincidence counters : no coincidence is detected.*

The amazing thing is of course that the source and the screen with the holes are the same for both experiments. We thus have the same "objects" passing through the screens. But in Fig. 1.b, we describe this object as a particle passing in only one hole, while in Fig. 1.a the object is described as a wave split between the two holes. This is the essence of wave-particle duality, on which we will comment later.

Now, a question arises : are the experiments of Fig. 1.a and Fig. 1.b only gedanken experiment, or are there real experiments corresponding to these setups ?

1.2. Experiments with massive particle

It is well known that interferences have been already observed with electrons [3] and with neutrons [4]. We can thus conclude that experiment of Fig. 1.a has been realized with objects (electrons, or neutrons) that we definitely consider as particles [5]. However, to our knowledge, nobody has tried an experiment such as the one of Fig. 1.b. Moreover, for most of the experiments that we know about, it is likely that the experiment of Fig. 1.b would not have had conclusive results, since the sources (electron gun, or neutron reactor) deliver a flux of particles with a certain probability of two particles being present simultaneously, and thus a non zero rate of coincidences is expected [6].

1.3. Experiments with light

The concept of wave-particle duality first emerged about light [7], and it is natural to look for such experiments in the domain of optics. The wave-like behaviour of light, even with extremely feeble sources, has been evidenced as early as 1909, and it has been confirmed repeatedly (Table I).

Table 1

Feeble light interference experiments. All these experiments have been realized with attenuated light from a usual source (atomic discharge).

Author	Date	Experiment	Detector	Photon flux (s^{-1})	Interferences
Taylor (a)	1909	Diffraction	Photography	10^6	Yes
Dempster et al. (b)	1927	(i) Grating	Photography	10^2	Yes
		(ii) Fabry Pérot	Photography	10^5	Yes
Janossy et al. (c)	1957	Michelson interferometer	Photomultiplier	10^5	Yes
Griffiths (d)	1963	Young slits	Image interferometer	2×10^3	Yes
Scarl et al. (e)	1968	Young slits	Photomultiplier	2×10^4	Yes
Donstov et al. (f)	1967	Fabry Pérot	Image intensifier	10^3	No
Reynolds et al. (g)	1969	Fabry Pérot	Image intensifier	10^2	Yes
Bozec et al. (h)	1969	Fabry Pérot	Photography	10^2	Yes
Grishaev et al. (i)	1969	Jamin interferometer	Image intensifier	10^3	Yes

(a) G.I. Taylor, Proc. Cambridge Philos. Soc. 15 (1909) 114.
(b) A.J. Dempster and H.F. Batho, Phys. Rev. 30 (1927) 644.
(c) L. Janossy and Z. Naray, Acta Phys. Hungaria 7 (1967) 403.
(d) H.M. Griffiths, Princeton University Senior Thesis (1963).
(e) G.T. Reynolds et al., Advances in electronics and electron physics 28 B (Academic Press, London, 1969).
(f) Y.P. Dontsov and A.I. Baz, Sov. Phys. JETP 25 (1967) 1.
(g) G.T. Reynolds, K. Spartalian and D.B. Scarl, Nuovo Cim. B 61 (1969) 355.
(h) P. Bozec, M. Cagnet and G. Roger, C.R. Acad. Sci. 269 (1969) 883.
(i) A. Grishaev et al., Sov. Phys. JETP 32 (1969) 16.

All these experiments consisted in the observation of interferences, or diffraction, with strongly attenuated light emitted by a usual source (thermal source, discharge lamp, laser). We can thus conclude that experiments of the type of Fig. 1.a have been realized with light.

When it comes to the particle-like behaviour of the light, one usually reads that the features of the photoelectric effect are a clear evidence of the existence of light quanta, as was first argued by Einstein [7]. In fact, a second thought to this question reveals that there is another possible interpretation for the photoelectric effect, in which light is not quantized [8]. In this interpretation, the light is taken as a classical electromagnetic wave, but the detector is quantized. More precisely, the detector is an atom with a stable ground state and a continuum of excited ionized states, separated from the ground state by a gap W_T (Fig. 2). The atom-light interaction is described by the hamiltonian

$$\hat{H}_I = \mathcal{E} . \hat{D} \tag{3.a}$$

with

$$\mathcal{E} = \mathcal{E}_0 \cos \omega t \tag{3.b}$$

the electric field of the wave, and $\hat{D}$ the electric dipole operator of the detector. All the well known features of the photoelectric effect (existence of a threshold, kinetic energy of the electron equal to $\hbar\omega - W_T, \dots$) are easily derived from this model. In this point of view, they are thus related to the quantization of the detector. This discussion shows that the existence of the photoelectric effect is certainly not sufficient to prove the particle like character of the light [9].

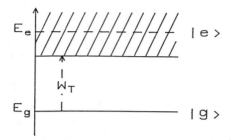

Fig. 2 . Model of detector for the photoelectric effect. The model considers an atom with a ground state and a continuum of excited ionized state. The interaction of the detector with a classical electromagnetic field (no photons involved) leads to all the known features of the photoelectric effect.

If one is really committed to demonstrating the particle-like character of the light, an experiment such as the one of Fig. 1.b is thus required. But there is no experiment of this type corresponding to the wave experiments of Table 1. We can thus conclude that the wave-like behaviour of light has been clearly evidenced (even with extremely attenuated light), but that the particle-like behaviour is far from having been so unquestionably demonstrated.

2. PARTICLE-LIKE BEHAVIOUR OF LIGHT : POSSIBILITY OF AN EXPERIMENT

2.1 Anticorrelation on a beam-splitter

The arrangement of Fig. 3, which is a straight forward modification of the scheme of Fig. 1.b, would be ideal to evidence a particle-like behaviour.

If light is really made of quanta, a single quantum should either be transmitted or be reflected by the beam splitter, but it should not be split. As a consequence, the coincidence counter should never register any joint detection. On the opposite, for a semi-classical model that describes the light as a classical wave, this wave is split on the beam splitter, and there is a non-zero probability of joint detection on both sides of the beam-splitter. The observation of an anticorrelation (zero coincidence) would thus be a convincing demonstration of the particle-like behaviour.

What do we expect if we send light on such an apparatus ? In fact, it all depends of the kind of light which is used.

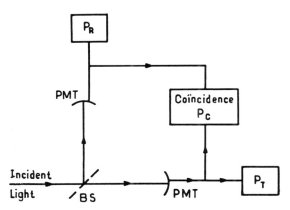

Fig. 3 . *Experiment for evidencing a particle — like behaviour.*
The light is detected on both sides of the beamsplitter by two
photomultiplier tubes : P_R *and* P_T *are probabilities of single*
photodetection ; P_C *is the probability of coincidence. This*
scheme is similar to Fig. 1.b.

2.2 One photon state versus quasi-classical state of the light

In order to make predictions on the issue of the experiment of Fig. 3, we resort to the Quantum Theory of Light [10] [11]. The result crucially depends on the quantum state of the light. For a one-photon state $|\psi\rangle = |n = 1\rangle$ (that is to say an eigenstate of the operator "number of photons", with the eigenvalue equal to 1) a complete anticorrelation is predicted. As a matter of fact, a single photon can only be detected once, and the probability of a joint detection is rigourously zero. One photon states would thus entail a particle-like behaviour.

But usual sources do not emit one-photon states. For instance, a pulsed laser emits a quasi-classical state (also called a "coherent state") [10] :

$$|\psi\rangle = |\alpha\rangle$$

For such a state, the probability to have n photodetections, $\mathcal{P}(n)$, is a Poisson distribution

$$\mathcal{P}(n) = \frac{|\alpha|^{2n}}{n!} e^{-|\alpha|^2} \tag{4}$$

This formula allows to interpret $|\alpha|^2$ as the average number of photons.

$$<n> = |\alpha|^2 \tag{5}$$

For such a state, it is clear that the probability of a double detection, $\mathcal{P}(2)$, is different from zero, so that coincidences will be observed and there is no particle-like behaviour. It is remarkable that this property remains true even for very feeble light, where the average photon number $|\alpha|^2$ is smaller than 1.

Now, all the usual sources (thermal lamps, discharge lamps, lasers) emit a mixture of quasi-classical states, even when they are strongly attenuated. There is thus no possibility to observe a particle-like behaviour with usual sources. It is interesting to notice that all the sources used for the experiments quoted in Table I are of this type. Had the authors of these experiments tried the experiment of Fig. 3, they would not have observed a particle-like behaviour.

In order to observe a particle-like behaviour, we thus need a special source, producing one-photon states. But another question then arises : how will we evidence the particle-like behaviour ? We need to be able to define the meaning of a zero coincidence rate (in experimental Physics, there is no absolute zero ; one has to define a threshold under which a quantity is taken null). In other words, we need a criterion to discriminate between a particle-like behaviour and a behaviour compatible with the semi-classical description.

2.3 Particle-like behaviour inequality

When contemplating the possibility of an experiment (even with ideal apparatus), it is clear that Fig. 3 is not definite enough. The notion of a coincidence is meaningfull only

50

if we define a gate : there will be a coincidence only if two detections happen during the same gate. Since on the other hand the experiment must be repeated in order to be able to define probabilities (of single or joint detections) we are led to the scheme of Fig. 4, based on a source emitting light pulses well separated in time. Each time a light pulse is emitted,

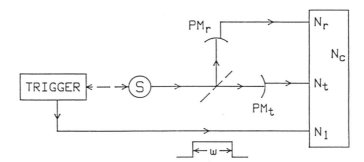

Fig. 4 . *Scheme of a more realistic experiment to evidence a particle — like behaviour of the light. The source s emits light pulses. The detections are allowed only during gates synchronized with the light pulses.*

a gate generator produces an electronic gate of duration w, which enables the counters to monitor a detection during the gate ; a coincidence is registered if both detectors are fired during the same gate.

Data are accumulated for a large number of light pulses. At the end of a run, the process has been repeated N_1 times (number of gates). One has monitored N_T counts on the transmitted arm, N_R counts on the reflected arm, and N_C coincidences. The relevant probabilities are immediately derived from these measurable numbers

$$P_R = \frac{N_R}{N_1} \qquad P_T = \frac{N_T}{N_1} \tag{6.a}$$

$$P_C = \frac{N_C}{N_1} \tag{6.b}$$

A semi-classical description of such an experiment predicts some coincidences, since the wave packet is split in two parts on the beam splitter. Both detectors may thus be fired simultaneously, and it is easy to show [12] that the probability of a joint detection obeys the inequality

$$P_C \geq P_R.P_T \tag{7}$$

In order to demonstrate this inequality, it is enough to make the following assumptions :
   - a wave packet is split on the beam splitter ;
   - the probability of a photodetection on a detector depends only on the intensity of the light impinging onto this detector : it is proportional to this intensity ;
   - the intensity of the light is a positive quantity, so that some Cauchy-Schwarz inequalities can be derived.

From (7) and (6), one can then derive an inequality for the numbers of counts [12]

$$\beta = \frac{N_1.N_C}{N_R.N_T} \geq 1 \tag{8}$$

This inequality must be satisfied if the experiment under consideration can be described by a semi-classical theory of the light. On the other hand, for a one photon state, we expect (if the Quantum Theory of the light is correct) no coincidence, that is to say a clear violation of this inequality. We thus have a way to characterize the particle-like behaviour of the light, namely the violation of inequality (8).

It is interesting to calculate the prediction of the Quantum Theory of light in the case of a quasi-classical state. Taking into account the Poisson distribution (eq. (4)), it is easy to show that in this case the inequality (7) (or (8)) is never violated. For a very attenuated pulse, with an average photon number smaller than 1, one finds the marginal value, i.e.

$$P_C = P_R.P_T \quad \text{or equivalently} \quad \beta = 1 \tag{9}$$

(very weak quasi-classical pulse).

The prediction (9) supports our claim that all the experiments of Table I, realized with very attenuated usual sources, would not have shown a particle-like behaviour. In order to observe such a behaviour, it has been necessary to use a special source emitting light in a state close to a one-photon state.

3. EXPERIMENTAL OBSERVATION OF A PARTICLE-LIKE BEHAVIOUR [12]

3.1. Source of one photon pulses

Suppose that an atom is brought at time t_0 into an excited resonance level, decaying to the ground state with a life time τ. Because of energy conservation, only one photon is emitted, and we get a one-photon pulse starting at time t_0 and decaying with a time constant τ. In a discharge lamp, a collection of atoms are excited at random times : when taking the average over t_0, one finds [13] that the light is now described by a density matrix which corresponds to a mixture of quasi-classical states. The one-photon character has disappeared. This is a usual source.

In order to keep the one-photon character, it is thus necessary to know the excitation time t_0 for each pulse, and to avoid the overlap of several pulses. The second condition is clearly achieved by attenuating the source, but the first one, which is essential, was overlooked in previous discussions.

We have been able to meet these requirements by use of a source designed to test Bell's inequalities [14]. This source is based on calcium atoms in a moderate density atomic beam [15], excited to the upper level of a two photon cascade (Fig. 5).

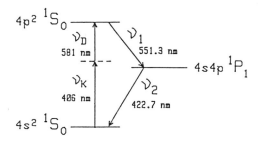

Fig. 5 . Radiative cascade in Calcium, used to produce the one − photon pulses. The atom is excited to its upper level by a two photon excitation with two lasers. It then reemits photons ν_1 and ν_2.

A first photon ν_1 is then emitted, and the atom in brought into the resonant excited level $|r\rangle$ at the time t_0 of emission of ν_1 : a one-photon pulse, corresponding to the photon ν_2, will then be emitted. This pulse decays with a time constant τ ; its starting time t_0 can be known by detection of ν_1.

3.2. Experiment with one-photon pulses

In fact, we have exactly followed the scheme of Fig. 4, the detection of ν_1 allowing to act the gate generator, and ν_2 being the one-photon light pulse. The gate duration w is taken equal to $w = 2\,\tau\;(\tau = 4.7\;ns)$ in order to have an almost complete overlap of the gate with the corresponding light pulse. If the excitation rate $\mathcal{N}_e$ of the cascades is kept much smaller than w^{-1}, we are in an almost ideal situation to realize the experiment of Fig 4. If $\mathcal{N}_e w$ is not very small compared to 1, there is some chance that two pulses are emitted during the same gate, and the state is no longer a pure one-photon state : the probability of having two photons is not exactly zero, and some joint detections are expected. However, for $\mathcal{N}_e w$ small enough, a clear violation of the inequality (7) is still predicted [11].

The results of the experiment are summarized in Fig. 6, which presents a plot of the quantity β, defined in eq. (8), as a function of the reduced excitation rate $\mathcal{N}_e w$. As expected, when $\mathcal{N}_e w$ is small enough, there is a very strong violation of the inequalities (7) and (8), that is to say that we observe a clear particle-like behaviour.

3.3. Experiment with a usual source

In order to support the discussion of § 2 on usual sources, and also to check the quite sophisticated detection system used in the experiment above, we have replaced the one photon source by a pulsed light emitting diode. The driving electric pulse was shaped in order to get a light pulse decaying in about 6 ns, and the gate generator of Fig. 4 was triggered by this electric pulse. We could then keep all the detection system (including the various delays, and gate duration) as it was in the above experiment. By adjustement of the pulse generator frequency, and by attenuation of the light pulse with a neutral density, we could achieve singles rates of the same order as in the above experiment. The coincidence rates have then been found much higher than in the experiment with one-photon states.

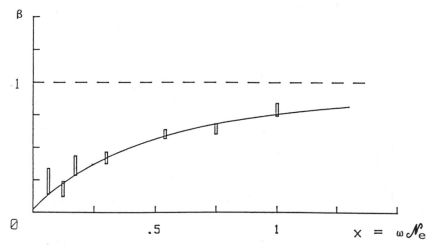

Fig. 6   *Correlation parameter β as a function of the reduced excitation rate $\mathcal{N}_e w$. A value of β smaller than 1 (anticorrelation) is the evidence of a particle − like behaviour. The solid line is the quantum theory prediction taking into account the possibility of two pulses being excited during the same gate. For a pure one − photon state, β would be exactly zero.*

More precisely, the coefficient β has always been found equal to 1, within one standard deviation.

We thus have an experimental confirmation that the light emitted by a usual source behaves according to the semi-classical description. In order to fully appreciate the meaning of this experimental result, it is interesting to note that in this experiment the average energy per pulse impinging on the photodetector is about 10^{-2} photon per pulse ! So, even when very attenuated, the light emitted by a usual source doesn't have a particle-like behaviour.

4. INTERFERENCES WITH A SINGLE PHOTON

We return now to the source delivering one-photon pulses, that have shown a particle-like behaviour (§ 3.2). It is then tempting to check whether these light pulses may also behave like a wave, i.e. allow to observe interferences. The quantum theory of light predicts indeed that interferences will happen, even with one photon pulses.

We have thus kept the same source and the same beam splitter as in Fig. 4, but the detectors on both sides of the beam splitter have been removed, and the two beams are recombined on a second beam splitter (Fig. 7).

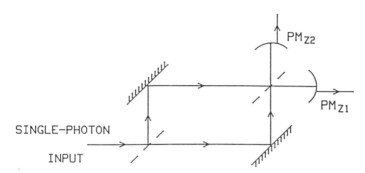

Fig. 7   Single photon interference experiment. The source and the beam − splitter are like in Fig. 4, but we have now a Mach−Zehnder interferometer. The detectors are gated as in Fig. 4, synchroneously with the light pulses.

We now have a Mach-Zehnder interferometer : the detection rates in the two outputs (1) and (2) are expected to be modulated as a function of the path difference in both arms of the interferometer. To guarantee that we are still working with one-photon pulses, the detectors $PM1$ and $PM2$ are gated synchroneously with the pulses, as they were in the experiment of § 3.2.

The interferometer has been carefully designed and built to give high visibility fringes with the large étendue beam produced by our source (about 0.5 mm^2 rad^2). The reflecting mirrors and the beam splitters are $\lambda/50$ flat on a 40 mm diameter aperture. A mechanical system driven by Piezzoelectric transducers permits to displace the mirrors while keeping their orientation exactly constant : this allows to control the path difference of the interferometer. Preliminary checks with usual light have then shown a strong modulation of the counting rates of PM_{Z1} and PM_{Z2} when the path difference is modified. For a source shaped as the one photon pulses source, the measured visibility is

$$V = 98.7 \% \pm 0.5 \% \tag{10}$$

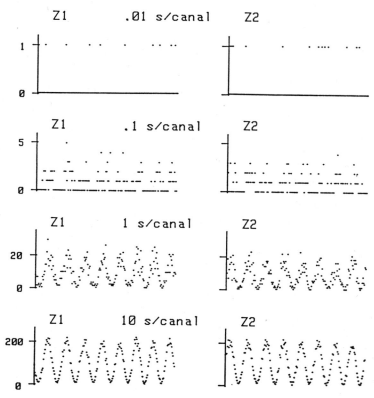

Fig. 8   *Number of detected counts in outputs* (1) *and* (2) *as a function of the path difference. The four sets of curves correspond to different counting times at each path difference. This experiment has been realized in a one — photon pulse regime* (β = 0.2). *Note that interferograms in outputs* (1) *and* (2) *are complementary.*

which is very close to the ideal value $V = 1$. We have then run this interferometer with the one-photon source.

Fig. 8 presents the results of such an experiment. The number of counts during a given time interval are measured as a function of the path difference. In the first curves, the counting time at each position was 0.01 s, while it was 10 s for the last recordings. This run was performed with the sources at a regime corresponding to an anticorrelation parameter $\beta = 0.2$ that is to say in the one photon regime. These recordings clearly show the interference fringes building up "one-photon at a time".

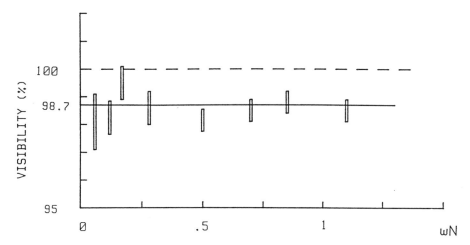

Fig. 9   *Observed visibility of the fringes as a function of the regime of the source. This visibility keeps close to 1 even in the almost pure one − photon regime.*

When data have been accumulated long enough, the signal to noise ratio is high enough to allow a measurement of the visibility of the fringes. We have repeated such measurements for various regimes of the source, corresponding to the different values of β shown on Fig. 6. The results, presented on Fig. 9, show that the visibility of the fringes keeps close to 1 -within the experimental uncertainties-even in a regime where the source emits almost pure one-photon pulses. As predicted by the quantum theory of light, single photon pulses do interfere. To our knowledge, this is the first experiment of this kind performed with light.

5. WAVE-PARTICLE DUALITY

We now rephrase the usual discussions on wave-particle duality, but we can do it about a real experiment instead of a gedanken experiment.

We first performed the experiment of Fig. 4, showing a clear anticorrelation on both sides of the beam splitter, and we claimed that it corresponds to a particle-like behaviour.

Indeed, if we want to visualize what happens in this experiment, the only possible image is that "something" is either reflected, or transmitted, on the beam-splitter, but it is not split : this corresponds to the behaviour of a classical particle.

The experiment of Fig. 7 showed interferences, and we claimed that it corresponds to a wave-like behaviour. Indeed, the detection rate in any output depends on the path difference between both arms. The only image [16] that we can find is that "something" is split on the first beam splitter, and recombined on the second one : this corresponds to the behaviour of a classical wave.

Now it has to be stressed that both experiments have been performed with the same source and the same beam spliter. In both experiments, we have the same light pulses impinging on the beam-splitter. But the images that we have to use for these light pulses are not compatible, and this is why we are led to say that there is *"duality"* : the light pulses are *wave and particle*.

Of course, this is just semantics, and giving a name to this incompatibility doesn't solve any problem. At this stage, we can however make a remark in the spirit of Bohr's complementarity : for a given experimental arrangement, only one behaviour will appear. For instance, with the device of Fig. 4, the light pulses assume a particle-like behaviour ; but with the apparatus of Fig. 7, they assume a wave-like behaviour. And the crucial point is that both apparatus are mutually incompatible : it is impossible to design an experiment in which one could test simultaneously the wave-like and the particle-like character of the light pulses. The selected behaviour depends on the selected apparatus.

This last statement may appear to be an issue to the problem. However, when taken litterally, new questions arise immediately : when does the system (the light pulse) "make the choice" to behave like a wave or like a particle ? Is it on the first beam splitter ? If it is so, this choice must depend on the kind of arrangement that will be encountered later on the path, and that might be changed afterwards ("delayed choice" experiment [17]). With such a description, we are immediately facing the problem of non-locality, and the image is certainly not very comfortable.

We do not know any image built from classical physics (and/or from our everyday experience) that can logically answer the question of a consistent description of the light pulses in these two experiments. On the other hand, we know that the quantum formalism describing the light pulses in the same for both experiments ; the mathematics are not ambiguous.

The problem then only arises when we ask the question of what image we must choose to describe the light (wave, or particle ?). Is it "a foolish question" [18] ? We do not think so, because our experience of Physics is that images are useful for imagining new situations. But we have to be extremely careful with images in quantum mechanics !

REFERENCES

[1]   R.P. Feynman, "Lectures on Physics, Quantum Mechanics", Addison-Weseley.
[2]   In that case, the fact that the experiment is dealing with particles is taken as evident, which is quite reasonable in some cases (with electrons or neutrons for instance). We however think that the discussion is more convincing when an experiment like (1.b) is presented (at the microscopic level, we do not see the particle, so how do we know that it is a particle ?).

[3]   See for instance O. Donati, G.F. Missiroli, and G. Pozzi, Am. J. Phys. $\underline{41}$, 639 (1973).

[4]   A.G. Klein, and S.A. Werner, Rep. Prog. Phys. $\underline{46}$, 259 (1983).

[5]   Wave effects (diffraction) have already been observed with atoms. However, no simple two-waves interference has yet been observed.

[6]   More precisely, it does not seem that these sources would have exhibited a particle-like behaviour in the sense of § 2.3, since the law of probability is close to a Poisson distribution.

[7]   A. Einstein, Annalen der Physik $\underline{17}$, 132 (1905).

[8]   W.E. Lamb, and M.O. Scully, "Polarisation, Matière et Rayonnement", volume in honour of Alfred Kastler, Presses Universitaires de France, Paris (1969).

[9]   We do not mean to imply that Einstein's interpretation of the photoelectric effect is not a good one. It is indeed a very clear and convincing interpretation. But we want to insist that other interpretations exist in which there is no need to consider the light as made out of quanta.

[10] R. Loudon, "The Quantum Theory of Light", Clarendon Press, Oxford (1983). C. Cohen-Tannoudji, J. Dupont-Roc, G. Grynberg, "Photons and Atoms", Wiley (1989).

[11] More detailed demonstrations of statements admitted in the present paper can be found in A. Aspect, P. Grangier, and G. Roger, J. Optics (Paris) $\underline{20}$, 119 (1989).

[12] P. Grangier, G. Roger, and A. Aspect, Europhys. Lett. $\underline{1}$, 173 (1986).

[13] P. Grangier, Thèse d'Etat, Paris (1986).

[14] A. Aspect, P. Grangier, and G. Roger, Phys. Rev. Lett. $\underline{47}$, 460 (1981) ; Phys. Rev. Lett. $\underline{49}$, 91 (1982) ; A. Aspect, J. Dalibard, and G. Roger, Phys. Rev. Lett. $\underline{49}$, 1804 (1982).

[15] A. Aspect, and P. Grangier, Lett. Nuovo Cimento $\underline{43}$, 345 (1985).

[16] For the discussion of the role of images, see A. Miller, in this volume.

[17] J.A. Wheeler, in "Quantum Theory and Measurement", Princeton University Press, Princeton (1989). T. Hellmuth, H. Walther, A. Zajonc, and W. Schleich, Phys. Rev. $\underline{A35}$, 2532 (1989).

[18] H. Feshbach, and V.F. Weisskopf, Physics Today $\underline{41}$, n° 10, 9 (Oct. 89).

"NICHT SEIN KANN WAS NICHT SEIN DARF," OR THE PREHISTORY OF EPR, 1909-1935:

EINSTEIN'S EARLY WORRIES ABOUT THE QUANTUM MECHANICS OF COMPOSITE SYSTEMS*

Don Howard

Department of Philosophy
University of Kentucky
Lexington, Kentucky

1. INTRODUCTION

The story of Einstein's misgivings about quantum mechanics and about his debate with Bohr has been told many times--by the participants themselves,[1] by their colleagues and contemporaries,[2] and by historians and philosophers of science of later generations.[3] So the question arises: Why tell the story yet again? The answer is that there is more to be said. I will argue that the standard histories have overlooked what was from early on the principal reason for Einstein's reservations about quantum mechanics, namely, the non-separability of the quantum mechanical account of interactions, something ultimately unacceptable to Einstein because it could not be reconciled with the field-theoretic manner of describing interactions.[4] Showing the significance of this issue for Einstein is important not only for the sake of setting right the historical record, but also because it makes Einstein's critique of quantum mechanics far more interesting--from the point of view of the physics involved--than if we see it resting merely on a stubborn old man's nostalgic attachment to classical determinism.

---

*The quote used in the title is taken from a letter of Wolfgang Pauli to Werner Heisenberg, 15 June 1935 (Pauli 1985, p. 402), in which Pauli takes issue with the EPR argument. Pauli himself took the quote from a poem by Christian Morgenstern, "Die unmögliche Tatsache," reprinted in the collection, "Alle Galgenlieder" (Berlin, 1932), p. 163.
[1] See Bohr 1949 and Einstein 1946.
[2] See, for example, Ehrenfest to Goudsmit, Uhlenbeck, and Dieke, 3 November 1927 (quoted in Bohr 1985, p. 38); see also Rosenfeld 1967.
[3] The accounts by Harvey Brown (1981), Arthur Fine (1979), Clifford Hooker (1972), Max Jammer (1974, 1985), Abraham Pais (1982), and John Stachel (1986) are those most highly to be recommended. Though he is not a historian, Bernard d'Espagnat has written insightfully about the Bohr-Einstein controversy, displaying an especially good understanding of the technical issues involved in Einstein's critique of the quantum theory and his dispute with Bohr; see d'Espagnat 1976, 1981.
[4] To my knowledge, Fine (1986) is the only author who has so far hinted at the importance of this worry in Einstein's thinking about quantum mechanics prior to 1935.

Sixty-Two Years of Uncertainty
Edited by A. I. Miller
Plenum Press, New York, 1990

Acccording to the standard accounts, Einstein's critique of the quantum
theory first took the form of doubts about its correctness. More specifically, he is supposed to have sought through a series of thought experiments
to exhibit violations of the Heisenberg uncertainty relations. Contemporary
witnesses and later commentators describe dramatic encounters between Einstein and Bohr at the 1927 and 1930 Solvay meetings, where, one by one, Bohr
found the flaws in Einstein arguments, culminating in his stunning refutation of Einstein's "photon box" experiment, a refutation that turned, ironically, upon Bohr's showing how a relativistic correction overlooked by
Einstein saves the day for the uncertainty relations. In this version of
history, it was only after Bohr had beaten down these attacks on the correctness of the quantum theory that Einstein reformulated his critique in
terms of doubts about the theory's completeness, the mature version of this
latter critique being found in the 1935 Einstein-Podolsky-Rosen (EPR) paper.

There is, of course, some truth to the standard history, even though it
was written by the victors, for Einstein did at one time have doubts about
the uncertainty relations. But it is far from being the whole story, and in
many crucial ways it is just plain wrong. It is not true that Einstein began to doubt the theory's completeness only after Bohr had parried his attempts to prove it incorrect. Einstein expressed public worries about incompleteness as early as the spring of 1927, and there are hints of such
worries earlier still. But more importantly, from a very early date, at
least 1925, Einstein was pondering the curious failure of classical assumptions abouut the independence of interacting systems made vivid in the new
Bose-Einstein statistics. Earlier still, certainly by 1909, Einstein had
recognized that the Planck formula for black-body radiation cannot be derived if one assumes that light quanta behave like the independent molecules
in the gases described by classical statistical mechanics. And by spring
1927, Einstein had recognized that quantum mechanics (or at least Schrödinger's wave mechanics) fails to satisfy the kind of separability principle
that he regarded as a necessary condition on any adequate physical theory, a
condition clearly satisfied by field theories like general relativity.

Einstein did worry as well about the failure of determinism, about the
peculiar consequences of indeterminacy, and about the curious nature and
role of measurement in quantum mechanics. But these were not, for Einstein,
fundamental problems. They were, instead, symptoms corollary to the one
basic problem of the quantum mechanical denial of the independence of interacting systems. And the main purpose of the famous series of thought experiments devised by Einstein, at least by the time of the 1930 photon-box
thought experiment, was to show that the non-separable quantum theory necessarily yields an incomplete description of physical events if one seeks to
apply it to systems assumed to satisfy a strict separability principle.

There is obvious irony in the circumstance that Einstein could not accept the non-separability of the quantum theory, because quantum non-separability is the almost inevitable issue of a line of development initiated
by Einstein's recognition that the Planck formula cannot be derived from the
assumption of mutually independent light quanta and furthered essentially by
Einstein's elaboration in 1924-1925 of Bose-Einstein statistics, where the
necessary denial of the independence of interacting systems emerges with
special clarity. The history of quantum mechanics up to 1926, which is often described as a search for a way consistently to marry the wave and particle aspects of light quanta and material particles, is, I think, better
described as a search for a mathematically consistent and empirically correct way of denying the mutual independence of interacting quantum systems.
Particles are naturally imagined as satisfying the separability principle,
and hence as being mutually independent. So too the waves familiar to us
from hydrodynamics, acoustics, and electrodynamics, but not the kind of
"waves" that interfere in the manner necessary to generate the right quantum

statistics, the "waves" that Schrödinger discovered must be located in configuration space, "waves" whose chief virtue is that the "wave" function for a joint system need not be decomposible into separate "wave" functions for the component systems. Einstein opened the line of research that led to Schrödinger's "wave" mechanics, but he could not accept the conclusion, for it was incompatible with his own deep commitment to the separable manner of describing interactions implicit in field theories like general relativity.

The first hints that something is seriously wrong with the standard histories of Einstein's critique of quantum mechanics emerged from a reexamination of the EPR argument initiated by Arthur Fine and since pursued by myself and others. This re-examination revealed that Einstein did not write the EPR paper, did not like the argument it contained, and from the summer of 1935 on espoused a rather different argument for incompleteness, one that turns crucially upon the just-mentioned, characteristically field-theoretic assumption about the independence of interacting systems, the assumption Einstein himself here dubs the "Trennungsprinzip" [separation principle].

Elsewhere I have written at length about Einstein's real argument for the incompleteness of quantum mechanics, about some of the systematic questions raised by the problem of the compatibility of quantum mechanics and field theory, and about Einstein's views on this question after the appearance of the EPR paper in 1935. Here I want to fill in the story for the period before the EPR paper. I am quite deliberate in seeking to do so with the benefit of hindsight, that is to say that, knowing how central the issue of the separability or independence of interacting systems became in Einstein's later discussions of quantum mechanics, I use that insight as a heuristic in trying to understand his earlier struggles with the problem, my working hypothesis being that the worry was similar from early to late.

In what follows, I will first review briefly what I have elsewhere written about Einstein's post-EPR critique of the quantum theory. Then I will turn to a careful retelling of the story of Einstein's worries about quantum mechanics from 1905 to 1935. I will start with Einstein's tantalizing remarks about the failure of separability at the time of his papers on Bose-Einstein statistics. I will then explore the background to these remarks in his earliest papers on the quantum hypothesis, from 1905 to 1909. Returning to the 1920s, I will outline Einstein's growing misgivings about the new quantum mechanics from 1925 to 1927, culminating in his first explicit criticism of the failure of separability in wave mechanics in the spring of 1927. The paper concludes with a review of the history of Einstein's famous Gedankenexperimente critical of quantum mechanics, my aim being to show that from the start his principal goal was to demonstrate how a non-separable quantum mechanics is necessarily incomplete when applied to systems assumed to be separable.

2. EINSTEIN ON LOCALITY AND SEPARABILITY AFTER EPR

The Einstein-Podolsky-Rosen (1935) paper is still commonly taken to represent the definitive statement of Einstein's mature misgivings about the quantum theory. In brief, the argument found there is this. First, a completeness condition is asserted as a necessary condition that must be satisfied by any acceptable scientific theory: "every element of the physical reality must have a counterpart in the physical theory" (EPR 1935, p. 777). Then a sufficient condition for the existence of elements of physical reality (the famous EPR reality criterion) is laid down: "If, without in any way disturbing a system, we can predict with certainty (i.e. with probability equal to unity) the value of a physical quantity, then there exists an element of physical reality corresponding to this physical quantity" (Einstein, Podolsky, and Rosen 1935, p. 777). And then, finally, by means of a rather complicated argument, it is shown that in an EPR-type thought experiment involving previously interacting systems, elements of physical reality exist corresponding to both of two conjugate parameters for one of the two interacting systems, since the value of either could have been predicted with certainty and without physically disturbing the system on the basis of measurements carried out on the other system. But quantum mechanics holds that conjugate parameters, like position and linear momentum along a common axis, cannot have simultaneously definite values. Quantum mechanics is, thus, incomplete, since it fails to satisfy the completeness condition.

That is the standard account of Einstein's incompleteness argument. But that account is seriously wrong. Einstein did think quantum mechanics incomplete, but for reasons significantly different from those advanced in the EPR paper. He repudiated the EPR argument within weeks of its publication; and from 1935 on, all of his discussions of incompleteness take a quite different form from that found in the EPR paper. He continued to be concerned with the peculiar way in which quantum mechanics describes interacting systems; but he never invoked the EPR completeness condition, he never invoked the reality criterion, and he never invoked the uncertainty relations. Moreover, what he does say makes far clearer than the EPR paper the connection between his critique of quantum mechanics, on the one hand, and his commitments to field theories and realism, on the other.

Einstein's own incompleteness argument first appears in correspondence with Erwin Schrödinger in June of 1935, barely one month after the publication of the EPR paper; it was repeated and refined in a series of papers and other writings between 1936 and 1949.[5] In outline, it is this. A complete theory assigns one and only one theoretical state to each real state of a physical system.[6] But in EPR-type experiments involving spatio-temporally separated, but previously interacting systems, A and B, quantum mechanics assigns different theoretical states, different "psi-functions," to one and the same real state of A, say, depending upon the kind of measurement we choose to carry out on B. Hence quantum mechanics is incomplete.

The crucial step in the argument involves the proof that system A possesses one and only one real state. This is held to follow from the conjunction of two principles that I (not Einstein himself) call the locality and separability principles. Separability says that spatio-temporally separated systems possess well-defined real states, such that the joint state of the composite system is wholly determined by these two separate states. Locality says that such a real state is unaffected by events in regions of space-time separated from it by a spacelike interval.[7] Einstein argues that both principles apply to the separated systems in the EPR-type experiment (if they are allowed to separate sufficiently before we perform a measurement on B). It follows that system A has its own well-defined real state from the moment the interaction between A and B ceases, and that this real state is unaffected by anything we do in the vicinity of B. But quantum mechanics, again, assigns different states to A depending upon the parameter

[5] The principal published texts are Einstein 1936, 1946, 1948, and 1949; another important source is Born 1969. For detailed references, see Howard 1985 or 1989.

[6] This is a curious conception of completeness, more akin to what is called in formal semantics "categoricity." For more on the background to the concept of the categoricity or "Eindeutigkeit" of theories in Einstein's work prior to the development of general relativity in 1915, see Howard 1988. A future paper will explore the issue in the years 1915 to 1935.

[7] What Einstein calls the "Trennungsprinzip" in his 1935 correspondence with Schrödinger combines both separability and locality. Einstein does not himself make the distinction clearly until 1946; see Howard 1985.

chosen for measurement on B. Thus, Einstein claims that the incompleteness of quantum mechanics--in the special sense of its assigning different theoretical states to one and the same real state--follows inevitably if we insist upon the principles of locality and separability.

Understanding that this was Einstein's real incompleteness argument is crucial to reconstructing the pre-history of the EPR experiment, and this for two reasons. First, because I want to argue that as early as 1927 and in virtually all of his later thought experiments critical of the quantum theory prior to 1935, it was the problem of non-separability that Einstein was really trying to articulate. And, second, because once we see that this was the real issue, we understand at last why Einstein's commitment to the program of field theories forced him to repudiate quantum mechanics. For as Einstein himself later explained, both locality and separability, but especially the latter, are built into the ontological foundations of field theories. The argument is simple. In a field theory, the fundamental ontology, the reality assumed by the theory, consists of the points of the space-time manifold and fundamental field structures, such as the metric and stress-energy tensors, assumed to be well defined at each point of the manifold.[8] Implicitly, therefore, any field theory assumes (i) that each point of the manifold, and by extension any region of the manifold, possesses its own real state, say that represented by the metric tensor, and (ii) that all interactions are to be described in terms of changes in these separate real states, which is to say that joint states are exhaustively determined by combinations of the relevant separate states, just as the separability principle demands. If this is correct (and I think it is), and if the quantum mechanical account of interactions denies separability, then there can be no reconciliation of the two. Moreover, Einstein had not inconsiderable (if not ultimately compelling) arguments--methodological, epistemological, and metaphysical--for retaining both locality and separability, which helps to explain his dogged commitment to the field theory program as an alternative to quantum mechanics.

For what follows, the point about the explanation of interactions in accordance with the separabilty principle bears elaboration. In one sense, two interacting systems even under a classical description are not independent of one another, since various correlations (if only momentum and energy conservation) are called into being by the interaction. But if the two systems are separable, always possessing well-defined separate states that exhaustively determine any joint properties--as is the case in classical mechanics, electrodynamics, and general relativity--then they are independent in the sense that each possesses its own separate "reality," if you will. And this independence manifests itself in the fact that all of the correlations between them can be explained in terms of their separate states. In the interesting case of statistical correlations of the kind to be considered below, this means that all joint probabilities for measurement outcomes, given the joint state of the two systems, always factorize as the product of separate probabilities for the individual measurement outcomes on the two systems, given, for each system, its own separate state.[9] The non-separability of the quantum mechanical account of interactions manifests itself precisely in the fact that joint probabilities do not thus factorize.

---

[8] It is important to note, however, that on Einstein's understanding of a field-theoretic ontology (at least that of general relativity), the points are not given independently of the structures defined upon them. The legacy of his wrangling with the "hole argument" ("Lochbetractung") was his regarding the points of the manifold as being only implicitly defined as the intersections of world lines. For details, see Stachel 1989.

[9] For more detail, see Howard 1989, pp. 239-241.

3. BOSE-EINSTEIN STATISTICS AND THE BOHR-KRAMERS-SLATER THEORY: 1924-1925

The full story of Einstein's struggle with the quantum goes back to 1900, when, as a student, he first read Planck's papers on irreversible radiation processes and began to think about the manner in which light and matter interact. And it was in 1909 that Einstein first asserted in print that the quantum hypothesis is incompatible with classical assumptions about the independence of interacting systems. But I want to start with what was happening at the beginning of 1925, when Einstein for all intents and purposes ceased contributing to the development of the quantum theory, and took on the role of the theory's chief critic.

A few months earlier, in June of 1924, Einstein received from the Bengali physicist Satyendra Nath Bose a letter and an accompanying manuscript with a strikingly new derivation of the Planck radiation law. What was novel in Bose's derivation—Einstein called it "an important advance" (Einstein 1924a, p. 181)—was that it made no explicit use of the wave-theoretical arguments until then standard, proceeding instead on the assumption that a volume filled with light quanta can be treated by methods standard in the kinetic theory of gases, except that a new kind of statistics is required, statistics fundamentally different from classical Boltzmann statistics. Einstein was so impressed that he translated Bose's paper himself and arranged for its publication in the Zeitschrift für Physik. Bose's approach made it possible for the first time to understand how, in calculating the probabilities, W, that enter the Boltzmann equation, $S = k \cdot \log(W)$, the quantum approach makes different assumptions about equiprobable cases than are made classically. Not that all of this was immediately apparent. For Einstein wrote to Ehrenfest on 12 July about Bose's paper: "Derivation elegant, but essence remains obscure" (EA 10-089). But the essence was soon to become clearer when Einstein applied Bose's idea not to a photon gas, but to a quantum gas of material particles.

Einstein went on to write three papers on the subject; they represent his last great substantive contribution to quantum mechanics. What is not now realized is that what they showed him about quantum mechanics may have forever dulled his enthusiasm for the topic. The first of these papers was presented to the Prussian Academy on 10 July 1924 (Einstein 1924b), the second, containing the prediction of the low-temperature phase transition since known as "Bose-Einstein condensation," was presented on 8 January 1925 (Einstein 1925a), and the third on 29 January (Einstein 1925b). The significance of all three is limited, for spin was not yet clearly understood, the exclusion principle had yet to be articulated by Pauli, and it would take two more years before the respective roles of Fermi-Dirac and Bose-Einstein statistics were clearly distinguished. But such limitations are not immediately relevant to the story of Einstein's doubts about the quantum theory.

What is relevant is a question raised by Ehrenfest. Section §7 of the second paper is titled: "Comparison of the Gas Theory Developed Here with That Which Follows from the Hypothesis of the Mutual Statistical Independence of the Gas Molecules." It begins thus:

> Bose's theory of radiation and my analogous theory of ideal gases have been reproved by Mr. Ehrenfest and other colleagues because in these theories the quanta or molecules are not treated as structures statistically independent of one another, without this circumstance being especially pointed out in our papers. This is entirely correct. If one treats the quanta as being statistically independent of one another in their localization, then one obtains the Wien radiation law; if one treats the gas molecules analogously, then one obtains the classical equation of state for ideal gases, even if one otherwise proceeds exactly as Bose and I have. (Einstein 1925a, p. 5)

After showing how, following Bose's method, one counts the number of "complexions" corresponding to a given macrostate, that is to say how one distributes particles over the cells of phase space, Einstein adds:

> It is easy to see that, according to this way of calculating, the distribution of molecules among the cells is not treated as a statistically independent one. This is connected with the fact that the cases that are here called "complexions" would not be regarded as cases of equal probability according to the hypothesis of the independent distribution of the individual molecules among the cells. Assigning different probability to these "complexions" would not then give the entropy correctly in the case of an actual statistical independence of the molecules. Thus, the formula [for the entropy] indirectly expresses a certain hypothesis about a mutual influence of the molecules—for the time being of a quite mysterious kind—which determines precisely the equal statistical probability of the cases here defined as "complexions." (Einstein 1925a, p. 6)

Exactly what Einstein meant by his comment about the connection between the failure of statistical independence and "a quite mysterious kind" of "mutual influence" of one molecule upon another is spelled out in a letter to Schrödinger of 28 February 1925 (evidently written before Schrödinger had seen Einstein's second gas theory paper):

> In the Bose statistics employed by me, the quanta or molecules are not treated as being <u>independent of one another</u>. . . . A complexion is characterized through giving the number of molecules that are present in each individual cell. The number of the complexions so defined should determine the entropy. According to this procedure, the molecules do not appear as being localized independently of one another, but rather they have a preference to sit together with another molecule in the same cell. One can easily picture this in the case of small numbers. [In particular] 2 quanta, 2 cells:

Bose-statistics		
	1st cell	2nd cell
1st case	••	—
2nd case	•	•
3rd case	—	••

independent molecules		
	1st cell	2nd cell
1st case	I II	-
2nd case	I	II
3rd case	II	I
4th case	-	I II

> According to Bose the molecules stack together relatively more often than according to the hypothesis of the statistical independence of the molecules. (EA 22-002)

And in a P.S., Einstein adds that the new statistics are really not in conflict with those employed in his 1916 papers on transition probabilities, where the standard Maxwell-Boltzmann distribution was employed (Einstein 1916a, 1916b), because it is really only in relatively dense gases where the difference between the statistics of independent particles and the Bose-Einstein statistics will be noticeable: "There the interaction between the molecules makes itself felt,— the interaction which, for the present, is accounted for statistically, but whose physical nature remains veiled."

In many modern textbooks and histories of the subject, the principal innovation embodied in Bose-Einstein statistics is described in terms at first glance quite different from those we have just found Einstein using. The new statistics are said to be those appropriate to "identical" or "indistinguishable" particles. What is meant is clear. In the two-particle, two-cell case cited by Einstein we cannot tell which of the two particles is which, that is to say, we cannot keep track of their individual identities, as we can in classical Boltzmann statistics; hence, cases two and three in the classical statistics must be regarded as just one case (case two) in Bose-Einstein statistics, weighted equally with the other two remaining cases. But the "identical particles" vocabulary is misleading, for in the important case two in Bose-Einstein statistics, the two particles are by no means identical: they occupy different cells of phase space and so differ in position or momentum. They are arguably identical in cases one and three, since they occupy the same cell. But these cases have their counterparts in the Boltzmann statistics. The interesting difference appears in just those cases where the particles are not identical. What is important is the fact that we cannot track the individual identities of Bose-Einstein particles. We cannot say, as we could classically, "Here is particle A" at time t_0, and "Here is particle A," at some later time, t_1; the particle observed at t_1 might just as well be particle B. Classically, we can track individual identities, which possibility leads to Boltzmann statistics. (Notice how Einstein uses numerical labels, I and II, to suggest the separate indentifiability of the classical particles, representing the Bose-Einstein particles by unlabeled dots.) It is equally misleading to speak here of "indistinguishable" particles. For even in Bose-Einstein statistics we know that in case two there are different particles, we just cannot tell which is which.

Another common way of characterizing the novelty of Bose-Einstein statistics is to say that such statistics are appropriate for material particles evincing the wave-like aspect shortly before suggested in de Broglie's dissertation (1924). As we shall see, it is wrong to credit the idea of material particles possessing simultaneously a wave-like aspect wholly to de Broglie, since Einstein was well-known even at the time to have toyed with such ideas since at least 1921, motivated by considerations of symmetry and unity--if massless photons have a dual nature as both waves and particles, then massive particles should as well. But otherwise this charactization of the innovation represented by Bose-Einstein statistics is not incorrect, inasmuch as the novel way of counting complexions in Bose-Einstein statistics can be regarded as necessitated by the possibility of interference between the particles (the particles interfere precisely because we cannot tell which is which), such interference being perhaps most easily visualized with wave-theoretical models. Einstein himself pointed to this way of conceiving Bose-Einstein statistics in his second gas theory paper (Einstein 1925a, pp. 9-10); and in an important preliminary to his own development of wave mechanics, Schrödinger later elaborated this suggestion in an attempt to find a plausible wave-theoretical physical interpretation of the statistics (Schrödinger 1926a). Still, it is striking that Einstein himself did not emphasize this way of viewing the new statistics. He preferred to emphasize the fact that the particles are not treated as statistically independent systems and that such a failure of statistical independence is a symptom of a physically mysterious interaction between the particles.

Why did Einstein prefer this way of characterizing what was novel in his new statistics? Of course he understood the connection between his work and deBroglie's ideas, a connection equally obvious to most of his contemporaries. What point was he trying to make by stressing instead the failure of statistical independence and the existence of mysterious interactions? Might his way of characterizing the situation even tell us something about his understanding of the significance of wave-theoretical models?

An important clue to Einstein's thinking is provided in a talk entitled "On the Ether" that Einstein gave to the Schweizerische Naturforschende Gesellschaft in September 1924, after he had received and assimilated Bose's paper. At the end of his talk he turned to Bose's work. After explaining that Bose had replaced the customary wave-theoretical derivations of the Planck radiation law with a derivation employing the methods of statistical mechanics, Einstein remarked: "Then the question obtrudes whether or not diffraction and interference phenomena can just be connected to the quantum theory in such a way that the field-like concepts of the theory merely represent expressions of the interactions between quanta, in which case the field would no longer be ascribed any independent physical reality" (Einstein 1924c, p. 93). What is interesting here, aside from Einstein's scepticism regarding the reality of matter waves (and even the wave nature of photons!), is his suggestion that the effects commonly regarded as symptoms of a system's having a wave-like nature, that is, diffraction and interference, are really better understood as reflecting interactions between quanta. Thus, where others see waves, Einstein sees evidence of the physically mysterious interactions between quantum systems that he believed underlie classically unexpected statistical correlations between such systems. For Einstein, it is quanta, both light quanta and material particles, together with their curious interactions, that are real. The device of wave-theoretical representations is merely an artifice, a convenient tool, a vivid image, for helping us to think clearly about quantum interactions and statistical correlations.

One additional idea that will later loom large for Einstein had not yet come to the fore in his remarks about Bose-Einstein statistics, which is that the kinds of statistical dependence evinced in Bose-Einstein statistics can obtain even between spacelike separated systems or events. But there is other evidence that this problem too was already on Einstein's mind, as the concluding paragraph of the just-quoted talk indicates. For in a seemingly abrupt shift, Einstein turns back to the main topic of the talk, the ether, by which he meant the space-time manifold plus metric, remarking that even if the quantum theory develops into a real theory, "we will not be able to dispense with the ether in theoretical physics, that is, with the continuum endowed with physical properties; for the general theory of relativity, to whose fundamental aspects physicists will indeed always cling, excludes an immediate distant action, but every local-action theory assumes continuous fields, and thus the existence of an 'ether'" (Einstein 1924c, p. 93).

Recall how a continuous field theory like general relativity incorporates the principle of local action. In effect, such a theory treats every point in the field, every point of the space-time manifold in the case of general relativity, as a separable, independent system, possessing its own physical state represented by the fundamental field parameter, which would be the metric tensor in general relativity. Within this framework, action is explained in terms of a change in the fundamental parameter being propagated from point to point across the field, which is to say that the value of the fundamental parameter at any point is always wholly determined by the field equations and by the values of that parameter at all immediately adjacent points. What is not allowed is for the value of the fundamental parameter at one point to be immediately functionally dependent upon values at distant points. It is the restriction to local action so conceived that Einstein had in mind when he said that all "local-action theories" assume continuous fields. General relativity, through its incorporation of the first-signal principle, is even more restrictive in this regard than classical field theories, like Maxwellian electrodynamics, that impose no upper bound on signal velocities. For in general relativity, even the admissible varieties of local action are constrained to occur only between points of the manifold that are timelike separated.

It is important to keep in mind Einstein's basic commitment to the separable field-theoretic ontology and its associated locality constraints, because it helps to understand why the Bose-Einstein statistics would appear puzzling to Einstein. For the field-theoretic way of explaining interactions requires us to assign separate states to spatially separated systems. These states would determine separately the probabilities for each system's behavior, and it would follow that joint probabilities would have to be determined wholly by these separate probabilities, which is to say that the joint probabilities would have to factorize. But that does not happen in Bose-Einstein statistics, which is why Einstein found them so mysterious.

Einstein's gas theory papers were not the first investigations to make acute various questions about the statistical correlations that obtain between interacting systems. In fact, Einstein had been worrying about the general problem of probability relations between interacting systems for a long time. Such concerns had most recently come to the fore in his reaction to the Bohr-Kramers-Slater (BKS) theory (Bohr, Kramers, and Slater 1924). The English version of the BKS paper appeared in April 1924, the German version on 22 May. We remember it today for its use of virtual fields determining the probabilities of individual atomic emissions (and absorptions), and for its suggestion that, in consequence of the merely probabilistic determination of transition events, energy and momentum are conserved only on average, over large numbers of quantum events, and not in individual events. Einstein, of course, opposed the BKS theory because of its abandonment of strict energy-momentum conservation, but that is far from the whole story.

As we will see, there is irony here. Einstein turns out eventually to repudiate quantum mechanics in part because of its denial of the statistical independence of distant systems. But one of the main things that troubled him about the BKS theory was precisely its assumption of the statistical independence of atomic transitions (absorption or emission of energy quanta) in distant systems, or rather its failure to assume correlations sufficient to guarantee strict energy-momentum conservation in individual events.

In the BKS theory, each atom is assumed to be the source of a virtual radiation field with components corresponding to all of that atom's possible transitions. The radiation field serves two purposes. First, it determines the probabilities for emissions and absorptions by the atom from which the field originates, that is to say, the transition probabilities introduced by Einstein in his 1916 quantum theory papers (Einstein 1916a, 1916b). Second, it serves as the vehicle through which that atom communicates with surrounding atoms. It accomplishes this by helping to determine the probabilities for absorption and induced emission in these other atoms, depending upon whether or not it interferes constructively or destructively with the virtual radiation field emanating from each of the latter. But as BKS themselves stress, the correlations engendered by this communication between atoms are quite weak:

> In fact, the occurrence of a certain transition in a given atom will depend on the initial stationary state of this atom itself and on the states of the atoms with which it is in communication through the virtual radiation field, but not on the occurrence of transition processes in the latter atoms. . . . As regards the occurrence of transitions . . . we abandon . . . any attempt at a causal connexion between the transitions in distant atoms, and especially a direct application of the principles of conservation of energy and momentum, so characteristic for the classical theories. (Bohr, Kramers, and Slater, p. 165)

Or again,

By interaction between atoms at greater distances from each other, where according to the classical theory of radiation there would be no question of simultaneous mutual action, we shall assume an independence of the individual transition processes, which stands in striking contrast to the classical claim of conservation of energy and momentum. Thus we assume that an induced transition in an atom is not directly caused by a transition in a distant atom for which the energy difference between the initial and the final stationary state is the same. On the contrary, an atom which has contributed to the induction of a certain transition in a distant atom through the virtual radiation field conjugated with the virtual harmonic oscillator corresponding with one of the possible transitions to other stationary states, may nevertheless itself ultimately perform another of these transitions. (p. 166)

And, finally, they add, in an interesting comment: "But it may be emphasized that the degree of independence of the transition processes assumed here would seem the only consistent way of describing the interaction between radiation and atoms by a theory involving probability considerations" (pp. 166-167). But, of course, this is wrong, as the later development of quantum mechanics was to show.

Consider more carefully the kind of coupling that BKS were assuming. The probability of a transition in a given atom, A, is determined by its associated virtual radiation field. This virtual radiation field can be altered by the effects of a radiation field propagating, subluminally, from another atom, B, and since the virtual field radiating from B is determined by B's current stationary state, the probability of a transition in A can depend upon the state (the virtual field) of B, which is to say that the probability of a transition at A can depend upon the _probabilities_ of various transitions at B. On the other hand, the probability of a transition at A is statistically independent of the actual occurrence of a transition at B. The first kind of dependence is wholly consistent with classical, _local_, field-theoretic models of interactions, since the changes in A's state (virtual field) induced by B's state (virtual field) are propagated subluminally. But dependence of the latter kind threatens classical models of local interaction, with prohibitions on "distant action"; it was general relativity's exclusion of such "Fernwirkungen" that Einstein cited in late 1924 as the main reason why general relativity would never be abandoned.

Einstein's objections to the BKS theory are recorded in at least three different places. Einstein gave a colloquium on the BKS theory in Berlin on 28 or 29 May, within days of the paper's German publication.[10] What may be a list of objections to the theory prepared for that occasion survives in the Einstein Archive (EA 8-076) under the title "Bedenken inbezug auf Bohr-Cramers." It begins as follows: "1) Strict validity of the energy principle in all known elementary processes. Assumption of the invalidity in distant actions unnatural." A similar list of objections is contained in a letter to Ehrenfest of 31 May 1924 (EA 10-087); it begins in the same vein: "1) Nature appears to adhere strictly to the conservation laws (Frank-Hertz, Stokes's rule). Why should distant actions be excepted?"

Perhaps the most interesting record of Einstein's objections, however, interesting because of its intended audience, is a letter from Pauli to Bohr of 2 October 1924, in which Pauli reports the contents of a conversation about the BKS theory that Pauli had with Einstein during the Innsbruck

---

[10]Rudolf Ladenburg to Kramers, 8 June 1924, as quoted in Bohr 1984, p. 27, gives the date as 28 May. But Wigner (1980, p. 461) reports that the colloquia took place regularly on Thursdays, which would make the date 29 May.

Naturforscherversammlung in late September (it was Pauli's first meeting with Einstein). The very first of Einstein's objections, as reported by Pauli, is this: "1. By means of fluctuation arguments one can show that, in the case of the statistical independence of the occurrence of elementary processes at spatially distant atoms, a system can, in the course of time, display systematic deviations from the first law, in that, for example, the total kinetic energy of a radiation-filled cavity with perfectly reflecting walls can, in the course of time, assume arbitrarily large values. He finds this dégoûtant (so he says)" (Pauli 1979, p. 164). What Einstein is pointing to in this example, also mentioned in the list of "Bedenken" for his Berlin colloquium and in the cited letter to Ehrenfest, is not the failure of energy and momentum to be conserved in individual events, which comes in objection 2, but rather the existence of systematic deviations from energy conservation even on the average; to Ehrenfest he describes this as a matter of the "constantly increasing Brownian motion" of a "mirror-box" (EA 10-087). In fact, the fluctuations turn out to be significant only in certain limiting cases (Schrödinger 1924), but that is of no consequence here. What is important is the clue that this and the other quoted remarks provide as to Einstein's real reservations about the BKS theory. Specifically, Einstein believed that any adequate quantum theory would have to incorporate at a basic level some kind of strong statistical dependence of spatially-separated systems, in order to secure strict energy-momentum conservation. And what he was searching for with his "mirror-box" thought experiment was a vivid way to show the consequences of the BKS theory's failure to do this.

Spatially separated systems are statistically independent in the BKS theory because it assigns a separate virtual wave field to each (spatially separated) atomic system. In this regard, the BKS theory resembles Einstein's own earlier speculations about "ghost fields" ["Gespensterfelder"] or "guiding fields" ["Führungsfelder"], which he had introduced to try to explain the interference effects between quantum systems, be they light quanta or material particles. And his reasons for objecting to the BKS theory are similar to the reported reason for his never having published his own ideas along this line; in his letter to Ehrenfest of 31 May 1924 he says of the BKS theory: "This idea is an old acquaintance of mine, but one whom I do not regard as a respectable fellow."

Here is how Wigner recalls Einstein's reasoning about this matter in his University of Berlin physics colloquium:

Yet Einstein, though he was fond of it [the "Führungsfeld" idea], never published it. He realized that it is in conflict with the conservation principles: at a collision of a light quantum and an electron for instance, both would follow a guiding field. But these guiding fields give only the probabilities of the directions in which the two components, the light quantum and the electron, will proceed. Since they follow their directions independently, it may happen that in one collision the light quantum is strongly deflected, the electron very little. In another collision, it may be the other way around. Hence the momentum and the energy conservation laws would be obeyed only statistically—that is, on the average. This Einstein could not accept and hence never took his idea of the guiding field quite seriously. (Wigner 1980, p. 463; emphasis mine)

The dilemma that Einstein faced here was that some kind of wave aspect had to be associated with light quanta and material particles to explain diffraction and interference, wave-like interference even between material particles being suspected by many at least since the discovery of the Ramsauer effect in 1920. And these wave-aspects—call them "ghost fields," "guiding fields," "virtual fields," or whatever—can at best determine probabilistically the motions of individual particles or the transitions in individual atoms. But as long as the "guiding" or "virtual fields" are assigned separately, one to each particle or atom, one cannot arrange both for the merely probabilistic behavior of individual systems and for correlations between interacting systems sufficient to secure strict energy-momentum conservation in all individual events. As it turned out, it was only Schrödinger's relocation of the wave fields from physical space to configuration space that made possible the assignment of joint wave fields that could give the strong correlations needed to secure strict conservation, and the even stronger correlations evinced in Bose-Einstein statistics. But as we shall see, the price to be paid for Schrödinger's innovation was a degree of non-separability between interacting systems that Einstein found intolerable because inconsistent with the field-theoretic manner of representing interactions.

Einstein had himself believed for some time that an adequate quantum theory would have to incorporate some kind of strong coupling between distant systems in order to secure strict energy-momentum conservation. Many other physicists, were still not sure about this matter as late as fall 1924, when Pauli wrote to Bohr, in the above-cited letter (2 October 1924): "And if you were to ask me what I believe about the statistical dependence or independence of quantum processes in spatially distant atoms, then I must answer honestly: I do not know. The Geiger experiment, which I hear is already being started, will indeed quite soon decide this question experimentally. It suits me equally well if it turns out one way or the other" (Pauli 1979, p. 165). But the mentioned Bothe-Geiger experiment (Bothe and Geiger 1924, 1925a, 1925b) and the Compton-Simon experiment (Compton and Simon 1925a, 1925b, 1925c) were soon to persuade most everyone that energy and momentum are strictly conserved in individual atomic events. Writing to Einstein on 9 January 1925, Ehrenfest put the matter thus: "If Bothe and Geiger find a 'statistical independence' of electron and scattered light quantum, that proves nothing. But if they find a dependence, that is a triumph for Einstein over Bohr. -- This time (by way of exception!) I believe firmly in you and would thus be pleased if dependence were made evident" (quoted from Bohr 1984, p. 77). However, the issue had already been decided, as Einstein explained to Lorentz on 16 December 1924: "Geiger and Bothe have carried out an experiment that speaks in favor of strict light quanta and against the views that Bohr-Cramers-Slater have recently developed. They showed that in the Compton effect the deflected radiation and the electron thrown out toward the other side are events statistically dependent upon one another. But, nevertheless, the energy-momentum principle appears to hold strictly and not only statistically" (EA 16-575).

Of course the statistical dependence demonstrated by Bothe-Geiger and Compton-Simon does not involve the kind of correlation that surfaces in Bose-Einstein statistics. One can explain energy-momentum conservation quite naturally in terms of a model positing distinguishable particles, systems whose separate identities can be tracked throughout their interactions, which is precisely how Einstein preferred to think of his light quanta and material particles. In more modern language, the correlations evinced in the Bothe-Geiger and Compton-Simon experiments can be explained in terms of common causes; there is here no threat of non-locality or non-separability.

But while Einstein preferred to think of light quanta and material particles as independent, distinguishable systems, he really already knew better. For one thing, there was the obvious problem that a simple corpuscular model is powerless to explain interference and diffraction, which is part of what drove Einstein to the unsuccessful "Führungsfeld" idea in the first place. And, more importantly, Einstein's own earlier work on the quantum hypothesis, in particular, his efforts to understand the relationship between his light quantum hypothesis and Planck's radiation law, had already taught him that light quanta do not, in fact, behave like the independent particles of classical statistical mechanics. In other words, already at the time of the BKS theory, Einstein had good reason to expect that an adequate quantum theory would require correlations between interacting systems beyond those needed to secure strict energy-momentum conservation.

4. EINSTEIN'S EARLIEST REMARKS ON THE INDEPENDENCE OF QUANTA: 1905-1914

Recall that what primarily distinguished Einstein's point of view from Planck's in 1905 is that, whereas Planck wanted to quantize only the process of a resonator's absorbing or emitting energy, Einstein wanted to introduce light quanta or photons as carriers of that energy even between elementary events of emission or absorption. That is to say, Einstein wanted to quantize the electromagnetic radiation field itself, arguing that Maxwell's equations should be regarded as describing merely the average behavior of a large number of light quanta (Einstein 1905, p. 132; 1906, p. 203). But these light quanta are not yet the photons or light quanta of the mature quantum mechanics of the late 1920s, and this for one crucial reason. Remember the following oft-quoted remark from Einstein's 1905 paper: "Monochromatic radiation of low density (within the domain of validity of Wien's radiation formula) behaves from a thermodynamic point of view as if it consisted of mutually independent energy quanta of the magnitude $R\beta\nu/N$" (Einstein 1905, p. 143; emphasis mine). I have deliberately emphasized the words whose import we usually do not appreciate when reading this passage.

In what sense did Einstein mean these quanta to be independent of one another? He was quite explicit on this point. The quanta are independent in the sense that the joint probability for two of them occupying specific cells in phase space is the product of the separate probabilities. After writing the relation, $W = W_1 \cdot W_2$ ("W" standing for probability, "Wahrscheinlichkeit"), Einstein comments: "The last relation says that the states of the two systems are mutually independent events" (Einstein 1905, p. 141). He had a good reason for postulating such independence. If one defines entropy according to Boltzmann's principle, $S = k \cdot \log(W)$, as Einstein thought one must, then the factorizability of the probability is a necessary and sufficient condition for the additivity of the entropy, itself a necessary condition in Einstein's eyes (Einstein 1905, p. 140).

Einstein never retreated from his belief in the existence of photons, but by 1909 it had become clear to him (if it was not already clear in 1905) that quanta conceived as independent particles, the quanta of 1905, are not the whole story about radiation. An explicit statement of this point first found its way into print in March 1909 in Einstein's masterful survey paper, "Zum gegenwärtigen Stand des Strahlungsproblems" (Einstein 1909a).

The context was yet another attempt to understand the relationship between his own light quantum hypothesis, which by itself was found to yield a formula for black-body radiation valid only in the Wien regime (ν/T large), and the kind of energy quantization implicit in Planck's radiation law. The method was that of fluctuation arguments, an approach that had served Einstein well in the past. He first asked what would be the mean-square fluctuations in the energy of a radiation-filled cavity, and, second, what would be the mean-square fluctuations in the radiation pressure, as manifested by fluctuations in the motion of a mirror suspended in the cavity. Both calculations led directly from Planck's radiation formula to a similar result, namely, an expression for the fluctuations that can be divided into two terms, the first of which Einstein interprets as arising from mutually independent light quanta, the second from interference effects of the kind to be expected were the radiation completely described by Maxwell's electrodynamics. Thus, with regard to the expression for energy fluctuations, Einstein says that this first term, $(R/Nk)\nu h n_0$, were it alone present, would yield fluctuations "as if the radiation consisted of pointlike quanta of energy $h\nu$ that move independently of one another" (Einstein 1909a, p. 189). And about the expression for radiation pressure fluctuations, he says: "According to the current theory [Maxwell's electrodynamics], the expression must reduce to the second term (fluctuations due to interference). If only the first term were present, then the fluctuations in radiation pressure could be completely explained through the assumption that the radiation consists of slightly extended complexes of energy $h\nu$ that move independently of one another" (Einstein 1909a, p. 190). A complete account of cavity radiation entails, however, the presence of both terms. And so it follows that a complete theory cannot assume only mutually independent light quanta; it must allow for some means whereby localized, pointlike quanta can, mysteriously, interfere with one another.

At the time Einstein wrote this survey paper (received 23 January 1909, published 15 March), he was still rather sanguine about the prospects for finding a theoretical model of radiation embodying both the existence of quanta and the possibility of their interfering, this without departing significantly from existing theoretical conceptions. Near the end of the paper he says that what is apparently needed is "a modification of our current theories," not "a complete abandonment of them" (Einstein 1909a, p. 192). But he was clearly struggling to understand how localized quanta could possibly interfere with one another.

This issue came to the fore in an exchange of letters between Einstein and Lorentz in May of 1909, shortly after Einstein read Lorentz's influential lecture on the radiation problem delivered to the 1908 International Congress of Mathematicians in Rome (Lorentz 1908a).[11] Lorentz had by this time reluctantly accepted Planck's radiation formula, instead of his preferred Rayleigh-Jeans formula (see Lorentz 1908b), but in a letter to Einstein of 6 May 1909 (EA 16-418), he pressed Einstein to explain how localized, mutually independent quanta could explain interference and diffraction. Einstein replied on 23 May, speaking first to the question of independence: "I am not at all of the opinion that one should think of light as being composed of mutually independent quanta localized in relatively small spaces. This would be the most convenient explanation of the Wien end of the radiation formlula. But already the division of a light ray at the surface of refractive media absolutely prohibits this view. A light ray divides, but a light quantum indeed cannot divide without change of frequency" (EA 16-419). Then he goes on to suggest how he really views the situation, introducing for the first time (as far as I can determine) the progenitor of his later "ghost" or "guiding" field idea:

> As I already said, in my opinion one should not think about constructing light out of discrete, mutually independent points. I imagine the situation somewhat as follows: . . . I conceive of the light quantum as a point that is surrounded by a greatly extended vector field, that somehow diminishes with distance. Whether or not when several light quanta are present with mutually overlapping fields one must imagine a simple superposition of the vector fields, that I cannot say. In any case, for the determination of events, one must have equations of motion for the singular points in addition to the differential equations for the vector field. (EA 16-419)

The point is, of course, that these vector fields will mediate the interactions among light quanta.

---

[11] See Einstein to Lorentz 13 April 1909 (EA 70-139); Einstein read the 1909 reprinting in the Revue génerale des sciences (Lorentz 1909).

Einstein's vector field idea first made its way into print in his second great survey paper of 1909, this his lecture "Über die Entwickelung unserer Anschauungen über das Wesen und die Konstitution der Strahlung" (Einstein 1909b), delivered to the Salzburg Naturforscherversammlung on 21 September. Einstein first reviews the radiation pressure fluctuation argument from the previous paper, and the interpretation of the two terms in the resulting expression for the fluctuations as quantum and interference terms respectively. But his growing realization that the quanta cannot be regarded as independent is reflected in his observation that the view of quanta as localized particles moving through space and being reflected independently of one another--the model that is the focus of his 1905 light quantum hypothesis paper--is "the crudest visualization of the light quantum hypothesis" (Einstein 1909b, p. 498). Einstein then introduces the vector field idea broached in the letter to Lorentz, but with the difference that the fields are here portrayed as "force fields" having the character of "plane waves." He concludes by noting that in introducing this idea, not yet an exact theory, he "only wanted to make it clear . . . that the two structural characteristics (undulatory structure and quantum structure), both of which should belong to radiation according to Planck's formula, are not to be viewed as irreconcilable with one another" (Einstein 1909b, p. 500).

The customary gloss on this last remark is that it is an anticipation of the notion of wave-particle duality. That is true, but it puts the emphasis in the wrong place. As we have seen, what was really going on here was, first, Einstein's coming to grips with the fact that photons or light quanta cannot be invested with the kind of independence from one another standardly assumed for the systems of particles to which classical statistical mechanics applies, and, second, his search for a theoretical model of quanta that would accomodate this lack of independence without compromising the principle that, at root, radiation has an atomistic structure.

Between 1909 and 1925, many investigations were inspired by Einstein's writings on light quanta, the principal aim being to understand more clearly the difference between Einstein's conception of independent light quanta and Planck's conception of quantized resonators. Several people theorized that the independence assumption had to be modified, and the conviction slowly gained force that the classical manner of counting complexions had to be modified after the manner of Planck's counting rule, though the theoretical foundations of the latter remained obscure. It was really only the papers of Bose and Einstein in 1924-1925 that began to clarify these matters. There is, however, one individual whose now almost entirely forgotten work on light quanta is of special interest because of the unexpected light it throws on Einstein's thinking about the independence problem during the 1910s. This is Mieczysław Wolfke, a young Polish physicist who took a degree under Otto Lummer at Breslau in 1910 and became a Privatdozent at the ETH in 1913. He moved to the University of Zurich, again as Dozent, in 1914, where he remained until assuming a professorship at the Warsaw Polytechnic in 1922. He was thus a colleague of Einstein's in Zurich for about eighteen months in 1913-1914; that relationship is important for our story.

Starting in late 1913, Wolfke published a series of papers developing a derivation of the Planck radiation formula starting from the assumption of what he termed "light atoms," which were conceived as being in some respects similar to Einstein's light quanta (Wolfke 1913a, 1913b, 1914a). Pressed by G. Krutkow (1914) to explain the difference between "light atoms" and "light quanta," especially to explain why Einstein's mutually independent light quanta lead to Wien's law whereas Wolfke's "light atoms" lead to Planck's law, Wolfke published in March of 1914 in the Physikalische Zeitschrift a short paper elaborating the different independence assumptions made by him and by Einstein. The crucial § 3 of his paper, entitled "The Decisive Presuppositions," reads as follows:

Mr. Einstein has personally drawn my attention to the difference in principle between the Einsteinian light quantum theory and the foregoing argument [deriving Planck's law from light atoms].

The definition of the independence of the light atoms from one another that one presupposes in the probability considerations is alone decisive for the derived radiation formulas.

In the above derivation . . . of the Planck radiation formula only this general assumption is used, namely that the light atoms are mutually independent with regard to their existence, in other words, it is assumed that the probability for the existence of a light atom of a specific frequency is independent of how many atoms of the same frequency are simultaneously present in the volume under consideration. Nevertheless, in my derivation no limiting assumptions were established regarding the spatial distribution of the light atoms.

However, in opposition to this, the Einsteinian light quantum theory presupposes the special case that light atoms are also spatially independent of one another, i.e, that the probability for a specific position of a light atom is independent of the simultaneous position of the other light atoms of the same frequency.

In consequence of this, the Einsteinian light quantum theory leads to the Wien radiation law, which, as is well known, can be regarded as a special case of the Planck radiation formula. (Wolfke 1914b, p. 309)

How much of this is Einstein and how much Wolfke is hard to say; such evidence must by handled with care. But certainly nothing in the foregoing analysis is inconsistent with what Einstein had earlier said.

The assumption that Wolfke was accurately reporting Einstein's views is strengthened by Wolfke's reply to Krutkow's further demand that he give a more formal characterization of the two kinds of independence (Krutkow 1914b). For Wolfke adverts precisely to Einstein's 1905 characterization of independence, namely, the factorizability of the associated probabilities:

In fact the Einsteinian light quanta behave like the individual, mutually independent molecules of a gas . . . . However, the spatial independence of the Einsteinian light quanta comes out even more clearly from Einstein's argument itself. From the Wien radiation formula Einstein calculates the probability $\underline{W}$ that all $\underline{n}$ light quanta of the same frequency enclosed in a volume $\underline{v_0}$ find themselves at an arbitrary moment of time in the subvolume $\underline{v}$ of the volume $\underline{v_0}$. The expression for this probability reads:

$$\underline{W} = (\underline{v}/\underline{v_0})^{\underline{n}}.$$

This probability may be interpreted as the product of the individual probabilities $\underline{v}/\underline{v_0}$ that an individual one of the light quanta under consideration lies in the subvolume $\underline{v}$ at an arbitrary moment of time. From the fact that the total probability $\underline{W}$ is expressed as the product of the individual probabilities $\underline{v}/\underline{v_0}$, one recognizes that it is a matter of individual mutually independent events. Thus we see that, according to Einstein's view, the fact that a light quantum lies in a specific subvolume is independent of the position of the other light quanta. (Wolfke 1914c, pp. 463-464)

What Einstein is represented as asserting is a more careful analysis of the type of independence that must be denied to light quanta in an adequate quantum theory. Both Einstein's original 1905 light quanta and the kind of quanta that would have to be assumed to derive the Planck radiation law are held to be independent from the point of view of their existence, which is to say that the probability for the existence of a light quantum of some specific frequency is independent of the number of other light quanta of

that frequency already in existence. But Einstein's quanta are independent
of one another also in the spatial sense, which is to say that the joint
probability for two quanta of the same frequency to occupy specific locations is factorizable as the product of separate probabilties for each to
occupy its own location. It is the same kind of spatial independence assumed in classical statistical mechanics, but it must be denied in order to
derive the Planck formula.[12]

In a later paper, Wolfke interpreted this failure of spatial independence as a matter of the quanta of a specific frequency tending to join together in complexes that he called "light molecules," separate light quanta
being designated "light atoms" (Wolfke 1921). The analogy is of course
strained, but it is interesting when one recalls how in 1925 Einstein characterized the novelty of the Bose-Einstein statistics by saying that the
molecules "have a preference to sit together with another molecule in the
same cell" of phase space and tend "to stack together" (Einstein to Schrödinger, 28 February 1925, EA 22-002).

Another path back to Einstein's 1924-1925 gas theory papers also leads
through Einstein's characterization of the independence of light quanta in
terms of the factorizability of joint probabilities and the associated
additivity of entropies for composite systems. For all that the additivity
principle was accorded fundamental importance by most of those who attended
to the foundations of statistical mechanics, there was a puzzle about additivity that had been known to physicists since the publication in 1902 of
Gibbs's Elementary Principles in Statistical Mechanics. In the final paragraph, Gibbs enunciated the paradox that was to come to be known by his name
(Gibbs 1902, pp. 206-207). He considered a chamber divided into two halves
by an impermeable barrier, each half filled by a gas; the entropy of the
whole system is the sum of the entropies of the two components. When the
barrier is removed, allowing the gases to mix, the total entropy will increase if the two gases are different in kind, whereas the total entropy
will stay the same if the two gases are of the same kind. But that should
not happen if the additivity principle is universally valid, because in both
cases the previously separated volumes diffuse throughout the whole chamber
in the same way, which should lead to an increase in the entropy of each
previously separated component; and then if the entropies of these components still add in the normal way, the total entropy after mixing in both
cases should go up. That it does not when the originally separated volumes
are identical in kind must be connected in some way to a failure of the additivity principle in the case of indistinguishable particles.

Curiously, Gibbs's own reaction to this paradox is rarely noted. It
was to infer that the paradox forces us for most purposes to use statistical
measures of entropy and other thermodynamic quantities calculated on the basis of what he called the "generic phase," rather than the "specific phase"
(Gibbs 1902, p. 207). Gibbs's conclusion is pertinent to the later history
of the paradox in quantum mechanics because of the way he defines "generic
phase." The "specific phase" is the phase as we normally conceive it in
classical statistical mechanics—a point in the standardly defined $6n$-
dimensional phase space for a system of n particles. The "generic phase" is
defined, in effect, as an equivalence class of specific phases differing
only through exchanging the positions of otherwise indistinguishible parti-

---

[12] Since we are concerned with the probability of a system's occupying a
given cell of phase space, Einstein must by the same logic be asssuming an
independence with respect to the instantaneous momenta or velocities of the
systems in question. Remember Einstein's glossing the necessary independence assumption in 1909 as the assumption that the quanta of energy "move
independently of one another" (Einstein 1909a, p. 189; emphasis mine).

cles. This is almost Bose-Einstein statistics, except for the way Gibbs weighted the points (cells) in his "generic phase" space, namely, as the sum of the weights of the specific phases related to the given generic phase by exchange of indistinguishable particles.[13] But that the rules for counting phases or cells must be modified in the case of indistinguishable particles was clearly recognized by Gibbs.

Various authors puzzled over the Gibbs paradox in succeeding years. The realization that its solution is somehow connected to the curious statistics of indistinguishable particles began to emerge with an important study by Planck (1922). But it was really only in Einstein's gas theory papers of 1924 and 1925 that the problem found its definitive solution. At the end of the first of these papers, Einstein raises the problem (Einstein 1924b, p. 267), and then in the second paper he offers this solution:

> These considerations thow light upon the paradox that was pointed to at the end of first paper. In order for two wave trains to interfere noticeably, they must agree with regard to V [phase velocity] and ν [frequency]. Moreover . . . it is necessary that v as well as m nearly agree for both gases. The wavefields associated with two gases of noticeably different molecular mass thus cannot noticeably interfere with one another. From this one can conclude that, according to the theory presented here, the entropy of a gas mixture is additively composed out of those of the components of the mixture, exactly as in the classical theory, at least as long as the molecular weights of the components diverge from one another somewhat. (Einstein 1925a, p. 10)

The solution, in other words, is that the additivity associated with the factorizability of probabilities, and hence with the classical conception of the independence of interacting systems, fails precisely in those cases where interference is possible, interference being the other symptom of the failure of independence.

Einstein's solution of the Gibbs paradox comes at the end of a section largely devoted to a calculation of the mean square fluctuation in the number of particles with energies falling within a given infinitesimal range. The resulting expression is a sum of two terms that Einstein interprets in a manner analogous to the interpretation he gave in 1909 to the two terms in his expression for fluctuations in radiation pressure in black-body radiation, only now, of course, we are talking about massive particles rather than massless photons. Thus the first term is said to represent the fluctuations that would arise were the particles composing the gas statistically independent of one another. The second is the interference term (Einstein 1925a, p. 9). So, with regard to their relative independence and their capacity to interfere, material particles behave just like photons; or in the less helpful if more standard gloss on of this result, wave-particle duality is extended finally to material particles as well as to photons.

5. THE TURNING AWAY: 1925-1927

The publication of Einstein's three gas theory papers marked the end of his substantive contributions to the development of quantum theory. From

---

[13]Apart from the vocabulary, Gibbs's distinction between generic and specific phase is virtually identical to the distinction introduced in Planck 1925 between so-called "Quantenzellen" and "Urzellen," except, of course, that Planck knew, in effect, how to weight his "quantum cells" properly, even if he did not know why this is the proper weighting. See Mehra and Rechenberg 1982, p. 616, for further discussion.

this time on, with but few exceptions, Einstein's time and energy were devoted mainly to the search for a unified field theory that would accomodate empirically well-established quantum phenomena within the field-theoretic framework, but not by incorporating wholesale the formal apparatus of the developing quantum theory, an apparatus that Einstein gradually came to regard as fundamentally inadequate. And my hypothesis is that his reason for so regarding the quantum theory, his reason for turning away from active work on it and turning back to unified field theory, was his finally coming to grips with the fact that the quantum theory's way of describing interacting systems is incompatible with the assumptions of separability, locality, and independence that are a necessary part of the field-theoretic approach as he understood it. The clear articulation of this insight was to take most of the rest of his life; but it was clear enough already in 1925 to turn Einstein away from further substantive work on the quantum theory.

In the history of the development of modern quantum mechanics, events began to move rapidly in the spring of 1925. The paper containing Pauli's enunciation of the exclusion principle (Pauli 1925) was published in March. Two months later the first results of the Bothe-Geiger experiments were announced (Bothe and Geiger 1925a), a complete account coming in June (Bothe and Geiger 1925b). The Bothe-Geiger experiment (and the Compton-Simon experiment, the results of which were published in September—Compton and Simon 1925c) convinced most physicists that the particle-like light quanta Einstein had advocated for years would have to be taken seriously, and that Bohr, Kramers, and Slater were wrong in asserting the statistical independence of transition processes in distant atoms. But at the same time, Einstein was arguing in his gas theory papers that material particles as well as light quanta exhibit wave-like interference effects, of the kind recently predicted in de Broglie's thesis (de Broglie 1924), and hence that they cannot be independent, distinguishable particles of the kind posited in classical mechanics, though Einstein himself may have wanted them to be that way. So Einstein was arguing that both light and matter have wave- and particle-like properties. How was this situation to be understood?

What convinced many physicists that a wave-like character of material particles would have to be taken just as seriously as the particle-like character of light was Walter Elsasser's wave-theoretical interpretation of the Ramsauer effect (Ramsauer 1920, 1921a, 1921b) as an interference phenomenon (Elsasser 1925). Ramsauer claimed to have demonstrated experimentally that the mean-free path of electrons passing through certain noble gases goes to infinity (the scattering cross-section goes to zero) as the velocity of the electrons declines. In effect, the atoms of the gas become invisible to the electrons. Remember, these are material particles that Ramsauer was studying. The result was so shocking that many physicists literally did not believe it. Born's reaction is typical. In a letter to Einstein of 29 November 1921 he characterized Ramsauer's claim as "simply insane" (Born 1969, p. 93). But in his note published in July of 1925, in which he cites Einstein's gas theory papers and de Broglie's dissertation, Elsasser showed that the Ramsauer effect could be interpreted quite straightforwardly as a result of interference between the electrons and the atoms in the gas.

One important figure had himself been thinking independently about the Ramsauer effect in much the same way as Elsasser. Here is what Bohr wrote to Hans Geiger on 21 April 1925 in response to Geiger's report of a new experiment by Bothe refuting another implication of the BKS theory (Einstein's reaction to this experiment is discussed below):

> I was quite prepared to learn that our proposed point of view about the independence of the quantum process in separated atoms would turn out to be wrong. . . . Not only were Einstein's objections very disquieting; but recently I have also felt that an explanation of col-

lision phenomena, especially Ramsauer's results on the penetration of slow electrons through atoms, presents difficulties to our ordinary space-time description of nature similar in kind to the those presented by the simultaneous understanding of interference phenomena and a coupling of changes of state of separated atoms by radiation. In general, I believe that these difficulties exclude the retention of the ordinary space-time description of phenomena to such an extent that, in spite of the existence of coupling, conclusions about a possible corpuscular nature of radiation lack a sufficient basis. (Bohr 1984, p. 79)

On the same day Bohr wrote much the same thing to James Franck (director of the institute where Elsasser worked in Göttingen): "It is, in particular, the results of Ramsauer concerning the penetration of slow electrons through atoms that apparently do not fit in with the assumed viewpoint. In fact, these results may pose difficulties for our customary spatio-temporal description of nature that are similar in kind to a coupling of changes of state in separated atoms through radiation. But then there is no more reason to doubt such a coupling and the conservation laws generally" (Bohr 1984, p. 350). What Bohr means here by "customary space-time description" and similar terms is precisely a description like that afforded by classical field theories or classical mechanics, where spatially separated systems are assumed to be separable. What is important is that the Ramsauer effect was seen as evidence for distant correlations between material particles of the kind then commonly represented by wave-theoretical interference.

Many physicists were impressed by Einstein's gas theory papers but puzzled about the new statistics, which they struggled to understand and reinterpret (see, for example, Planck 1925, Schrödinger 1925). Most puzzling was the physical significance of the denial of independence in Bose-Einstein statistics. How could spatially localized material particles fail to be independent? How could they interfere with one another? Encouraged by Elsasser's note and by Einstein's own nod toward de Broglie, several of these thinkers, most notably Landé (1925) and Schrödinger (1926a), sought to develop consistent wave-theoretical interpretations of the new statistics, thinking this the only way to understand the non-independence of interacting systems. Indeed, Elsasser himself, in proposing his interpretation of the Ramsauer effect as a wave-like interference phenomenon, wrote of Einstein's "detour through statistics" (Elasasser 1925, p. 711).

Schrödinger's paper is an important first step toward the development of wave mechanics. It begins as follows:

> In the new gas theory recently developed by A. Einstein, this surely counts, in general, as the essential point, namely, that an entirely new kind of statistics, the so-called Bose statistics, are to be applied to the movements of gas molecules. One's natural instinct rightly resists viewing this new statistics as something primary, incapable of further explanation. On the contrary, there seems to be disguised within it the assumption of a certain dependence of the gas molecules upon one another, or an interaction between them, which nevertheless in this form can only be analyzed with difficulty.
> One may expect that a deeper insight into the real essence of the theory would be obtained if we were able to leave as it was the old statistical method, which has been tested in experience and is logically well founded, and were to undertake a change in the foundations in a place where it is possible without a sacrificium intellectus. (Schrödinger 1926a, p. 95)

Schrödinger goes on to observe that what yields Einstein's gas theory is the application to molecules of the kind of statistics which, applied to "light atoms" (cf. Wolfke 1914b, 1921), gives the Planck formula. But he notes that we can derive the latter using the "natural" statistics if only we apply these statistics to the "ether resonators," that is to the degrees of freedom of radiation. Schrödinger then suggests that the same trick will work with gas molecules, if we simply interchange the concepts, "manifold of energy states" and "manifold of carriers of these states." And he comments:

> Thus one must simply fashion our model of the gas after the same model of cavity radiation that corresponds to what is still not the extreme light quantum idea; then the natural statistics--basically the convenient Planck method for summing states--will lead to the Einstein gas theory. That means nothing else than taking seriously the de Broglie-Einstein undulation theory of moving corpuscles, according to which the latter are nothing more than a kind of "foamy crest" on wave radiation that constitutes the underlying basis of everything. (Schrödinger 1926a, p. 95)

The idea is, as Einstein himself had suggested, to try to understand the failure of independence in Bose-Einstein statistics by means of the wave-theoretical conception of interference. But what this means is that one of the primary motivations behind Schrödinger's development of wave mechanics was the desire to explain the curious statistics of interacting systems, and in particular the failure of probabilities to factorize, that had come to the fore in the Bose-Einstein statistics.

Einstein followed all of these developments closely, as one can best tell from his correspondence with Ehrenfest and Schrödinger. Indeed, the collaboration between Einstein and Schrödinger was such that Schrödinger tried unsuccessfully to persuade Einstein to be listed as coauthor of a paper on degeneracy in a Bose-Einstein gas (Schrödinger 1926b).[14] Through most of 1925 and into early 1926, Einstein was primarily concerned with such investigations of the new Bose-Einstein statistics. He was also following the controversy over the Pauli exclusion principle and Fermi-Dirac statistics for spin-$\frac{1}{2}$ particles like the electron, being at first quite sceptical.

Late in 1925, however, Einstein also turned his attention to the new matrix formalism. The fundamental papers of Heisenberg, Born, and Jordan appeared in the Zeitschrift für Physik between 18 September 1925 and 4 February 1926. From the start, Einstein was critical of this approach. Thus, in a letter to Michele Besso of 25 December 1925, he wrote: "The most interesting thing that theory has yielded recently is the Heisenberg-Born-Jordan theory of quantum states. A real multiplication sorcery in which infinite determinants (matrices) take the place of cartesian coordinates. Most ingenious and so greatly complicated as sufficiently to protect it against a proof of incorrectness" (Speziali 1972, pp. 215-216). In a letter to Ehrenfest of 12 February 1926, he wrote: "I have still busied myself much with Heisenberg-Born. Though with all manner of admiration for the idea, I incline more and more to the view that it is incorrect" (EA 10-130). On 13 March he said much the same in a letter to Lorentz: "Though with all manner of admiration for the spirit that resides in these works, my instinct struggles against this way of conceiving things" (EA 16-594). One month later, on 12 April, he offered this more detailed criticism in a letter to Ehrenfest and praises Schrödinger's wave mechanics by comparison:

> The Born-Heisenberg thing will certainly not be right. It appears not to be possible to arrange uniquely the correspondence of a matrix

---

[14] See Schrödinger to Einstein, 5 November 1925 (EA 22-005); Einstein to Schrödinger, 14 November (EA 22-009); and Schrödinger to Einstein, 4 December (EA 22-010). Einstein did present the paper at the 7 January meeting of the Berlin Academy, in whose proceedings it was published.

function to an ordinary one. Nevertheless, a mechanical problem is supposed to correspond uniquely to a matrix problem. On the other hand, Schrödinger has constructed a highly ingenious theory of quantum states of an entirely different kind, in which he lets the De Broglie waves play in phase space. The things appear in the Annalen. No such infernal machine, but a clear idea and -- "compelling" in its application. (EA 10-135)

And the initial enthusiasm for Schrödinger is echoed in a letter to Lorentz on the same day: "Schrödinger has, in press, a theory of quantum states, a truly ingenious carrying out of de Broglie's idea" (EA 16-600).

Schrödinger's first wave-mechanics paper had appeared in the Annalen on 13 March. Given the background to it explored above, namely, Schrödinger's employment of Einstein's own ideas on the wave nature of material particles in an attempt to understand the physical meaning of Einstein's new statistics, it is not surprising that Einstein was initially favorably inclined toward wave mechanics. But Einstein's enthusiasm was short lived. Schrödinger's proof of the equivalence of wave and matrix mechanics was published on 4 May. It may be coincidence, but the first hint of doubt about wave mechanics on Einstein's part appears at just this time. Thus, a few days earlier, on 1 May, he wrote to Lorentz: "Schrödinger's conception of the quantum rules makes a great impression on me; it seems to me to be a bit of reality, however unclear the sense of waves in n-dimensional q-space remains" (EA 16-604). By 18 June, the doubts about Schrödinger's approach began to crystallize, as Einstein explained in another letter to Ehrenfest: "Schrödinger's works are wonderful--but even so one nevertheless hardly comes closer to a real understanding. The field in a many-dimensional coordinate space does not smell like something real" (EA 10-138). And four days later he voiced similar doubts to Lorentz: "The method of Schrödinger seems indeed more correctly conceived than that of Heisenberg, and yet it is hard to place a function in coordinate space and view it as an equivalent for a motion. But if one could succeed in doing something similar in four-dimensional space, then it would be more satisfying" (EA 16-607).

Throughout the summer and fall of 1926, Einstein continued to worry about the significance of waves in coordinate space. On 21 August he wrote to Sommerfeld: "Of the new attempts to obtain a deeper formulation of the quantum laws, that by Schrödinger pleases me most. If only the undulatory fields introduced there could be transplanted from the n-dimensional coordinate space to the 3 or 4 dimensional! The Heisenberg-Dirac theories compel my admiriation, but to me they don't smell like reality" (Hermann 1968, p. 108). On 28 August he wrote to Ehrenfest:

> Admiringly--mistrustfully I stand opposed to quantum mechanics. I do not understand Dirac at all in points of detail (Compton effect). . . . Schrödinger is, in the beginning, very captivating. But the waves in n-dimensional coordinate space are indigestible, as well as the absence of any understanding of the frequency of the emitted light. I have already written to you that the canal ray experiments have turned out entirely in the sense of the undulatory theory. Hic waves, hic quanta! both realities stand rock solid. Aber der Teufel macht einen Vers darauf (der sich _wirklich_ reimt). (EA 10-144)[15]

And as late as 29 November the same combination of admiration and growing distrust is still evident in a letter to Sommerfeld: "The successes of

---

[15]This last remark is untranslatable without loss of meaning. Literally, it means: "But the devil makes a poem about this (that _really_ rhymes). Figuratively it means something like, "only the devil can make sense of it."

Schrödinger's theory make a great impression, and yet I do not know whether it is question of anything more than the old quantum rules, i.e., a question of something corresponding to an aspect of the real events. Has one really come closer to a solution of the riddle?" (EA 21-356). But surely Einstein's most famous remark from this period is the one found in his letter to Born of 4 December: "Quantum mechanics very much commands attention. But an inner voice says that that is still not the real thing. The theory delivers much, but it hardly brings us closer to the secret of the old one. In any case, I am convinced that He does not play dice. Waves in 3n-dimensional space, whose velocities are regulated by potential energy (e.g., rubber bands) . . ." (Born 1969, pp. 129-130). Remember that it was Born himself who earlier in the summer of 1926 had first introduced the probabilistic interepretation of the Schrödinger wave function (Born 1926a, 1926b).

After the turn of the year, the doubts about wave mechanics finally hardened into conviction. Thus Einstein wrote on 11 January 1927 to Ehrenfest: "My heart is not warmed by the Schrödinger business—it is noncausal and altogether too primitive" (EA 10-152). And the same hardening of Einstein's opinion is clear from a letter to Lorentz of 16 February 1927: "The quantum theory has been completely Schrödingerized and has much practical success from that. But this can nevertheless not be the description of a real process. It is a mystery" (EA 16-611). Within two months this conviction was to evolve into a sharp and penetrating critique.

What was Einstein doing during this period that might help to explain his growing disenchantment with the new quantum mechanics of Heisenberg, Dirac, Born, Jordan, and Schrödinger, a theory whose development owed so much to the stimulus of Einstein's own investigations, including most recently his gas theory papers? On the one hand, Einstein—always the good empiricist—was paying careful attention to new experimental developments that might shed light on the quantum puzzles, especially those probing distant correlations and wave-like interference phenomena. Thus, for example, in the above-cited 13 March letter to Lorentz he reports with delight the results of an experiment by Bothe (Bothe 1926) yielding another refutation of the BKS theory and a vindication of his own light quantum hypothesis.

In Bothe's experiment, a piece of copper foil is weakly irradiated with x-rays, producing flouresence radiation (approximately 2 events per second), nearly all of which is captured in two oppositely situated Geiger counters perpendicular to the incident x-rays. According to the BKS theory, the radiation emitted from an atom is represented by a nearly spherical wave emanating from the atom; the wave determines the probability of subsequent absorptions in all parts of space reached by the wave. There would therefore be a nonvanishing probability that radiation will be absorbed simultaneously in each chamber, that is to say a small but significant probability that coincidences will be detected. But according to Einstein's light quantum hypothesis, emission is a directed process; if the emitted quantum is absorbed in one chamber, there is no chance of a simultaneous absorption in the other chamber. Thus, no coincidences should be detected, except those few arising accidentally from nearly simultaneous emissions. What Bothe found was a coincidence rate far lower than would be expected on the BKS theory, a rate very close to that predicted by the light quantum hypothesis. Einstein reported the results to Lorentz as Bothe described them: "He found complete statistical independence of the absorption events" (EA 16-594). But as Einstein was surely aware, the result could as well be described as showing a strong statistical dependence between events in the two chambers, a perfect correlation between detection in one chamber and non-detection in the other.

On 16 March, three days after reporting to Lorentz the results of the Bothe experiment, and three days after Schrödinger's first wave mechanics paper (Schrödinger 1926c) appeared in the Annalen, Einstein submitted a note to _Die Naturwissenschaften_ (Einstein 1926a) in which he himself proposed another experiment. Something about the old quantum theory that had long troubled Einstein was the fact that the frequency of radiation emitted from an atom is not related to any intrinsic periodicity of the atom itself, such as periodic mechanical motions of charge carriers (electrons) within the atom, as it should be according to classical Maxwellian electrodynamics. Instead, the frequency is related only to the difference in energy between two stationary states. But then, so Einstein seems to have reasoned, radiation of identical frequency emitted by different atoms should not cohere. Einstein's experiment was designed to exhibit such coherence, if it existed, in the transverse radiation emitted by separate atoms in an atomic beam (canal ray). There was controversy about whether the experimental design was sufficient to yield an unambiguous decision,[16] but when the findings were finally reported (Rupp 1926), Einstein at least took them as confirming coherence, even if he was not entirely happy with that result, preferring, as he did, a world of independent light quanta and particles.[17] Here then we have another experiment, this one initiated by Einstein himself, which interested Einstein primarily for the light it shed on the peculiarities of the quantum mechanical account of interacting systems. In this case it was a matter of the peculiar interference exhibited by systems that would have to be regarded, from a classical point of view, as wholly independent. Ironically, Einstein was yet again contributing to showing that the world may not be the way he wanted it to be.[18]

There is other evidence that Einstein was brooding about the strange quantum mechanics of composite systems in the spring of 1926, and this specifically in connection with Schrödinger's wave mechanics. One month after his proposal for the canal ray interference experiment, on 16 April, Einstein wrote to Schrödinger that the new theory showed "true genius," but that he was troubled by one feature of it:

> With justified enthusiasm, Herr Planck has shown me your theory, which I too have then studied with the greatest interest. In the course of this study, one doubt has occurred to me, which you can hopefully banish for me. If I have two systems that are not at all coupled with one another, and E_1 is a possible value of the energy of the first according to quantum mechanics, with E_2 such a value for the second system, then $E_1 + E_2 = E$ must be such a value for the total system composed of the two. But I do not see how your equation

$$\text{div grad } \phi + \frac{E^2}{h^2(E - \Phi)} \phi = 0$$

> should express this property.
> So that you will see what I mean, I set down another equation that would satisfy this requirement:

$$\text{div grad } \phi + \frac{E - \Phi}{h^2} \phi = 0$$

---

[16] See, for example, Schrödinger to Einstein, 23 April 1926 (EA 22-014), and Einstein to Schrödinger, 26 April (EA 22-018); see also Joos 1926, and Bohr to Einstein, 13 April 1927 (Bohr 1985, pp. 418-421).

[17] See Einstein 1926b; see also Einstein to Ehrenfest, 18 June 1926 (EA 10-138), and 28 August (EA 10-144), as well as Einstein to Lorentz, 22 June 1926 (EA 16-607).

[18] In his letter to Ehrenfest of 24 November 1926 (EA 10-148), Einstein alludes to yet another experiment that must have been concerned with the same cluster of problems: "Our experiment on the Compton effect is still not ready, but it will certainly succeed." I have been able to determine neither the details of the experiment, nor who Einstein's collaborator was.

Then the two equations

$$\text{div grad } \phi_1 + \frac{E_1 - \Phi_1}{h^2} \phi_1 = 0 \text{ (valid for the phase space of the 1st system)}$$

$$\text{div grad } \phi_2 + \frac{E_2 - \Phi_2}{h^2} \phi_2 = 0 \ (\ ''\quad ''\quad ''\quad ''\quad ''\quad ''\quad '' \text{ 2nd } '' \ )$$

have as a consequence

$$\text{div grad } (\phi_1 \phi_2) + \frac{(E_1 + \Phi_1) - (E^2 + \Phi_2)}{h^2}(\phi_1 \phi_2) = 0 \quad \text{(valid in combined q-space)}$$

One requires for the proof only to multiply the equations with ϕ_1 or ϕ_2 and then add. $\phi_1 \phi_2$ would thus be a solution to the equation for the combined system belonging to the energy value $E_1 + E_2$.

I have tried in vain to establish a relation of this kind for your equation. (EA 22-012)

In fact, Einstein had misremembered the Schrödinger equation, which was pre-cisely the one Einstein proposed as possessing the desired additivity prop-erty. Einstein noted the error himself in a postcard to Schrödinger of 22 April (EA 22-013), which must have crossed in the mail Schrödinger's letter of 23 April (EA 22-014) pointing out the same thing. But as slips of memory go, this one is interesting for what it reveals about Einstein's concerns. And that the description of composite systems was his concern is made evi-dent in his next postcard to Schrödinger, dated 26 April, after receipt of Schrödinger's letter of the 23rd. After again apologizing for his error, Einstein writes: "I am convinced that you have found a decisive advance with your formulation of the quantum condition, just as I am convinced that the Heisenberg-Born path is off the track. There the same condition of system-additivity is not fulfilled" (EA 22-018). Clearly, Einstein had not yet seen Schrödinger's proof of the equivalence of wave and matrix mechanics (Schrödinger 1926e).

It would take Einstein another year to pinpoint the difficulty with wave mechanics that he was just beginning to sniff out. Here in the spring of 1926 he was arguing that if there exist two solutions of the Schrödinger equation, ϕ_1 and ϕ_2, for two different systems, then $\phi_1 \cdot \phi_2$ should also be a solution for the joint system, as it indeed is. What he was to argue in the spring of 1927 was that if ϕ is any solution of the Schrödinger equation for a joint system, then it ought to be equivalent to a product, $\phi_1 \cdot \phi_2$, of sepa-rate solutions for the separate subsystems, which is generally not the case.

His growing doubts about quantum mechanics are apparently what led Ein-stein in late summer 1926 to turn his attention back to a problem connected with general relativity that he had neglected since 1916. The problem was that of deriving the equations of motion for a test particle from the gener-al relativistic field equations.[19] Einstein himself had dealt with the issue in an approximate way in 1916, showing that particles follow geodesic paths (at least if the particle's own gravitational field is neglected), and it had in the meantime been investigated by a number of others, including Hermann Weyl and Arthur Eddington, who established exact results. Curious-ly, however, Einstein seems to have been largely unaware of this work when

---

[19]The first mention of the problem of motion that I can find in Ein-stein's correspondence from this period is in a letter to Besso of 11 August 1927, apparently written while Einstein was visiting in Zurich. It may thus be the case that his interest in returning to the problem was stimulated by conversations with Schrödinger, who was himself then teaching at the Univer-sity of Zurich.

he tackled the problem in 1926 as if it were still an open matter. He eventually submitted two papers on the subject to the Berlin Academy--in January (Einstein and Grommer 1927) and November (Einstein 1927d) 1927--which largely reproduce the results of the earlier work.[20]

Exactly what Einstein hoped to achieve is not clear, though he seems to have hoped that deriving the equations of motion for elementary particles from general relativity would lead to progress in the quantum theory, an earlier attempt to tie gravitation to quantum mechanics by means of an overdetermined set of field equations having failed (Einstein 1923). Thus, in summarizing the results of the first paper, Einstein and Grommer write: "The progress achieved here lies in the fact that it is shown for the first time that a field theory can include a theory of the mechanical behavior of discontinuities. This can be of significance for the theory of matter, or the quantum theory" (Einstein and Grommer 1927, p. 13). And in a letter to Weyl of 26 April 1927 (this in reply to Weyl's letter of 3 February complaining about Einstein's neglect of Weyl's own earlier work on the subject), Einstein writes:

> I attach so much importance to the whole issue because it would be very important to know whether or not the field equations are to be seen as refuted by the facts of the quanta. One is indeed naturally inclined to believe this and most do believe it. But until now still nothing appears to me to have been proved about this.
> The new results in the quantum domain are really impressive. But in the depths of my soul I cannot reconcile myself to this head-in-the-sand conception of the half-causal and half-geometrical. I still believe in a synthesis of the quantum and wave conceptions, which I feel is the only thing that can bring about a definitive solution. (EA 24-088)

More specifically, Einstein seems from the start to have hoped that some of the characteristically non-classical features of the quantum theory might result; at least so it seems from slightly despairing negative remarks, like this in his letter to Ehrenfest of 24 November 1926: "The equations of motion of singularities can really be derived relativistically. But it appears that absolutely nothing 'unclassical' is to be obtained thereby" (EA 10-148). The despair changed to hope early in 1927, after the publication of the first paper on the equations of motion. Thus, on 11 January 1927, he writes to Ehrenfest: "The problem of motion has become pretty, even if there is still a slight snag in it. In any case, it is interesting that the field equations can determine the motion of singularities. I even think that this will once again determine the development of quantum mechanics, but the way there is still not to be perceived" (EA 10-152). And on 5 May he writes, again to Ehrenfest: "I published the paper on the relativistic dynamics of the singular point indeed a long time ago. But the dynamical case still has not been taken care of correctly. I have now come to the point where I believe that results emerge here that deviate from the classical laws of motion. The method has also become clear and certain. If only I would calculate better! . . . It would be wonderful if the accustomed differential equations would lead to quantum mechanics; and I do not regard it as being at all out of the question" (EA 10-162). The hope was not realized, but as late as November 1927, when he published his second note on the problem of motion, Einstein continued to regard the question as an open one.

Einstein was looking to general relativity to provide equations of motion for elementary particles, because he thought it one of the principal

---

[20] For the history of this problem, see Havas 1989, upon which I have relied extensively.

shortcomings of Schrödinger's wave mechanics that it failed to do so. In effect, what was lacking in the new quantum mechanics, as was soon made vivid by Heisenberg's enunciation of the uncertainty relations (Heisenberg 1927), was the concept of a world-line or a trajectory, establishing which is the aim of the problem of motion. Thus, the criticism of Schrödinger in the above-quoted letter of 11 January 1927 to Ehrenfest concludes: "I do not believe that kinematics must be discarded." Moreover, given his preference for a view of nature in which the separability of interacting systems is assumed, Einstein would naturally look to field theories, like general relativity, to supply the want in quantum mechanics.

Exactly how a field theory might accomplish the end of saving the notion of independent systems while reproducing the empirically established quantum facts was a question Einstein did not prejudge. He seems ready to consider a number of alternatives, but all within the larger framework of the field-theoretic way of individuating systems. So it is no accident, for example, that in mid-1926 and early 1927 he begins to show renewed interest in, and later genuine enthusiasm for five-dimensional Kaluza-Klein theories, a subject he had touched upon four years before (Einstein and Grommer 1923), and which had been revived by Oskar Klein in April 1926 as a way of trying to unify quantum mechanics and general relativity (Klein 1926). In a letter to Ehrenfest of 18 June 1926, for example, Einstein concludes a paragraph criticizing Schrödinger, with the remark: "I am curious about what Herr Klein has found; give him my best" (EA 10-138); and on 3 September, after the July publication of Klein's paper, he remarks to Ehrenfest: "Klein's paper is beautiful and impressive" (Pais 1982, p. 333). On 16 February 1927, at about the time of two short notes of his own on the Kaluza-Klein theory (Einstein 1927a), he commented in a letter to Lorentz: "It turns out that the unification of gravitation and Maxwell's theory by means of the five-dimensional theory (Kaluza-Klein-Fock) is accomplished in a completely satisfactory way" (EA 16-611). And on 5 May he writes Ehrenfest: "The last paper by O. Klein pleased me very much; he really appears to be a level-headed fellow" (EA 10-162). By early the following year, Einstein had given up entirely the hope that equations of motion derived from general relativity would solve the quantum problem, looking now exclusively to Kaluza-Klein theories, as he explained to Ehrenfest on 21 January 1928: "I think I told you that the derivation of the law of motion according to the rel. theory has finally succeeded. But it simply comes out classically. I think that Kaluza-Klein have correctly indicated the way to advance further. Long live the 5th dimension" (EA 10-173; as quoted in Havas 1989, p. 249).

On 23 February 1927, shortly after the presentation of his first paper on Kaluza-Klein theories, Einstein gave a talk at the University of Berlin under the title, "Theoretisches und Experimentelles zur Frage der Lichtentstehung" (Einstein 1927b). The only significance of the talk is that it gives us yet another clue to the issues that were claiming Einstein's attention in the spring of 1927. He singles out for attention a recent experiment of Bothe's that is, in a way, a progenitor of the two-slit diffraction experiment (Bothe 1927). Bothe arranged for radiation from a single x-ray source to be divided into two beams, each of which impinges on a paraffin block, P_1 and P_2, respectively. Part of the scattered radiation from each block is then allowed to scatter a second time from a third paraffin block, S, placed midway between the other two, and one then measures the magnitude of the Compton effect in the resulting twice-scattered radiation. Bothe's declared aim was to use the experiment to decide between two different ways in which light quanta may be associated with a wave field. In the first, the entire energy and momentum of the field is taken to be concentrated in a single "super" light quantum. In the second, each quantum has its normal energy, $h\nu$, and momentum, $h\nu/c$, the individual quanta being associated with partial waves, the total wave field being regarded as a product of the activity of many quanta. Bothe argued that the first conception would lead to

an anomalous Compton effect, whereas the second, because of constructive interference between the two coherent partial waves incident upon S, would
yield the same Compton effect as if S were irradiated by only a single beam.
The results decisively favored the second point of view.

Einstein's report of these results is interesting because of how it
differs from Bothe's. Einstein says merely that the results support the
view that "light has a particle-like character, and is thus corpuscular"
(Einstein 1927b, p. 546). This is true, inasmuch as Bothe claimed to have
demonstrated that energy and momentum are associated with light quanta after
the fashion of Einstein's own corpuscular conception of the quanta. But
Bothe went on to point out that the interference crucial to the experiment's
outcome is incompatible with a radically corpuscularian conception of light:
"It thus turns out that the spatio-temporal localization of the quanta does
not go so far as to permit one to speak, generally, of a continuous 'motion.' In this we glimpse the principal result of the investigation" (Bothe
1927, p. 342). And in order to make this point even more vividly, Bothe
published a schematic diagram of the experiment, to help show that a simple
conception of light quanta as strictly localized particles cannot explain
the alternating light and dark bands in a typical interference pattern. Of
course, Einstein too understood that interference phenomena ruled out a radical corpuscularian conception of light. Indeed, he immediately follows his
characterization of the Bothe experiment with the remark: "But other characteristics of light, the geometrical characteristics and the interference
phenomena, cannot be explained by the quantum conception" (Einstein 1927b,
p. 546), and he ended his talk thus: "What nature demands of us is not a
quantum theory or a wave theory, instead nature demands of us a synthesis of
both conceptions, which, to be sure, until now still exceeds the powers of
thought of the physicists" (Einstein 1927b, p. 546). But Einstein's not
mentioning that Bothe's experiment itself dramatically revealed this very
duality suggests that his instinctive sympathies still lay with the radical
corpuscularian view.

It is against this background that we must assess what is assuredly
Einstein's most interesting critical comment on the quantum theory from this
period. At the meeting of the Berlin Academy on 5 May 1927, Einstein presented a paper entitled "Does Schrödinger's Wave Mechanics Determine the
Motion of a System Completely or Only in the Statistical Sense?" ["Bestimmt
Schrödinger's Wellenmechanik die Bewegung eines Systems vollständig oder nur
im Sinne der Statistik?"]. As word of the talk spread, it evidently aroused
considerable interest, as witness Heisenberg's letter to Einstein of 19 May,
where he writes: "In a roundabout way, through Born, Jordan, I learned that
you had written a paper in which you put forward the same points that you
advanced in the recent discussion, namely, that it would still be possible
to know the paths of corpuscles more exactly than I would like. Now I naturally have a burning interest in this. . . . I do not know whether you would
find it very immodest if I might ask you for any proofs of this work?" (EA
12-173). And, at least initially, Einstein thought he had established a secure result, writing to Ehrenfest on 5 May: "I have also now carried out a
little investigation concerning the Schrödinger business, in which I show
that, in a completely unambiguous way, one can associate definite movements
with the solutions, something which makes any statistical interpretation un-
necessay" (EA 10-162). For reasons that we will explore shortly, the work
was never published, but a manuscript version survives in the Einstein Archive (EA 2-100).

Einstein's idea for associating definite movements with any solution of
the Schrödinger equation was the following. Given any solution, Ψ, for definite total energy E and potential energy Φ, it is possible to express div
grad Ψ as a sum of n terms Ψ_{ab}, to each of which we can associate a definite
"direction" in n-dimensional configuration space; the n "directions" will be

defined separately for the value of Ψ at each point in configuration space. Following Schrödinger, Einstein styled div grad Ψ a "metric" in configuration space, calling the Ψ_{ab} the "tensor of Ψ-curvature" and div grad Ψ the scalar of this tensor. He then showed that the total kinetic energy of the system can also be expressed as a sum of terms, one corresponding to each of the "directions" in configuration space, and that, having thus decomposed the kinetic energy, one can associate with each term in the expression for the total kinetic energy a "velocity" in the corresponding "direction." Whether one can make physical sense out of these "velocities" is not clear; but an answer to the question is not essential to our story.

What is essential is the reason Einstein himself gave for abandoning this effort. The copy in the Archive has attached to it an extra sheet headed "Nachtrag zur Korrektur" ["Added in Proof"]. Pais reports "that the paper was in print when Einstein requested by telephone that it be withdrawn" (Pais 1982,p. 444). One might guess that the addition helps to explain the withdrawal:

Added in proof. Herr Bothe has in the meantime calculated the example of the anisotropic, two-dimensional resonator according to the schema indicated here and thereby found results that are surely to be rejected from a physical standpoint. Stimulated by this, I have found that the schema does not satisfy a general requirement that must be imposed on a general law of motion for systems.
Consider, in particular, a system Σ that consists of two energetically independent subsystems, Σ_1 and Σ_2; this means that the potential energy as well as the kinetic energy is additively composed of two parts, the first of which contains quantities referring only to Σ_1, the second quantities referring only to Σ_2. It is then well known that

$$\Psi = \Psi_1 \cdot \Psi_2 \ ,$$

where Ψ_1 depends only on the coordinates of Σ_1, Ψ_2 only on the coordinates of Σ_2. In this case we must demand that the motions of the composite system be combinations of possible motions of the subsystems.
The indicated scheme does not satisfy this requirement. In particular, let μ be an index belonging to a coordinate of Σ_1, ν an index belonging to a coordinate of Σ_2. Then $\Psi_{\mu\nu}$ does not vanish. . . .
Herr Grommer has pointed out that this objection could be taken care of by means of a modification of the stated schema, in which we employ not the scalar Ψ itself, but rather the scalar $\lg \Psi$ for the definition of the principal directions. The elaboration of this idea should occasion no difficulty, but it will only be presented when it has been shown to work in specific examples. (EA 2-100)

My guess is that the article was withdrawn when Einstein realized that Grommer's suggested route around the non-separability problem failed to work.

This is a crucial text for my argument that the separability problem was all along at the forefront of Einstein's worries about the shortcomings of quantum mechanics. He says here, simply, that separability is a necessary condition on any theory aiming to describe the motions of physical systems. And while the specific instance of non-separability discussed in the "Nachtrag" concerns Einstein's own refinement of Schrödinger's wave mechanics, he surely realized that exactly the same problem infects Schrödinger's original theory. Remember that just one year earlier he had wrongly criticized that theory for failing to satisfy the converse condition (that the product, $\phi_1 \cdot \phi_2$, of any two solutions for the separate systems, ϕ_1 and ϕ_2, should also be a solution for the joint system). Now he has found what is, for him, the right criticism. For Schrödinger's theory fails to satisfy the requirement that any solution of the Schrödinger equation for the composite system be expressible as a product of solutions for the separate subsystems. Quantum mechanics implies interference effects between the two subsystems that make no sense from the point of view of the classical model of independent particles. This is the same problem that first presented itself to Einstein in his 1909 papers on radiation and surfaced again in the curious quantum statistics for material particles in Einstein's 1924-1925 gas theory papers. And it is the same problem that lay behind Einstein's own argument for the incompleteness of quantum mechanics first elaborated in his correspondence with Schrödinger in the summer of 1935, right after the publication of the EPR paper, where the separability problem had been obscured.

6. THE 1927 SOLVAY MEETING

Most of us know about the sequence of thought experiments by which Einstein sought to convince Bohr and others of the inadequacies of quantum mechanics through Bohr's account of his dispute with Einstein (Bohr 1949). On the whole, it is an accurate account, confirmed in some points of detail by other contemporary evidence. But it is not the whole story, and on at least one crucial point—the aim of Einstein's famous "photon-box" thought experiment, it is seriously in error, as are other standard accounts deriving in part from it, such as Jammer's history of Einstein's objections to quantum mechanics (Jammer 1974, 1985). And, even more importantly, most readers of Bohr's review article are unlikely to realize that non-separability was the main issue over which Bohr and Einstein were really arguing. Einstein, of course, understood perfectly well what the issue was (the "problem" is EPR):

Of the "orthodox" quantum theoreticians whose position I know, Niels Bohr's seems to me to come nearest to doing justice to the problem. Translated into my own way of putting it, he argues as follows: If the partial systems A and B form a total system which is described by its ψ-function $\psi(AB)$, there is no reason why any mutually independent existence (state of reality) should be ascribed to the partial systems A and B viewed separately, not even if the partial systems are spatially separated from each other at the particular time under consideration. The assertion that, in this latter case, the real situation of B could not be (directly) influenced by any measurement taken on A is, therefore, within the framework of quantum theory, unfounded and (as the paradox shows) unacceptable. (Einstein 1949, pp. 681-682)

What I want now to do is to review the history of Einstein's Gedankenexperimente with the explicit aim of showing how, through it all, non-separability was the real issue that Einstein was trying to bring to the fore.

The first of the famous Gedankenexperimente dates from the 1927 Solvay meeting, held in Brussels from 24 through 29 October. As reported by Bohr (1949, pp. 211-218), Ehrenfest (letter to Goudsmit, Uhlenbeck, and Dieke, 3 November 1927, reprinted in Bohr 1985, pp. 415-418), and others (for example, Heisenberg 1967, pp. 107-108), most of the interesting discussion between Bohr and Einstein took place outside of the organized conference sessions, at breakfast and during walks between the hotel and the meeting. What exactly Bohr and Einstein discussed is not as clear as it might be, because the records left by them differ in crucial ways. They agree in placing at the center of those discussions the precursor of what we now call the single-slit diffraction experiment. But what that experiment was supposed to show, and what else the dicussion touched upon is not clear.

Consider first the published version of Einstein's contribution to the general discussion, which seems to have taken place on Friday, 28 October, just before the close of the conference (Einstein 1927c). Since the discussion took place near the end of the conference, and since Einstein submitted the written version of his remarks a month later,[21] and thus with ample time for reflection, it seems safe to assume that this version represents in some sense the culmination of Einstein's thinking on that occasion.

After sketching the single-slit experiment, Einstein remarks that there are two ways of interpreting the quantum theory in such a context. What he calls "Interpretation I" is essentially the ensemble interpretation that he insisted in later years is the only tenable way of understanding the quantum theory. On this view, the wave function is associated with a large collection of similar systems; in the present case Einstein says it refers to an "electron cloud" corresponding to "an infinity of elementary processes" (p. 101). "Interpretation II," presumably that favored by defenders of the theory like Heisenberg and Schrödinger, regards the quantum theory as "a complete theory of the individual processes" (p. 101), meaning that a wave function is associated with each individual electron, providing a maximally complete description of its behavior. Having remarked that it is only "Interpretation II" that permits us to explain energy-momentum conservation in individual events and thus results such as those found in the Bothe-Geiger experiment, Einstein says that he nevertheless wants to make some criticisms of this interpretation:

> The scattered wave moving towards P does not present any preferred direction. If $|\psi|^2$ was simply considered as the probability that a definite particle is situated at a certain place at a definite instant, it might happen that one and the same elementary process would act at two or more places of the screen. But the interpretation according to which $|\psi|^2$ expresses the probability that this particle is situated at a certain place presupposes a very particular mechanism of action at a distance which would prevent the wave continuously distributed in space from acting at two places of the screen. In my opinion one can only counter this objection in the way that one does not only describe the process by the Schrödinger wave, but at the same time one localizes the particle during the propagation. I think that de Broglie is right in searching in this direction. If one works exclusively with the Schrödinger waves, interpretation II of $|\psi|^2$ in my opinion implies a contradiction with the relativity postulate.
>   I would still like briefly to indicate two arguments which seem to me to speak against viewpoint II. One is essentially connected with a multidimensional representation (configuration space) because only this representation makes possible the interpretation of $|\psi|^2$ belonging to interpretation II. Now, it seems to me that there are objections of principle against this multidimensional representation. In fact, in this representation two configurations of a system which only differ by the permutation of two particles of the same kind are represented by two different points (of configuration space), which is not in agreement with the new statistical results. Secondly, the peculiarity of the forces of acting only at small spatial distances finds a less natural expression in the configuration space than in the space of three or four dimensions. (Einstein 1927c, pp. 102-103)

This text should be better known, if only for Einstein's having here raised, for the first time that I know, the problem of the non-relativistic character of wave-packet collapse. (This criticism echoes the problem of distant

---

[21] The German manuscript of a fragment of Einstein's remarks (EA 16-617), included in a letter of Einstein to Lorentz (the conference organizer) of 21 November 1927 (EA 16-615), carries a notation in an unknown hand in the upper left-hand corner: "Allg. Disk Freitag" ("Gen[eral] Disc[ussion] Friday").

correlations between events of detection and non-detection explored in Bothe's 1926 experiment, the one whose results Einstein excitedly reported to Lorentz. And on at least one earlier occasion, Einstein had worried that matrix mechanics might not be generally covariant, though for entirely different reasons; see Einstein to Ehrenfest, 12 February 1926, EA 10-130). But what interests me more is his next criticism, the one having to do with the new statistics.

Recall Einstein's letter to Schrödinger of 28 February 1925, written right after publication of his second and third gas theory papers, where he so clearly explained to Schrödinger both the structure of the two-particle state space presupposed by Bose-Einstein statistics and how this structure is related to the failure of statistical independence between two Bose-Einstein particles. Though he did not say so explicitly--there was no need for one master of statistics to remark on such a triviality to another master--one could have made the point about the failure of independence by noting that the two-particle state space is not simply a product of the two one-particle state spaces, as would be the case with classical, two-particle Boltzmann statistics. What Einstein is now saying is that something is fundamentally wrong with Schrödinger's employment of configuration space, because the two-particle configuration space is the product of the two one-particle configuration spaces, contrary to what must be the case in order to derive Bose-Einstein statistics. Of course Einstein is mistaken here, but not about the structure of the two-particle configuration space. His error is his not understanding that the state space of Schrödinger's (and Heisenberg's) quantum mechanics is not configuration space, but instead a rather differently structured Hilbert space (in the now-standard representation), and that in the two-particle Hilbert space, only a single ray (vector), and hence a single quantum state, is associated with the two mentioned configurations, so that the derivation of the novel statistics proceeds without difficulty. But the fact that Einstein was thus mistaken is less important for our purposes than the fact that he was still brooding about the quantum mechanics of composite systems.

One thing puzzles me about Einstein's thinking at this time. Earlier in 1927, in May, he had identified non-separability as a principal failing, from his point of view, of Schrödinger's wave mechanics. But now, in October, he is still committed, apparently, to the new statistics he had helped to introduce, and clearly aware that the novelty of these statistics is connected to the failure of traditional assumptions about the statistical independence of systems, which is to say the failure of the probabilities to factorize. From Born's statistical interpretation of the wave function, with which Einstein was well-acquainted, it is but a short step to making the connection between the non-factorizability of the probabilities in Bose-Einstein and Fermi-Dirac statistics and the non-separability of two-particle wave functions. Why Einstein seems not yet to have made that connection is a mystery to me, all the more so since the derivation of the new statistics from wave-mechanical fundamentals had already been accomplished by Dirac (1926).

Surprisingly, virtually none of Einstein's published objections to the quantum theory at the 1927 Solvay meeting are reported in Bohr's well-known account (Bohr 1949), and this in spite of the fact that Bohr footnotes that publication (Bohr 1949, p. 212). Bohr does allude to the wave-packet collapse problem: "The apparent difficulty, in this description, which Einstein felt so acutely, is the fact that, if in the experiment the electron is recorded at one point A of the plate, then it is out of the question of ever observing an effect of this electron at another point (B), although the laws of ordinary wave propagation offer no room for a correlation between two such events" (Bohr 1949, pp. 212-213). But he quickly moves on to emphasize a different cluster of issues:

Einstein's attitude gave rise to ardent discussions within a small circle, in which Ehrenfest . . . took part in a most active and helpful way. . . . The discussions . . . centered on the question of whether the quantum-mechanical description exhausted the possibilities of accounting for observable phenomena or, as Einstein maintained, the analysis could be carried further and, especially, of whether a fuller description of the phenomena could be obtained by bringing into consideration the detailed balance of energy and momentum in individual processes. (Bohr 1949, p. 213)

And then, while claiming to "explain the trend of Einstein's arguments," Bohr goes on to introduce the familiar refinements--a movable diaphragm and the addition of a second diaphragm with two slits--all by way of elaborating his own complementarity interpretation that precludes measurements more accurate than those permitted by the uncertainty relations on the grounds that the requisite experimental arrangements would be mutually exclusive.

We have no detailed independent record of the 1927 discussions between Einstein and Bohr, so for all we know this may well be an accurate account. It is true that Einstein had doubts about the uncertainty relations, and it is true that he held out hope for a more complete fundamental theory. Moreover, the only piece of contemporary evidence of which I know, namely, Ehrenfest's letter to Goudsmit, Uhlenbeck, and Dieke of 3 November 1927, largely confirms Bohr's account:

It was delightful for me to be present during the conversations between Bohr and Einstein. Like a game of chess. Einstein all the time with new examples. In a certain sense a sort of Perpetuum Mobile of the second kind to break the UNCERTAINTY RELATIONS. Bohr from out of the philosophical smoke clouds constantly searching for the tools to crush one example after another. Einstein like a jack-in-the-box: jumping out fresh every morning. Oh, that was priceless. But I am almost without reservation pro Bohr and contra Einstein. (Quoted in Bohr 1985, p. 38)

Still, my instincts tell me that something is not right about the Bohr (and Ehrenfest) account of the Bohr-Einstein discussion; at the very least they put the emphasis in the wrong place.

There is no reason to doubt that Einstein offered Gedankenexperimente aiming (at least in part) to exhibit violations of the uncertainty relations; it would have been the kind of intellectual game that Einstein so enjoyed. Moreover, he had a special reason to dispute the uncertainty relations with Bohr in particular, because six months earlier, in a letter to Einstein of 13 April, Bohr had deployed the uncertainty relations in disputing Einstein's interpretation of the Rupp experiment as favoring, unambiguously, a wave-like conception of radiation (see Bohr 1985, pp. 418-421). But if the uncertainly relations really were the main sticking point for Einstein, why did Einstein not say so in the published version of his remarks, or anywhere else for that matter in correspondence or in print in the weeks and months following the Solvay meeting? My guess is that it is because any doubts Einstein had about the validity of the uncertainty relations were secondary to his deeper worries about the way quantum mechanics describes composite (interacting) systems.[22]

---

[22]Harvey Brown has also noted that Einstein's published remarks at the 1927 Solvay meeting are not directed toward questioning the uncertainty relations, as Bohr claims, and that these remarks instead anticipate the 1935 EPR argument; see Brown 1981, p. 61.

One clue that supports my interpretation is provided by Bohr's own account of the discussion. As Bohr tells it, the general drift of the discussion between himself and Einstein during the week of the 1927 Solvay meeting was toward an ever more careful consideration of the "detailed balance of energy and momentum in individual processes," meaning, as he explains, an ever more careful consideration of the interaction between the electron and the diaphragm. Einstein evidently argued that by measuring the recoil momentum of the diaphragm, one could predict accurately the lateral component of the electron's momentum, and that when this information was combined with the particle's position, as defined by the aperture in the diaphragm, one could thus predict the precise position at which the electron would hit the screen, whereas the quantum theory yields just probabilities for its hitting various points of the screen. Bohr replied with the standard complementarity argument, namely, that in order to measure the diaphragm's recoil momentum one would have to detach it from its mount so that it could move freely in the lateral direction, but that in doing so one thereby loses all precise knowledge of the particle's position when it passes through the slit, since the slit's location is now indefinite.

What this means is that as the discussion between Bohr and Einstein progressed, what may have begun as doubts about the implications of the uncertainty relation for the description of the individual electron evolved into a discussion of the quantum mechanical two-body problem, the two bodies being the electron and the diaphragm. In Bohr's own words: "As regards the quantum-mechanical description, we have to deal here with a two-body system consisting of the diaphragm as well as of the particle" (Bohr 1949, p. 216). And later on, after describing how the situation is made even more vivid by consideration the two-slit diffraction experiment, Bohr remarks, in words reminiscent of his reply to EPR: "We . . . are just faced with the impossibility, in the analysis of quantum effects, of drawing any sharp separation between an independent behaviour of atomic objects and their interaction with the measuring instruments which serve to define the conditions under which the phenomena occur" (Bohr 1949, p. 218). There is a very good reason why the discussion may have taken these turns: first from a consideration of the adequacy of the wave function as a description of an individual electron to the validity of the uncertainty relations, and then from the uncertainty relations to the quantum mechanical two-body problem and non-separability.

Remember that Einstein had been worrying since at least the early 1920s about how to reconcile a probabilistic description of individual systems with the conservation laws. Einstein's 1916 papers on radiative transformations had shown him that an element of chance would likely have to enter an adequate future quantum theory; and Einstein knew as well that some kind of wave-like character had to be associated with both light quanta and material particles to explain interference effects. Einstein's "ghost fields" or "guiding fields" accomplished both ends. But recall Wigner's report of Einstein's own reasons for never having pushed the idea of "ghost fields" or "guiding fields". It was that if each of two interacting systems is guided independently (in the statistical sense) by a separate "guiding field" one cannot guarantee energy-momentum conservation, and it was his insistence on strict energy-momentum conservation in individual events that determined his opposition to the BKS theory. This problem was solved by Schrödinger's shifting the wave function from physical space to configuration space, but at the price of non-separability, a failure of independence that Einstein knew from his gas theory papers had to be part of the quantum theory but that he still found too bitter a pill to swallow when confronted by it in wave mechanics. Another expression of Einstein's desire that the behavior of systems such as electrons be determined independently by their own states is his insistence that these systems be represented as localized, particle-like systems. As we saw above, the desire for localization was strengthened by Einstein's worry that a non-relativistic action-at-a-distance would be implied by our taking the wave function itself as real. And remember that, as early as 1924, Einstein was arguing that the wave-like aspects of quantum systems should be seen not as something real, but merely as a convenient device for representing interactions between systems.

So what Einstein wanted was an ontology of (1) independently controlled, localized systems, but also (2) systems that satisfy strict energy-momentum conservation. Classical mechanics and classical field theories, including general relativity, manage to reconcile these two desiderata. Quantum mechanics does not.

One can imagine Bohr responding to Einstein's published 1927 Solvay discussion remarks by emphasizing just this point. Einstein said that "Interpretation II," which associates the wave function with individual systems rather than ensembles, is the only acceptable interpretation because it is necessary in order to secure strict energy-momentum conservation. But then he insists that we understand the system to which the wave function is associated to be strictly localized. Bohr would have pointed out that, according to quantum mechanics, one cannot have both. He had already argued this very point forcefully in his address at the Volta Congress in Como one month earlier: "The very nature of the quantum theory thus forces us to regard the space-time co-ordination and the claim of causality, the union of which characterizes the classical theories, as complementary but exclusive features of the description" (Bohr 1927, p. 580).[23] And: "According to the quantum theory a general reciprocal relation exists between the maximum sharpness of definition of the space-time and energy-momentum vectors associated with the individuals. This circumstance may be regarded as a simple symbolical expression for the complementary nature of the space-time description and the claims of causality" (Bohr 1927, p. 582). The key to understanding passages like this is realizing that when Bohr talks about "space-time coordination" or "space-time description," he means the kind of description in which quantum systems such as electrons are regarded as localized, the kind of description Einstein preferred, the kind of description that strongly suggests, if it does not actually entail, the separability of such localized systems. And when Bohr talks about the "claims of causality," he means--as he explains himself--strict conservation of sharply defined energy and momentum.

The discussion would have turned from the uncertainty relations to the quantum mechanics of interacting systems, focussing on the interaction between the electron and the diaphragm, because uncertainty intrudes only when one severs conceptually the physical link between the two interacting systems, that is to say, only when one pretends that two really non-separable systems are separable. Consider the position-momentum uncertainty relationship for, say, the x-axis in connection with two interacting systems, the case made famous in the EPR paper. The total linear momentum after the interaction, $p_1 + p_2$, and the relative separation, $x_1 - x_2$, are compatible observables; both can be defined with arbitrary sharpness. It is only the individual momenta and positions, p_1, x_1 and p_2, x_2, that are incompatible, subject to the uncertainty relations. It is only the pure case, non-factorizable joint state that contains all of the correlations necessary to preserve the link between the positions and momenta of the two systems. If

---

[23] This is quoted from the version published in Nature in April of 1928, but the progenitor of this specific remark can be found in a manuscript dated as early as 12-13 October, and it is otherwise wholly consistent with the argument of even the earlierst suriving manuscripts of the Como talk. For more on the history of the various manuscripts, see Bohr 1985.

one pretends that the two systems are separable, that means employing a mixture over factorized joint states, not a pure case. Such a mixture can preserve the momentum correlations or the position correlations, but not both. If you choose the former, you get strict momentum conservation, but no spatial localizability; if you choose the latter, you get localizability, but no momentum conservation.

Thus, the reason Einstein cannot get both localizability and energy-momentum conservation is because of the non-separability of interacting systems in quantum mechanics. The point deserves emphasis. Non-separability is the basic phenomenon that distinguishes quantum physics from classical physics; uncertainty is merely a symptom. Uncertainty intrudes only when one pretends to describe the properties of an independent system, there being no really independent systems. As Bohr himself was wont to say, "isolated material particles are abstractions, their properties on the quantum theory being definable and observable only through their interaction with other systems" (Bohr 1927, p. 581). In a separable universe, there need be no uncertainty. So even where uncertainty seems to be the issue, quantum nonseparability is the real heart of the matter.

Whether or not the discussion between Einstein and Bohr actually proceeded in this fashion is impossible to say. I offer this scenario as a reconstruction that at least reconciles the otherwise rather different seeming records published by Einstein and Bohr. But I do think it a plausible scenario, one that helps us to make better sense of the later history of Einstein's objections to the quantum theory. And one further piece of documentary evidence strengthens my conviction that worries about non-separability really lay behind the October 1927 controversy with Bohr. Remember that, for Einstein, it is field theories that provide the clearest embodiment of a separable ontology, each point of the underlying manifold being regarded as endowed with its own, separate, well-defined state, say in the form of a metric tensor. Just one month after the Solvay meeting, in late November, Einstein returned again to the problem of motion in general relativity, the problem he had begun exploring with Grommer late in 1926, taking up now specifically the equation of motion for elementary particles like the electron. His second note on this subject was presented to the Berlin academy on 24 November. In the introduction, he says the following about the results of his investigations:

> This result is of interest from the point of view of the general question whether or not field theory stands in contradiction with the postulates of the quantum theory. The majority of physicists are indeed today convinced that the facts of the quanta rule out the validity of a field theory in the customary sense of the word. But this conviction is not grounded in a sufficient knowledge of the consequences of the field theory. For that reason, the further tracing of the consequences of the field theory with regard to the motion of singularities seems to me, for the time being, still to be imperative, this in spite of the fact that a thorough command of the numerical relationships has been accomplished, in another way, by quantum mechanics. (Einstein 1927d, p. 235)

The whole point of trying to derive the equation of motion from field-theoretic first principles is to show that the motion of a particle is wholly determined by the values of the fundamental field parameters, such as the metric tensor, at points of the manifold immediately adjacent to the particle's trajectory. That is to say that successfully deriving such an equation of motion for elementary particles would rule out any non-separability between interacting particles, since the field would mediate the interaction in a purely local fashion.

7. THE PHOTON-BOX AND BEYOND: 1930-1935

After the 1927 Solvay meeting, Einstein's interest in quantum mechanics dropped off markedly. He devoted himself ever more single-mindedly to developing a unified field theory, hoping, but always in vain, to find thereby a deeper field-theoretic foundation for quantum mechanics. Characteristic of his resigned attitude during this time is a remark in a letter to Ehrenfest of 23 August 1928: "I believe less than ever in the essentially statistical nature of events and have resolved to apply the tiny capacity for work that is still given to me according to my own taste, in a manner independent of the contemporary goings on" (EA 10-186). What little he had to say about quantum mechanics was confined to attempts to articulate yet more clearly why he did not think it to be the final word in fundamental physics.

His next major contribution along these lines, the famous "photon-box" Gedankenexperiment, came at the 1930 Solvay meeting, held in Brussels the week of 20-25 October. Our only record is found in Bohr's later recollection (Bohr 1949) and in manuscript notes for a talk Bohr gave at the University of Bristol a year later, on 5 October 1931 (Bohr 1931). Einstein's only recorded comment in the proceedings of the conference, an inconsequential remark after a talk by Pierre Weiss (Solvay 1930, p. 360), has nothing to do with foundational problems.

The details are well-known. A radiation-filled cavity has in its side a shutter controlled by a clock. The shutter opens for an instant at a definite time, allowing the escape of one photon. Weighing the box before and after the release, we can determine the energy of the emitted photon with arbitrary accuracy, and when we combine this result with the known time of emission, we supposedly have a violation of the energy-time uncertainty relation. Bohr's ironic refutation of the experiment is also well known. He pointed out that the weighing requires that the box be accelerated in a gravitational field, and that this affects the rate of the clock just enough to secure agreement with the uncertainty relations. The irony, of course, is that relativity is here invoked to save quantum mechanics.

On the face of it, this is merely another attempt to find a violation of the uncertainty relations, which is indeed all it might be. But there is evidence that, here again, Einstein's real aim may well have been to bring out the peculiariities, from a classical point of view, of the quantum mechanical account of interactions. The evidence is a letter from Ehrenfest to Bohr of 9 July 1931, written immediately after Ehrenfest had visited Einstein in Berlin. According to Jammer (Jammer 1974, pp. 171-172; 1985, pp. 134-135), from whom most of us have learned about the letter, Ehrenfest reported that Einstein no longer wanted to use the photon-box thought experiment to disprove the uncertainty relations, but "for a totally different purpose" (Jammer 1985, p. 134), the implication being that disproving the uncertainty relations had been the original intention behind the photon-box thought experiment. But Jammer has misread the letter. What Ehrenfest really wrote to Bohr is this: "He said to me that, for a very long time already, he absolutely no longer doubted the uncertainty relations, and that he thus, e.g., had BY NO MEANS invented the 'weighable light-flash box' (let us call it simply L-F-box) 'contra uncertainty relation,' but for a totally different purpose" (BSC-AHQP).[24] Einstein may have wanted to dispute the

---

[24] "Er sagte mir, dass er schon sehr lange absolut nicht mehr an die Unsicherheitsrelation zweifelt und dass er also z.B. den 'waegbaren Lichtblitz-Kasten' (lass ihn kurz L-W-Kasten heissen) DURCHAUS nicht 'contra Unsicherheits-Relation' ausgedacht hat, sondern fuer einen ganz anderen Zweck."

uncertainty relations in 1927, but as we see by his own testimony to Ehrenfest, that was not his purpose at the time of the 1930 Solvay meeting.

Ehrenfest goes on to explain Einstein's real intention. What Einstein
wanted, says Ehrenfest, is a "machine" that emits a projectile in such a way
that, _after_ the projectile has been emitted, an inspection of the machine
will enable the experimenter to predict _either_ the value of the projectile's
magnitude A _or_ the value of its magnitude B, these values then being measurable when the projectile returns after a relatively long time, it having
been reflected at some location sufficiently distant ($\frac{1}{2}$ light-year) to in-
sure that there will be a spacelike separation between the projectile and
the machine at the time we inspect the machine. According to Ehrenfest:
"It is thus, for Einstein, beyond discussion and beyond doubt, that, because
of the uncertainty relation, one must naturally choose between the either
and the or. But the [experimentor] can choose between them AFTER the projectile is already finally under way" (BSC-AHQP).[25] And: "It is interesting to get clear about the fact that the projectile, which is already flying
around isolated 'for itself,' must be prepared to satsify very different
'non-commutative' predictions, 'without knowing as yet' which of these predictions one will make (and test)" (BSC-AHQP).[26] The photon-box turns out
to satisfy all of the requirements for such a "machine," the two quantities,
A and B, being respectively, the time of the photon's return and its energy
or color (wavelength).

Jammer is quite right that we see here all of the ingredients of Einstein's later incompleteness arguments, but Jammer's interpretation is skew-
ed by his taking the published EPR argument as a correct guide to Einstein's
views, rather than the quite different version first presented in Einstein's
correspondence with Schrödinger from the summer of 1935, the version featur-
ing the separation principle. Since we know that separability was the main
issue in 1935, we should look for it here in 1931. It's not hard to find.

In fact, the logic of the 1931 version of the photon-box _Gedankenex-
periment_ is almost exactly that of Einstein's own 1935 incompleteness argument. The whole point of placing the reflector $\frac{1}{2}$ light-year away is to
assure a spacelike separation between the inspection of the photon-box and
the projectile. The argument works as a criticism of the quantum theory
only if one assumes that the projectile, when thus separated from the box,
is, in virtue of that separation and its therefore possessing its own independent reality, wholly unaffected by what we do to the box when we inspect it. But quantum mechanics makes a different assumption. It says
that, if we weigh the box, we can predict the color of the returning photon
exactly, but that its time of return will be indefinite, whereas if we check
the clock, we can predict the time of the photon's return exactly, its color
now being indefinite. In other words, quantum mechanics says that the state
we ascribe to the photon depends crucially on what we do to the box. What
Einstein is thus arguing is that classical assumptions about the separability of previously interacting systems lead to different results than the
quantum mechanical account of interactions, and that, if we adhere to these

---

[25] "Es steht also fuer Einstein ausser Discussion und ausser Zweifel,
dass man, wegen der Unsicherheitsrelation natuerlich zwischen dem entweder
und oder waehlen muss. Aber der Frager kann dazwischen waehlen, NACHDEM das
Projectil endgueltig schon unterwegs ist."
[26] "Es ist interessant sich deutlich zu machen, dass das Projectil, das
da schon isoliert 'fuer sich selber' herumfliegt darauf vorbereitet sein
muss sehr verschiedenen 'nichtcommutativen' Prophezeihungen zu genuegen,
'ohne noch zu wissen' welche dieser Prophezeihungen man machen (und pruefen)
wird."

assumptions of separability, as Einstein, the champion of field theories, clearly thought we must, then the quantum theory must be judged incomplete.

Essentially the same Gedankenexperiment reported to Bohr by Ehrenfest was presented by Einstein himself at a colloquium in Berlin on 4 November 1931 (Einstein 1931). The published report leaves the aim of the experiment unclear, but it is interesting because Einstein is said to have stressed, himself, that the two measurements on the box--the weighing or the reading of the clock--cannot both be performed, just as Ehrenfest had reported to Bohr, indicating again that disputing the uncertainty relation is not Einstein's aim. In Bohr's account of his controversy with Einstein, the demonstration that the two parameters cannot be measured simultaneously was precisely his (Bohr's) triumph over Einstein at the 1930 Solvay meeting. But how could Einstein so easily accomodate this point, if it were really such a devasting critique of his original idea. The answer is that Einstein never intended to assert that both measurements could be performed simultaneously on the box, or at least that such a possibility was never a crucial part of the experiment. Bohr, Jammer, and others have taken it to be crucial only because they wrongly believed that the uncertainty relations, rather than non-separability, was Einstein's real target.

That Bohr was still not clear about the real point of Einstein's argument is evident from the fact that in his 5 October 1931 talk at the University of Bristol (Bohr 1931), three months after Ehrenfest's 9 July letter informing him of how Einstein wanted to use the photon-box experiment, Bohr gave a quite different account of the experiment, essentially the same as in his later recollections (Bohr 1949), presenting it as an objection to the uncertainty relations. If Bohr misunderstood Einstein in this way in 1931, how do we know that he was not guilty of exactly the same misunderstanding in October 1930 and in his later recollections?[27]

Einstein spent three months (11 December to 4 March) in the United States in late 1930 and early 1931, mostly at Cal Tech. He evidently spent some of this time talking with his Cal Tech colleagues Richard C. Tolman and Boris Podolsky about his objections to quantum mechanics. On 26 February 1931 they submitted to the Physical Review a note entitled "Knowledge of Past and Future in Quantum Mechanics" (Einstein, Tolman, and Podolsky 1931), which Einstein himself apparently credited primarily to Tolman (see Einstein 1931, p. 23). The stated aim is to show that, contrary to what some had claimed (see, for example, Heisenberg 1930, p. 20), the past behavior of a particle cannot be known any more precisely than its future behavior; but the ETP Gedankenexperiment (to coin a designation) is of interest for our story because it involves a modification of the photon-box arrangement that permits the study of correlations not between the box and the emitted photon, but between two particles both emitted from the box.

The box is now fitted with two holes opened by the same shutter. One of the two emitted particles travels directly to an observer at 0; the other follows a different trajectory, reflected toward 0 at a great distance from the box. Weighing the box before and after the release of the particles allows us to determine their total energy. ETP argue that measuring the time

---

[27] Several other authors have questioned the cogency of Jammer's account of the photon-box thought experiment, arguing as I do (but without having examined Ehrenfest's letter to Bohr of 9 July 1931) that a proof of incompleteness was the real aim. See Hooker 1972, p. 78; Hoffmann 1979, pp. 187, 190; Fine 1979, p. 157; and Brown 1981, pp. 67-69. I highly recommend Harvey Brown's account of the matter for its careful consideration of technical matters, and I thank Brown for drawing my attention to the Hooker and Hoffmann references.

of arrival of the first particle at O, along with its momentum, would enable one to calculate the time when the shutter opened and thus to predict both the time of arrival and the energy of the second particle, contrary to the limitations of the uncertainty relation. That being ruled out by the quantum theory, ETP conclude that it must not be possible to measure both the time of the first particle's arrival and its pre-arrival momentum. This may have been how Tolman meant to use the arrangement. But it obviously lends itself to other uses that may have been of more interest to Einstein. Thus, if the particles are once again taken to be photons (whose velocity is a known constant), then one has the option of measuring either the time of arrival of the first photon or its energy and thus predicting either the time of arrival of the second photon, or its energy (color). But the geometry of the experiment (the great distance of the reflector from O) insures that measurements performed on the first photon cannot affect the second, if we assume separability and locality, and thus that both the time of the second photon's arrival at O and its energy correspond to independently real properties of the photon. Once again, the assumption of separability leads to results in conflict with the quantum theory.

The only direct account that Einstein himself ever gave of the history of these Gedankenexperimente was in an exchange of letters with the Cal Tech physicist, Paul S. Epstein in the latter part of 1945, following the publication of an article by Epstein, "The Reality Problem in Quantum Mechanics," in the June issue of the American Journal of Physics (Epstein 1945). Epstein had introduced his own thought experiment—a variation on the two-slit diffraction experiment—to illustrate the central point of the EPR argument. A beam of light, S, is split by a half-silvered mirror N-N', and each resulting beam, S_1 and S_2, is then reflected again by a perfect mirror, M_1 and M_2, respectively, after which the beams are recombined at a second half-silvered mirror O-O' producing two final beams, S_3 and S_4, each of which enters a detector. Consider a beam S of such low intensity that just one photon at a time passes through the apparatus. Epstein says that if the mirrors M_1 and M_2 are fixed, preventing us from determining, by the mirrors' recoil, which path a given photon travels, the reflected beams S_1 and S_2 are coherent and interfere at the second half-silvered mirror O-O', so that by suitably adjusting the geometry of the arrangement we can make all of the emerging photons go into one detector, say that corresponding to S_3. But if mirrors M_1 and M_2 are movable, so that we can tell which path each photon travels, the beams, S_1 and S_2, are incoherent, there is no interference at O-O', and equal numbers of photons show up, on average, in each detector.

Epstein's analysis of the experiment is somewhat confused. In his first, undated letter, Einstein points out that Epstein had spoken glibly of the ψ-function of one of the photons, ignoring its interaction with the mirror. Einstein explains that Epstein has ignored the non-separability of the joint photon-mirror system: "Now I do not understand the following in your treatment of the mirror example with a light quantum. If a mirror is movable (laterally), then the total system is a system with two types of coordinates (e.g. Q for the mirror, q for the quantum). There is then no ψ (q,t) at all, as long as no 'complete' observation of the mirror is at hand. Then, in terms of the theory, one cannot at all ask how ψ (q,t) is constituted as a function of the time t" (EA 10-581). But then, instead of continuing with an analysis of Epstein's experiment, Einstein sketchs his own preferred way of viewing the matter.

He considers two previously interacting particles "described as completely as possible in the sense of the quantum theory by ψ (Q,q,t)" (EA 10-581). After a sufficiently long time, the particles have separated, and now we ask "in what sense each individual particle corresponds to a real state of affairs" (EA 10-581). To learn something about q, we perform a measurement on Q, and we can arrange the measurement so that "the ψ (q,t) resulting from this measurement and from the $\psi\,(Q,q,t)$ has <u>either</u> a sharp position (for a given value of the time), <u>or</u> a sharp momentum" (EA 10-581). In order to decide what the "real state of affairs" is with regard to the second particle, we ask whether the measurement on the first particle has a real, physical influence on the second particle. If there were such an influence (and Einstein thinks Epstein prefers this view), it would mean the existence of superliminal effects, against which Einstein's "physical instinct" struggles, and it would be difficult to see how such effects could be incorporated in the quantum theory. But if the measurement on the first particle has no physical influence on the second particle, "then all of the determinations for the second particle that result from the possible measurements on the first particle must be true of the second particle, if no measurement at all were performed on the first particle" (10-581). It follows that quantum mechanics is incomplete. We recognize here yet another statement of Einstein's own (non-EPR) post-1935 incompleteness argument. In this there is nothing new. What is significant is what follows in the next letter.

Epstein responded on 5 November 1945 (EA 10-582), confessing that he never really understood the EPR paper, that he had in the meantime restudied it, but was still confused by some of the calculations in it. Einstein answered on 10 November (EA 10-583). He begins by declining to discuss the mathematical questions, saying that Schrödinger had settled them in a thorough treatment shortly after the publication of the EPR paper (Schrödinger 1935, 1936). He says, then, that it may be better if he shows Epstein how he himself first arrived at the incompleteness argument: "I myself first came upon the argument starting from a simple thought experiment. I think it would be best for us if I exhibited this to you" (EA 10-583). The arrangement is the following.

A photon-box can move freely in the x-direction. An observer rides with the box and has at his or her disposal various instruments, including a clock for timing the opening of the shutter and tools with which to measure the box's position. Before starting the experiment, the observer allows the box to come to rest, something that can be determined by means of light emitted from the box being reflected from a distant wall. Of course, knowing that the box is at rest, that is, that it has zero momentum in the x-direction means that its position is unknown. Now the experimenter opens and closes the shutter at a definite time, allowing one photon to emerge. At this point, the experimenter has an option to measure one of two things. He or she can <u>either</u> anchor the box to the reference frame, permitting a precise measurement of the box's position, and thus a prediction of the exact time when the emitted photon will be received at some distant location S, which means, says Einstein, a "sharp determination of the position of the photon." <u>Or</u> the experimenter can make a new measurement of the box's recoil momentum, in which case he or she can predict exactly the energy or color of the emitted photon. What does the experiment show? Einstein says:

> As soon as it has left the box B, the light quantum represents a certain "real state of affairs," about whose nature we must seek to construct an <u>interpretation</u>, which is naturally in a certain sense arbitrary.
> This interpretation depends essentially upon the question: should we assume that the subsequent measurement we make on B <u>physically</u> influences the fleeing light quantum, that is to say, the "real state of affairs" characterized by the light quantum?
> Were that kind of a physical effect from B on the fleeing light quantum to occur, it would be an action at a distance, that propagates with superluminal velocity. Such an assumption is of course logically possible, but it is so very repugnant to my physical instinct, that I am not in a position to take it seriously--entirely apart from the fact that we cannot form any clear idea of the structure of such a process.

Thus I feel myself forced to the view that the real state of affairs corresponding to the light quantum is independent of what is subsequently measured on B. But from that it follows: Every characteristic of the light quantum that can be obtained from a subsequent measurement on B exists even if this measurement is not performed. Accordingly, the light quantum has a definite localization and a definite color.

Naturally one cannot do justice to this by means of a wave function. Thus I incline to the opinion that the wave function does not (completely) describe what is real, but only a to us empirically accessible maximal knowledge regarding that which really exists. . . . This is what I mean when I advance the view that quantum mechanics gives an incomplete description of the real state of affairs. . . .

If one is of the view that a theory of the character of quantum mechanics is definitive for physics, then one must either completely renounce the spatio-temporal localization of the real, or replace the idea of a real state of affairs with the notion of the probabilities for the results of all conceivable measurements. I think that this is the view that most physicists currently have in mind. But I do not believe that this will prove to be the correct path for the long run. (EA 10-583)

Here we have all of the ingredients of Einstein's own post-1935 incompleteness argument, including, most importantly, the separability principle in the form of the assumption that an independent real state of affairs is associated with the light quantum from the moment it leaves the box.

By Einstein's own account, this Gedankenexperiment and, presumably, the indicated interpretation of it, was the starting point from which the incompleteness argument developed. However, Einstein does not say exactly when the experiment first occurred to him, so we cannot insert it at a definite place in our chronology of Einstein Gedankenexperimente on the basis of any direct evidence. Can indirect arguments be brought to bear? If the Bohr-Jammer account is correct, according to which sometime in the summer of 1931 Einstein changed his mind about how to deploy the photon-box thought experiment, then the "Epstein" experiment had to come later. But Ehrenfest's 9 July 1931 letter to Bohr shows that Einstein did not change his mind about the use to which the experiment was to be put, that he intended it from the start as showing the incompleteness of the quantum theory. If that is the case, then the "Epstein" version of the photon-box arguably came first, as Einstein says. And from one point of view, the 1930 Solvay photon-box is sufficiently simpler than the "Epstein" photon-box that it may be regarded as a refinement of the latter; for simply weighing the box is a lot easier than performing the complicated series of momentum measurements sketched in the letter to Epstein.

If the "Epstein" photon-box Gedankenexperiment goes first in the chronology then, we must revise our understanding of how the 1930 Solvay photon-box was to be deployed in line with the analysis given in Ehrenfest's letter to Bohr. The logic of the argument would have been the same as that in the "Epstein" photon-box. After the photon is emitted, the experimenter can measure either the energy of the box, by a second weighing, or the time of emission, and depending upon which measurement he or she makes, a different prediction can be made about the photon. All that Bohr's famous critique concerns is the question, inessential from Einstein's point of view, whether or not the two measurements on the box can be carried out simultaneously.

From this point on, the tendency of Einstein's thinking is clear. He wanted to show the quantum theory to be incomplete, and he wanted to do this by showing that the assumption of separability (plus the assumption of locality), is incompatible with the claim that the theory gives a complete de-

scription of individual systems. All that changes are the details of the
Gedankenexperimente intended to demonstrate this.

The first of these new experiments dates from April of 1932. When Einstein returned from his third annual visit to the United States, his ship
lay over for three days in Rotterdam, where Ehrenfest came from Leyden to
visit him on 4 April (see Jammer 1985, pp. 135-136). The next day, Einstein
wrote to Ehrenfest: "Yesterday you nudged me into modifying the 'box-
experiment' in such a way that it would employ concepts less foreign to the
wave theorists. I do this in the following, where I employ only such idealizations that I know will appear unobjectionable to you" (EA 10-231). The
experiment is a modified Compton scattering experiment, in which the scattered photon is assumed to move along the same axis that the scattering mass
m is free to move along, the photon being reflected at a distant mirror back
to an experimenter who sits near the mass m. Einstein assumes that the mass
m is initially at rest (zero momentum) which means that its location is in-
determinate. He then argues that when the scattered photon returns to the
experimenter, he or she can measure either the photon's momentum, enabling
the experimenter to deduce the momentum of the mass m, or the photon's time
of arrival, enabling the experimentor to deduce the time when the initial
scattering occurred and thus the precise position of m right after the scattering. (There is an obvious error here.) The important point is that we
can thus deduce either the position or the momentum of m, without in any way
disturbing m itself. Einstein's conclusion comes as no surprise: "Thus,
without any experiment on m, it is possible to predict, according to a free
choice, either the momentum or the position of m with in principle arbitrary
accuracy. This is the reason why I feel myself motivated to attribute objective reality to both. It is to be sure not logically necessary, that I
concede" (EA 10-231).

The last documented stage in the development of Einstein's Gedankenexperimente can be dated to sometime during the spring or summer of 1933, when
Einstein was staying in Le Coq sur Mer, Belgium after his return to Europe,
in late March, in the wake of Hitler's Machtergreifung, and before his final
departure for the United States in early September. Léon Rosenfeld was then
a lecturer at the University of Liège, and had just finished his famous
joint paper with Bohr on the measurability of field quantities in quantum
electrodynamics (Bohr and Rosenfeld 1933). He gave a lecture on the topic
in Brussels, which Einstein attended. After the talk, Einstein approached
Rosenfeld wanting to discuss not the topic of the lecture but the general
problem of completeness, about which he said he still felt a certain "un-
easiness" ["Unbehagen"]. Rosenfeld quotes Einstein as follows:

> What would you say of the following situation? Suppose two particles
> are set in motion towards each other with the same, very large, momen-
> tum, and that they interact with each other for a very short time when
> they pass at known positions. Consider now an observer who gets hold
> of one of the particles, far away from the region of interaction, and
> measures its momentum; then, from the conditions of the experiment, he
> will obviously be able to deduce the momentum of the other particle.
> If, however, he chooses to measure the position of the first particle,
> he will be able to tell where the other particle is. This is a per-
> fectly correct and straightforward deduction from the principles of
> quantum mechanics; but is it not very paradoxical? How can the final
> state of the second particle be influenced by a measurement performed
> on the first, after all physical interaction has ceased between them?
> (Rosenfeld 1967, pp. 127-128)

Through the haze of Rosenfeld's again not unbiased recollection we can recognize here the same logic that is by now quite familiar, but elaborated

against a _Gedankenexperiment_ that is growing ever more refined, toward the conceptual purity of the EPR-type experiment.

8. CONCLUSION

After 1935, Einstein's reasons for thinking quantum mechanics incomplete were intimately connected to his firm belief in the separability principle. It is the separability principle that licenses the crucial inference that the undisturbed system in EPR-type _Gedankenexperimente_ has its own unique separate state independently of any measurements we might carry out on the other system. And if, according to the quantum theory, measurements on the other system lead us to ascribe different states (different psi-functions) to the undisturbed system depending upon the kind of measurement we perform on the first system, then it follows that quantum mechanics does not yield a complete description of undisturbed system. But, of course, as Bohr pointed out, quantum mechanics denies the separability of previously interacting systems. Einstein understood quite well that it was this disagreement over separability that stood between him and Bohr, as evidenced by his remark to Schrödinger in the summer of 1935: "One cannot get at the talmudist [Bohr] if one does not make use of a supplementary principle: the 'separation principle'" (Einstein to Schrödinger, 19 June 1935, EA 22-047). But Einstein was committed to separability because of his deeper commitment to field theories, and their associated way of describing interactions in purely local terms, a description that rests fundamentally on the assumption that every system, indeed, every point of the space-time manifold, has its own separate state in the form of well-defined values of the fundamental field parameters like the metric tensor.

In this paper I have been arguing that Einstein's worries over the way quantum mechanics describes interacting systems did not begin in 1935. On the contrary, I have shown that the puzzling behavior of interacting quantum systems had been at the forefront of Einstein's concern from at least 1909 and that these worries began to crystallize in the mid-1920s into the belief that, because it regards interacting systems as non-separable, quantum mechanics would be fundamental inadequate. And I have argued, finally, that the real aim of the famous series of _Gedankenexperimente_ starting at the 1927 Solvay meeting was to bring out precisely this feature of the quantum theory and to exhibit the, to Einstein, unacceptable consequences to which it leads. That is to say, I argued that Einstein's concern over the uncertainty relations and the breakdown of strict causality in quantum mechanics was secondary to his deeper concern over the quantum mechanical account of interactions.

One important test of this reconstruction of Einstein's views would be to determine whether or not Einstein's contemporaries understood his reservations about the quantum theory in this manner. Let me show that this was the case by quoting from just two letters written right after the publication of the EPR paper.

The first is a letter from Pauli to Heisenberg of 15 June 1935, in which Pauli prodded Heisenberg into composing a "pedagogical" reply to EPR, a reply that, unfortunately, was never published.[28] Pauli writes:

---

[28] For the text of Heisenberg's reply, "Ist eine deterministische Ergänzung der Quantenmechanik möglich?", see Pauli 1985, pp. 409-418; Heisenberg enclosed a copy with his letter to Pauli of 2 Juli 1935. Another copy, in typescript, is in the Einstein Archive, EA 5-207.

Einstein has again expressed himself publicly on quantum mechanics, indeed in the 15 May issue of Physical Review (together with Podolsky and Rosen--no good company, by the way). As is well known, every time that happens it is a catastrophe. "Weil, so schließt er messerscharf--nicht sein kann was nicht sein darf" (Morgenstern). . . .

He now understands this much, that one cannot simultaneously measure two quantities corresponding to non-commuting operators and that one cannot simultaneously ascribe numerical values to them. But where he runs into trouble in this connection is the way in which, in quantum mechanics, two systems are joined to form a composite system. . . .

A pedagogical reply to [this] train of thought must, I believe, clarify the following concepts. The difference between the following statements:

a) Two systems 1 and 2 are not in interaction with one another (= absence of any interaction energy). . . .

b) The composite system is in a state where the subsystems 1 and 2 are _independent_. (Decomposition of the eigenfunction into a product.) . . . .

Quite independently of _Einstein_, it appears to me that, in providing a systematic foundation for quantum mechanics, one should _start_ more from the composition and separation of systems than has until now (with Dirac, e.g.) been the case. -- This is indeed--as Einstein has _correctly_ felt--a very fundamental point in quantum mechanics, which has, moreover, a direct connection with your reflections about the _cut_ and the possibility of its being shifted to an arbitrary place. . . .

NB Perhaps I have devoted so much effort to these matters, which are trivialities for us, because a short time ago I received an invitation to Princeton for the next winter semester. It would be fun to go. I will by all means make the Morgenstern motto popular there. (Pauli 1985, pp. 402-404)

Notice that Pauli uses the past perfect tense: "as Einstein has _correctly_ felt" ["wie Einstein _richtig_ gefühlt hat"]. What he is characterizing here is not what he has just learned from the EPR paper, in which the issue of separability is anyway almost totally obscured (see Einstein to Schrödinger, 19 June 1935, EA 22-047); instead, he is describing the view that he has long associated with Einstein.

The second letter is from Schrödinger to Pauli, sometime between 1 July and 9 July 1935. It was prompted by Arnold Berliner's having asked Schrödinger to write a reply to EPR for Die Naturwissenschaften and Berliner's having told Schrödinger that Pauli was quite agitated about the matter. Schrödinger portrays Einstein's fundamental view of the nature of reality, the view lying behind the incompleteness argument, as follows: "He has a model of that which is real consisting of a map with little flags. To every real thing there must correspond on the map a little flag, and vice versa" (Pauli 1985, p. 406). This is Schrödinger's marvelously vivid way of characterizing Einstein's view of the fundamental ontology of field theories, the ontology which gives the most radical possible expression to the separability principle. Schrödinger has rightly discerned that it is this fundamental commitment that animated Einstein's opposition to quantum mechanics.

ACKNOWLEDGEMENTS

Many individuals, institutions, and organizations deserve thanks for their contributions to this paper and to the research that stands behind it. John Stachel and Robert Cohen very generously offered the hospitality of, respectively, the Center for Einstein Studies and the Center for Philosophy and History of Science, both at Boston University, where much of the research for this paper was conducted; I want also to thank John Stachel for his guidance through the Einstein Archive and for his critical comments on this paper. Arthur Fine has offered helpful criticism of some of the ideas incorporated here, as have Harvey Brown, Simon Saunders, Andrew Elby, and Robert Clifton. Special thanks go to W. Gerald Heverly, of the Special Collections Department at Hillman Library, University of Pittsburgh, for providing timely help in securing some needed documentation. Support for various phases of the research was provided by the National Science Foundation (research grant no. SES-8421040), the American Philosophical Association, the Deutscher Akademischer Austauschdienst, and the University of Kentucky Research Foundation. I wish to thank the Hebrew University of Jerusalem, which holds the copyright, for permission to quote from unpublished letters and papers of Einstein, and the Museum Boerhaave, Leiden, for permission to quote from unpublished correspondence of Paul Ehrenfest; items in the Einstein Archive are cited by their numbers in the Control Index.

REFERENCES

Bohr, N., 1927, The Quantum Postulate and the Recent Development of Atomic Theory, Nature (Suppl.), 121 (1928):580.

Bohr, N., 1931, Space-Time Continuity and Atomic Physics, lecture, University of Bristol, England, 5 October 1931, in Bohr 1985, p. 363.

Bohr, N., 1935, Can Quantum-Mechanical Description of Physical Reality Be Considered Complete? Phys. Rev., 48:696.

Bohr, N., 1949, Discussion with Einstein on Epistemological Problems in Atomic Physics, in Schilpp 1949, p. 199.

Bohr, N., 1984, "Collected Works," E. Rüdinger, ed., vol. 5, "The Emergence of Quantum Mechanics (Mainly 1924-1926)," K. Stolzenburg, ed., North-Holland, Amsterdam.

Bohr, N., 1985, "Collected Works," E. Rüdinger, ed., vol. 6, "Foundations of Quantum Physics I (1926-1932)," J. Kalckar, ed., North-Holland, Amsterdam.

Bohr, N., Kramers, H. A., and Slater, J. C., 1924, The Quantum Theory of Radiation, Phil. Mag., 47:785; Über die Quantentheorie der Strahlung, Zeitschr. f. Phys., 24:69; page numbers for English version cited from reprinting in "Sources of Quantum Mechanics," B. L. van der Waerden, ed., Dover, New York, 1967, p. 159.

Bohr, N., and Rosenfeld, L., 1933, Zur Frage der Messbarkeit der elektromagnetischen Feldgrössen, Kgl. Dan. Vid. Selsk., Mat.-Fys. Medd. 12 (no. 8):1.

Born, M., 1926a, Zur Quantenmechanik der Stoßvorgänge. [Vorläufige Mitteilung.], Zeitschr. f. Phys., 37:863.

Born, M., 1926b, Quantenmechanik der Stoßvorgänge, Zeitschr. f. Phys., 38:803.

Born, M., ed., 1969, "Albert Einstein--Hedwig und Max Born. Briefwechsel 1916-1955," Nymphenburger, Munich.

Bose, S. N., 1924, Plancks Gesetz und Lichtquantenhypothese, Zeitschr. f. Phys., 26:178.

Bothe, W., 1926, Über die Kopplung zwischen elementaren Strahlungsvorgängen, Zeitschr. f. Phys., 37:547.

Bothe, W., 1927, Lichtquanten und Interferenz, Zeitschr. f. Phys., 41:332.

Bothe, W., and Geiger, H., 1924, Ein Weg zur experimentellen Nachprüfung der Theorie von Bohr, Kramers und Slater, Zeitschr. f. Phys., 26:44.

Bothe, W., and Geiger, H., 1925a, Experimentelles zur Theorie von Bohr, Kramers und Slater, Naturw., 13:440.

Bothe, W., and Geiger, H., 1925b, Über das Wesen des Comptoneffekts. Ein experimenteller Beitrag zur Theorie der Strahlung, Zeitschr. f. Phys., 32:639.

Brown, H., 1981, O Debate Einstein-Bohr sobre a Mecânica Quântica, Cadernos de História e Filosofia da Ciência, 2:51.

Compton, A. H., and Simon, A. W., 1925a, Measurements of β-rays Excited by Hard X-rays, Phys. Rev., 25:107.

Compton, A. H., and Simon, A. W., 1925b, Measurements of β-rays Associated with Scattered X-rays, Phys. Rev., 25:306.

Compton, A. H., and Simon, A. W., 1925c, Directed Quanta of Scattered X-rays, Phys. Rev., 26: 289-299.

de Broglie, L., 1924, "Recherche sur la théorie des quanta," Masson et Cie, Paris; reprinted in Ann. d. physique 3 (1925):22.

d'Espagnat, B., 1976, "Conceptual Foundations of Quantum Mechanics," 2nd ed., W. A. Benjamin, Reading, Massachusetts.

d'Espagnat, B., 1981, "A la Recherche du Réel: Le regard d'un physicien," 2nd ed., Gauthier-Villars, Paris; English edition: "In Search of Reality," Springer-Verlag, New York, 1983.

Dirac, P. A. M., 1926, On the Theory of Quantum Mechanics, Proc. Roy. Soc. (London), A 112:661.

Einstein, A., 1905, Über einen die Erzeugung und Verwandlung des Lichtes betreffenden heuristischen Gesichtspunkt, Ann. d. Phys., 17:132.

Einstein, A., 1906, Zur Theorie der Lichterzeugung und Lichtabsorption, Ann. d. Phys., 20:199.

Einstein, A., 1907, Die Plancksche Theorie der Strahlung und die Theorie der spezifischen Wärme, Ann. d. Phys., 22:180.

Einstein, A., 1909a, Zum gegenwärtigen Stand des Strahlungsproblems, Phys. Zeitschr., 10:185.

Einstein, A., 1909b, Über die Entwickelung unserer Anschauungen über das Wesen und die Konstitution der Strahlung, Verh. Dtsch. Phys. Ges., 11:482; reprinted in Phys. Zeitschr., 10 (1909):817.

Einstein, A., 1916a, Strahlungs-emission und -absorption nach der Quantentheorie, Verh. Dtsch. Phys. Ges., 18:318.

Einstein, A., 1916b, Zur Quantentheorie der Strahlung, Mitt. Phys. Ges. Zürich, 47; reprinted in Phys. Zeitschr., 18 (1917):121.

Einstein, A., 1923, Bietet die Feldtheorie Möglichkeiten für die Lösung des quantenproblems? Sitz. Preuss. Akad. d. Wiss., Phys.-math. Klasse, 359.

Einstein, A., 1924a, Anmerkung des Übersetzers (following Bose 1924), Zeitschr. f. Phys., 26:181.

Einstein, A., 1924b, Quantentheorie des einatomigen idealen Gases, Sitz. Preuss. Akad. d. Wiss., Phys.-math. Klasse, 261.

Einstein, A., 1924c, Über den Äther, Verh. Schweiz. naturf. Ges., 105 (part 2):85.

Einstein, A., 1925a, Quantentheorie des einatomigen idealen Gases. Zweite Abhandlung, Sitz. Preuss. Akad. d. Wiss., Phys.-math. Klasse, 3.

Einstein, A., 1925b, Quantentheorie des idealen Gases, Sitz. Preuss. Akad. d. Wiss., Phys.-math. Klasse, 18.

Einstein, A., 1926a, Vorschlag zu einem die Natur des elementaren Strahlungs-Emissionsprozesses betreffenden Experiment, Naturw., 14:300.

Einstein, A., 1926b, Interferenzeigenschaften des durch Kanalstrahlen emittierten Lichtes, Sitz. Preuss. Akad. d. Wiss., Phys.-math. Klasse, 334.

Einstein, A., 1927a, Zu Kaluzas Theorie des Zusammenhanges von Gravitation und Elektrizität. Erste und zweite Mitteilung, Sitz. Preuss. Akad. d. Wiss., Phys.-math. Klasse, 23.

Einstein, A., 1927b, Theoretisches und Experimentelles zur Frage der Lichtentstehung, Zeitschr. f. ang. Chem., 40:546 (editorial report of lecture, 23 February 1927, to Mathematisch-physikalische Arbeitsgemeinschaft, University of Berlin).

Einstein, A., 1927c, Contribution to Discussion générale des idées émises, in Solvay 1927, p. 253; quoted from English translation in Bohr 1985, p. 101.

Einstein, A., 1927d, Allgemeine Relativitätstheorie und Bewegungsgesetz, Sitz. Preuss. Akad. d. Wiss., Phys.-math. Klasse, 235.

Einstein, A., 1931, Über die Unbestimmtheitsrelationen, Zeitschr. f. ang. Chem., 45:23 (editorial report of colloquium).

Einstein, A., 1936, Physik und Realität, Journ. Franklin Institute, 221:313.

Einstein, A., 1946, Autobiographisches--Autobiographical Notes, in Schilpp 1949, p. 1; reprinted with corrections as "Autobiographical Notes: A Centennial Edition," P. A. Schilpp, trans. and ed. Open Court, La Salle and Chicago, Illinois, 1979.

Einstein, A., 1948, Quanten-Mechanik und Wirklichkeit, Dialectica, 2:320.

Einstein, A., 1949, Remarks concerning the Essays Brought together in this Co-operative Volume, in Schilpp 1949, p. 665.

Einstein, A., and Grommer, J., 1923, Beweis der Nichtexistenz eines überall regulären zentrisch symmetrischen Feldes nach der Feld-theorie von Th. Kaluza, Scripta Universitatis atque Bibliothecae Hierosolymitanarum: Mathematicae et Physica, 1 (no. 7).

Einstein, A., and Grommer, J., 1927, Allgemeine Relativitätstheorie und Bewegungsgesetz, Sitz. Preuss. Akad. d. Wiss., Phys.-math. Klasse, 2.

Einstein, A., Podolsky, B., and Rosen, N., 1935, Can Quantum-Mechanical Description of Physical Reality Be Considered Complete? Phys. Rev., 47:777.

Einstein, A., Tolman, R. C., and Podolsky, B., 1931, Knowledge of Past and Future in Quantum Mechanics, Phys. Rev., 37:780.

Elsasser, W., 1925, Bemerkungen zur Quantenmechanik freier Elektronen, Naturw., 13:711.

Epstein, P. S., 1945, The Reality Problem in Quantum Mechanics, Amer. Journ. Phys., 13:127.

Fine, A., 1979, Einstein's Critique of Quantum Theory: The Roots and Significance of EPR, in "After Einstein: Proceedings of the Einstein Centennial Celebration at Memphis State University, 14-16 March 1979," P. Barker and C. G. Shugart, eds., Memphis State University Press, Memphis, Tennessee, 1981, p. 147; reprinted in Fine 1986, p. 26.

Fine, A., 1986, "The Shaky Game: Einstein, Realism, and the Quantum Theory," University of Chicago Press, Chicago.

Gibbs, J. W., 1902, "Elementary Principles in Statistical Mechanics Developed with Especial Reference to the Rational Foundation of Thermodynamics," Charles Scribner's Sons, New York.

Gibbs, J. W., 1905, "Elementare Grundlagen der statistischen Mechanik," E. Zermelo, trans., Johann Ambrosius Barth, Leipzig.

Havas, P., 1989, The Early History of the 'Problem of Motion' in General Relativity, in Howard and Stachel 1989, p. 234.

Heisenberg, W., 1927, Über den anschaulichen Inhalt der quantentheoretischen Kinematik und Mechanik, Zeitschr. f. Phys., 43:172.

Heisenberg, W., 1930, "The Physical Principles of the Quantum Theory," C. Eckart and F. C. Hoyt, trans., University of Chicago Press, Chicago.

Heisenberg, W., 1967, Quantum Theory and Its Interpretation, in Rozental 1967, p. 94.

Hermann, A., ed., 1968, "Albert Einstein/Arnold Sommerfeld. Briefwechsel. Sechzig Briefe aus dem goldenen Zeitalter der modernen Physik," Schwabe & Co., Basel and Stuttgart.

Hoffmann, B., 1979, "Albert Einstein: Creator and Rebel" (with the collaboration of Helen Dukas), Viking Press, New York.

Hooker, C., 1972, The Nature of Quantum Mechanical Reality: Einstein Versus Bohr, in "Paradigms & Paradoxes: The Philosophical Challenge of the Quantum Domain," R. G. Colodny, ed., University of Pittsburgh Press, Pittsburgh, p. 67.

Howard, D., 1985, Einstein on Locality and Separability, Stud. Hist. Phil. Sci. 16:171.

Howard, D., 1988, Einstein and Eindeutigkeit: A Neglected Theme in the
  Philosophical Background to General Relativity, in "Einstein and
  the History of General Relativity II," J. Eisenstaedt and A. J.
  Kox, eds., Einstein Studies, vol. 3, Birkhäuser, Boston (forth-
  coming).
Howard, D., 1989, Holism, Separability, and the Metaphysical Implications of
  the Bell Experiments, in "Philosophical Consequences of Quantum
  Theory: Reflections on Bell's Theorem," J. T. Cushing and E.
  McMullin, eds., University of Notre Dame Press, Notre Dame,
  Indiana, p. 224.
Howard, D., and Stachel, J., eds., 1989, "Einstein and the History of
  General Relativity," Einstein Studies, vol. 1, Birkhäuser, Boston.
Jammer, M., 1974, "The Philosophy of Quantum Mechanics: The Interpretations
  of Quantum Mechanics in Historical Perspective," John Wiley &
  Sons, New York.
Jammer, M., 1985, The EPR Problem in Its Historical Context, in "Symposium
  on the Foundations of Modern Physics," P. Lahti and P. Mittel-
  staedt, eds., World Publishing Company, Singapore, p. 129.
Joos, G., 1926, Modulation und Fourieranalyse im sichtbaren Spektralbereich,
  Phys. Zeitschr., 27:401.
Klein, O., 1926, Quantentheorie und fünfdimensionale Relativitätstheorie,
  Zeitschr. f. Phys., 37:895.
Krutkow, G., 1914a, Aus der Annahme unabhängiger Lichtquanten folgt die
  Wiensche Strahlungsformel, Phys. Zeitschr., 15:133.
Krutkow, G., 1914b, Bemerkung zu Herrn Wolfkes Note: "Welche Strahlungs-
  formel folgt aus der Annahme der Lichtatome?" Phys. Zeitschr.,
  15:363.
Landé, A., 1925, Lichtquanten und Kohärenz, Zeitschr. f. Phys., 33:571.
Lorentz, H. A., 1908a, Le partage de l'énergie entre la matière pondérable
  et l'éther, Nuovo Cimento, 16:5.
Lorentz, H. A., 1908b, Zur Strahlungstheorie, Phys. Zeitschr., 9:562.
Lorentz, H. A., 1909, Le partage de l'énergie entre la matière pondérable et
  l'éther, Rev. gén. d. sci., 20:14.
Mehra, J., and Rechenberg, H., 1982, "The Historical Development of Quantum
  Theory," vol. 1, "The Quantum Theory of Planck, Einstein, Bohr and
  Sommerfeld: Its Foundation and the Rise of Its Difficulties, 1900-
  1925," Springer-Verlag, New York, Heidelberg, and Berlin.
Pais, A., 1982, "'Subtle is the Lord . . .': The Science and the Life of
  Albert Einstein," Clarendon, Oxford; Oxford University Press, New
  York.
Pauli, W., 1925, Über den Zusammenhang des Abschlußes der Elektronengruppen
  im Atom mit der Komplexstruktur der Spektren, Zeitschr. f. Phys.,
  31:765.
Pauli, W., 1979, "Wissenschaftlicher Briefwechsel mit Bohr, Einstein,
  Heisenberg u.a.," vol. 1, "1919-1929," A. Hermann, K. von Meyenn,
  and V. F. Weisskopf, eds., Springer-Verlag, New York, Heidelberg,
  and Berlin.
Pauli, W., 1985, "Wissenschaftlicher Briefwechsel mit Bohr, Einstein,
  Heisenberg u.a.," vol. 2, "1930-1939," K. von Meyenn, A. Hermann,
  and V. F. Weisskopf, eds., Springer-Verlag, Berlin, Heidelberg,
  New York, and Tokyo.
Planck, M., 1922, Über die freie Energie von Gasmolekülen mit beliebiger
  Geschwindigkeitsverteilung, Sitz. Pruess. Akad. d. Wiss., Phys.-
  math. Klasse, 63.
Planck, M., 1925, Zur Frage der Quantelung einatomiger Gase, Sitz. Pruess.
  Akad. d. Wiss., Phys.-math. Klasse, 49.
Ramsauer, C., 1920, Über den Wirkungsquerschnitt der Gasmoleküle gegenüber
  langsamen Elektronen, Phys. Zeitschr., 21:576.
Ramsauer, C., 1921a, Über den Wirkungsquerschnitt der Gasmoleküle gegenüber
  langsamen Elektronen, Ann. d. Phys., 64:513.

Ramsauer, C., 1921b, Über den Wirkungsquerschnitt der Edelgase gegenüber langsamen Elektronen, Phys. Zeitschr., 22:613.

Rosenfeld, L., 1967, Niels Bohr in the Thirties: Consolidation and Extension of the Conception of Complementarity, in Rozental 1967, p. 114.

Rozental, S., ed., 1967, "Niels Bohr: His Life and Work as Seen by His Friends and Colleagues," John Wiley & Sons, New York.

Rupp, E., 1926, Über die Interferenzfähigkeit des Kanalstrahllichtes, Sitz. Preuss. Akad. d. Wiss., Phys.-math. Klasse, 341.

Schilpp, P. A., ed., 1949, "Albert Einstein: Philosopher-Scientist," The Library of Living Philosophers, vol. 7, The Library of Living Philosophers, Evanston, Illinois.

Schrödinger, E., 1924, Bohrs neue Strahlungshypothese und der Energiesatz, Naturw., 12:720.

Schrödinger, E., 1925, Bemerkungen über statistische Entropiedefinition beim idealen Gas, Sitz. Pruess. Akad. d. Wiss., Phys.-math. Klasse, 434.

Schrödinger, E., 1926a, Zur Einsteinschen Gastheorie, Phys. Zeitschr., 27:95.

Schrödinger, E., 1926b, Die Energiestufen des idealen einatomigen Gasmodels, Sitz. Pruess. Akad. d. Wiss., Phys.-math. Klasse, 23.

Schrödinger, E., 1926c, Quantisierung als Eigenwertproblem. (1. Mitteilung.) Ann. d. Phys., 79:361.

Schrödinger, E., 1926d, Quantisierung als Eigenwertproblem. (2. Mitteilung.) Ann. d. Phys., 79:489.

Schrödinger, E., 1926e, Über das Verhältnis der Heisenberg-Born-Jordanschen Quantenmechanik zu der meinen, Ann. d. Phys., 79:734.

Schrödinger, E., 1935, Discussion of Probability Relations between Separated Systems, Proc. Cambr. Phil. Soc., 31:555.

Schrödinger, E., 1936, Probability Relations between Separated Systems, Proc. Cambr. Phil. Soc., 32:446.

Solvay, 1927, "Électrons et photons: Rapports et discussions du cinquième conseil de physique tenu a Bruxells du 24 au 29 Octobre 1927," Gauthier-Villars, Paris, 1928.

Solvay, 1930, "Le Magnétisme: Rapports et discussions du sixième conseil de physique tenu a Bruxells du 20 au 25 Octobre 1930," Gauthier-Villars, Paris, 1932.

Speziali, P., ed., 1972, "Albert Einstein-Michele Besso: Correspondance 1903-1955," Hermann, Paris.

Stachel, J., 1986, Einstein and the Quantum: Fifty Years of Struggle, in "From Quarks to Quasars: Philosophical Problems of Modern Physics," R. G. Colodny, ed., University of Pittsburgh Press, Pittsburgh, p. 349.

Stachel, J., 1989, Einstein's Search for General Covariance, 1912-1915, in Howard and Stachel 1989, p. 63.

Wigner, E. P., 1980, Thirty Years of Knowing Einstein, in "Some Strangeness in the Proportion: A Centennial Symposium to Celebrate the Achievements of Albert Einstein," H. Woolf, ed., Addison-Wesley, Reading, Massachusetts, p. 461.

Wolfke, M., 1913a, Zur Quantentheorie. (Vorläufige Mitteilung), Verh. Dtsch. Phys. Ges., 15:1123.

Wolfke, M., 1913b, Zur Quantentheorie. (Zweite vorläufige Mitteilung), Verh. Dtsch. Phys. Ges., 15:1215.

Wolfke, M., 1914a, Zur Quantentheorie. (Dritte vorläufige Mitteilung), Verh. Dtsch. Phys. Ges., 16:4.

Wolfke, M., 1914b, Welche Strahlungsformel folgt aus der Annahme der Lichtatome? Phys. Zeitschr., 15:308.

Wolfke, M., 1914c, Antwort auf die Bemerkung Herrn Krutkows zu meiner Note: "Welche Strahlungsformel folgt aus den Annahme der Lichtatome?" Phys. Zeitschr., 15:463.

Wolfke, M., 1921, Einsteinsche Lichtquanten und räumliche Struktur der Strahlung, Phys. Zeitschr., 22:375.

INTRODUCTION TO TWO-PARTICLE INTERFEROMETRY

Michael A. Horne, Abner Shimony, and Anton Zeilinger

Department of Physics
Stonehill College
North Easton, MA

Departments of Philosophy and Physics
Boston University
Boston, MA

Atominstitut der Österreichischen Universitäten
Vienna, Austria, and
Physik Department
Technische Universität München
Garching, Federal Republic of Germany

Ordinary interferometry employs beams of particles -- photons, electrons, neutrons, and possible other particles -- but the phenomena which it studies arise when two amplitudes associated with a _single_ particle combine at a locus. When the single particle is characterized by a quantum state, the two amplitudes have a definite phase relation. The variation of the relative phase as one or more parameters vary gives rise to the familiar interferometric "fringe" pattern, which characteristically is sinusoidal.

The phenomena of two-particle interferometry also arise from the combination of two amplitudes with a definite phase relation. The radical innovation is the employment of beams of _two-particle systems_, with each pair in an "entangled" state, that is, a state which cannot be expressed as a simple product of quantum states of the two particles separately. That quantum mechanics permits in principle the existence of pairs of spatially separated particles in entangled states has been known at least since the classical paper of Einstein, Podolsky, and Rosen (1935), and the actual existence of such pairs has been known since the analysis by Bohm and Aharonov (1957) of the experiment of Wu and Shaknov (1950). It is only in the last five years, however, that beams of entangled two-particle systems have been subjected to the traditional interferometric techniques of splitting, directing, and combination.[1]

In this lecture we shall analyze a schematic arrangement (Fig. 1) to show that when the particle pairs are appropriately prepared, then quantum mechanics predicts two-particle interference fringes and predicts at the same time the non-occurrence of single-particle fringes. We shall then illustrate the experimental potentialities of two-particle interferometry by showing how this arrangement makes possible a test of Bell's Inequality without polarization analysis.

Sixty-Two Years of Uncertainty
Edited by A. I. Miller
Plenum Press, New York, 1990

113

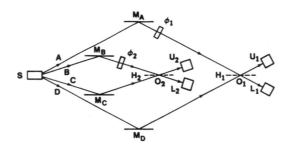

Fig. 1. An arrangement for two-particle in-
        terferometry with variable phase shif-
        ters.

In the arrangement of Fig. 1 an ensemble of particle pairs is emitted
from the source S into the beams A,B,C,D, each pair in the ensemble being in
the entangled quantum state

$$|\psi> = 2^{-\frac{1}{2}}(|A>_1|C>_2 + |D>_1|B>_2). \tag{1}$$

This state describes a coherent superposition of two distinct pairs of correlated paths for particles 1 and 2. In one of these, particle 1 enters beam
A and is reflected from mirror M_A to phase shifter $\emptyset_1$ en route to beam splitter H_1, from which it proceeds either into the upper channel U_1 or the lower
channel L_1; while particle 2 enters beam C and is reflected from mirror M_C
to beam splitter H_2, from which it proceeds either into the upper channel
U_2 or the lower channel L_2. In the other pair of correlated paths particle
1 enters beam D and proceeds to U_1 or L_1 via mirror M_D and H_1, while particle
2 enters beam B and proceeds to U_2 or L_2 via mirror M_B, phase shifter $\emptyset_2$,
and H_2. The beams A,B,C,D are assumed to be in a single plane, and their
directions ensure momentum conservation (i.e., the sum of the momenta of
particles 1 and 2 in A and C respectively equals the sum of the momenta of
particles 1 and 2 in D and B respectively). We wish to calculate the probabilities that the two particles will jointly enter each of the four possible
pairs of exit channels: (U_1,U_2), (U_1,L_2), (L_1,U_2), and (L_1,L_2). Quantum mechanically each of these probabilities is expressed as the absolute square
of a total probability amplitude, for instance,

$$P_\psi(U_1,U_2,|\emptyset_1,\emptyset_2) = |A_\psi(U_1,U_2|\emptyset_1,\emptyset_2)|^2, \tag{2}$$

where the dependence of this probability upon the initial quantum state
and upon the variable phase shifters $\emptyset_1$ and $\emptyset_2$ has been indicated explicitly.
There are two contributions to the probability amplitude A_ψ: one comes from
particle 1 entering beam A and eventually being reflected from H_1, while particle 2 enters beam C and eventually is transmitted through H_2; whereas the
other comes from particle 1 entering beam D and eventually being transmitted
through H_1, while particle 2 enters beam B and is reflected from H_2. In the
first contribution particle 1 encounters the phase shifter $\emptyset_1$, and in the
second particle 2 encounters the phase shifter $\emptyset_2$. We need to calculate the
relative phase of these two contributions.

A necessary preliminary to this calculation is the derivation of an equation governing the phase relations of reflected and transmitted rays from
a lossless beam-splitter, when two rays are incident symmetrically upon its
two faces, as indicated in Fig. 2. The rays correspond to quantum states
of definite linear momentum and are denoted by $|I>$ and $|J>$ respectively. If
the beam-splitter is symmetric, the moduli of the reflected and the trans-
mitted output from each incident ray are equal. Let $|I'>$ denote the total

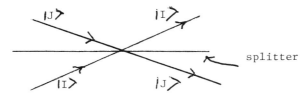

Fig. 2. Incident rays |I>, and |J'> reflected and
        transmitted from a symmetric, lossless
        beam-splitter.

output state from incident |I>, and |J'> denote the total output state from
incident |J>. Following Zeilinger (1981) we use losslessness to connect out-
put to input by a unitary operator U and use the symmetries to write

$$|I'> = 2^{-\frac{1}{2}}(e^{it}|I> + e^{ir}|J>) = U|I>, \tag{3a}$$

$$|J'> = 2^{-\frac{1}{2}}(e^{ir'}|I> + e^{it'}|J>) = U|J>, \tag{3b}$$

where the real numbers r and r' are the phase shifts due to reflection, and
the real numbers t and t' are the phase shifts due to transmission through
the beam-splitter. Because of the orthogonality of |I> and |J> and unitar-
ity, |I'> and |J'> are orthogonal, and hence

$$0 = <I'|J'> = \frac{1}{2}[e^{i(r'-t)} + e^{i(t'-r)}] \tag{4}$$

so that

$$r' - t = t' - r + \pi \pmod{2\pi}. \tag{5}$$

    We now return to Fig. 1 in order to calculate the probability amplitude
$A_\psi(U_1,U_2|\emptyset_1,\emptyset_2)$. Let r_1 and t_1 be the phase shifts of Eq. (3a) associated
with reflection and transmission of the ray incident upon beam-splitter H_1
from below, and r_1' and t_1' be the phase shifts of Eq. (3b) associated with
reflection and transmission of the ray incident upon H_1 from above. Let
r_2,t_2,r_2',t_2' have analogous meanings for beam-splitter H_2. Let s_1 be the
phase change associated with the upper path of particle 1 from S to H_1,
omitting $\emptyset_1$, and s_1' be the phase change associated with the lower path, via
beam D; likewise, let s_2 be the phase change associated with the upper path
of particle 2 from S to H_2, omitting $\emptyset_2$, and s_2' the phase change associated
with the lower path, via beam C. Finally we use the letters $\emptyset_1$ and $\emptyset_2$ not
only to designate the apparatus used for variable phase shifting, but also
for the amounts of these phase shifts -- an ambiguity of notation which will
cause no confusion. Using Eqs. (3a) and (3b) and collecting all these
phases we obtain

$$A_\psi(U_1,U_2|\emptyset_1,\emptyset_2) = \frac{2^{-\frac{1}{2}}}{2}[\exp i(s_1+\emptyset_1+r_1+s_2'+t_2') +$$
$$\exp i(s_1'+t_1'+s_2+\emptyset_2+r_2)]. \tag{6}$$

Hence,

$$P_\psi(U_1,U_2|\emptyset_1,\emptyset_2) = 1/4[1 + \cos(\emptyset_1 - \emptyset_2 + w)], \tag{7}$$

where w is a total fixed phase shift, independent of the variable phase
shifts $\emptyset_1$ and $\emptyset_2$, specifically,

$$w = s_1+r_1+s_2'+t_2'-s_1'-t_1'-s_2-r_2. \tag{8}$$

Likewise,

$$P_\psi(U_1,L_2|\emptyset_1,\emptyset_2) = \tfrac{1}{4}[1 + \cos(\emptyset_1 - \emptyset_2 + w')], \tag{9}$$

where

$$w' = s_1+r_1+s_2'+r_2'-s_1'-t_1'-s_2-t_2, \tag{10}$$

and expressions similar to Eqs. (7) and (9) can be given for $P_\psi(L_1,U_2|\emptyset_1,\emptyset_2)$ $P_\psi(L_1,L_2|\emptyset_1,\emptyset_2)$. In short, the probability of joint entrance of particles 1 and 2 into any of the four possible pairs of channels depends sinusoidally upon the difference $\emptyset_1-\emptyset_2$ of the variable phase shifts. Thus quantum mechanics predicts two-particle interference fringes in the experimental arrangement that has been described. What is extraordinary is that there are no one-particle interference fringes in this arrangement, as one can see by adding Eqs. (7) and (9) to obtain the probability that particle 1 will enter channel U_1, regardless of the behavior of particle 2:

$$P_\psi(U_1|\emptyset_1,\emptyset_2) = P_\psi(U_1,U_2|\emptyset_1,\emptyset_2) + P_\psi(U_1,L_2|\emptyset_1,\emptyset_2) =$$

$$\tfrac{1}{2} + \cos(\emptyset_1 - \emptyset_2 + w) + \cos(\emptyset_1 - \emptyset_2 + w') = \tfrac{1}{2}, \tag{11}$$

because by Eqs. (8), (10), and (5),

$$w' = w + (r_2'-t_2-t_2'r_2) = w + \pi\,(\mathrm{mod}\ 2\pi). \tag{12}$$

In fact, no matter what the values are of the variable phase shifts $\emptyset_1$ and $\emptyset_2$, the single-particle probabilities are the same, namely $\tfrac{1}{2}$. This result is at first very surprising, not only because of the sinusoidal behavior of the two-particle probabilities but also because in the arrangement of Fig. 1 each of the particles 1 and 2 seems to be subjected separately to a Mach-Zehnder interferometric experiment.

The quantum mechanical explanation for the absence of single-particle interference fringes is obtained by returning to the entangled state of Eq. (1) and inquiring what it implies about the state of particle 1 by itself and the state of particle 2 by itself. Neither 1 nor 2 is in a pure quantum state, but both can be described by statistical or density operators W_1 and W_2, as discussed, for example, by Beltrametti and Cassinelli (1981), 66, where

$$W_1 = \tfrac{1}{2}(|A\rangle\langle A| + |D\rangle\langle D|), \tag{13}$$

$$W_2 + \tfrac{1}{2}(|B\rangle\langle B| + |C\rangle\langle C|). \tag{14}$$

All predictions concerning particle 1 alone, neglecting correlations with particle 2, can be obtained from Eq. (13) and will be in exact agreement with those obtained from Eq. (1); and all predictions concerning particle 2 alone can be obtained from Eq. (14) and will agree with Eq. (1). Now W_1 is the statistical operator that would correctly describe an ensemble, of which half of the members are in quantum state $|A\rangle$ and half are in quantum state $|D\rangle$, though infinitely many other ensembles (so-called "mixtures") are correctly described by W_1. And likewise, W_2 is the statistical operator that would correctly describe the ensemble of which half are in state $|B\rangle$ and half are in state $|C\rangle$. Of course, neither of these ensembles would exhibit interference fringes, since each particle in each ensemble travels from source to output channel by only one path. Hence, neither ensemble takes advantage of the Mach-Zehnder interferometer to bring together contributions by two different paths, with definite phase relations, as required for single-particle interference fringes. Another way to put the matter is to say that the entangled state of Eq. (1) shows a definite phase relation between two two-particle states, namely $|A\rangle_1|C\rangle_2$ and $|D\rangle_1|B\rangle_2$, but no definite phase relations between single-particle states.

An obvious question is how one can know that the quantum state of a pair of particles emerging from the source has the form of Eq. (1). There are two ways to answer this question, one hard and one easy. The hard way is to describe quantum mechanically the process which gives birth to the two-particle pair and show that the sresulting quantum state of !+2 has the desired form. The (relatively) easy way is to do two-particle interferometry, in order to see whether two-particle interference fringes are exhibited, for it is straightforward to show that if the quantum state of each pair emerging from the source is a product of single-particle states, then the two-particle fringe behavior of Eqs. (7) and (9) will not be exhibited. So far, the only realizations of two-particle interferometry have used pairs of photons produced by the interaction of single photons with an appropriate crystal[1], and in these experiments the observation of two-particle interference fringes provides decisive evidence for the entangled state of the emerging two-photon system.

At the conclusion of the lecture "An Exposition of Bell's Theorem" in this volume it was noted that there is no intrinsic reason why a polarization experiment is necessary for the purpose of testing Bell's Inequality. Indeed, the arrangement of Fig. 1 of the present lecture is a special case of the schematic arrangement of Fig. 1 of that lecture and can be used to test an Inequality, when the following identifications are made: the outcomes of analysis of particle 1 are passage into channels U_1 and L_1, and the conventional values s_m assigned to these two outcomes are 1 and -1 respectively; likewise the outcomes of analysis of particle 2 are passage into channels U_2 and L_2, and the values t_n assigned to these are 1 and -1 respectively; and the variable parameters a and b are taken to be the variable phase shifts $\emptyset_1$ and $\emptyset_2$. Then Inequality (4) of "An Exposition of Bell's Theorem" can be rewritten as

$$-2 \leq E_w(\emptyset_1', \emptyset_2') + E_w(\emptyset_1', \emptyset_2'') + E_w(\emptyset_1'', \emptyset_2') - E_w(\emptyset_1'', \emptyset_2'') \leq 2. \tag{15}$$

The quantum mechanical expectation value of the products of outcomes, when the variable phase shifts are $\emptyset_1$ and $\emptyset_2$, is

$$\begin{aligned} E_\psi(\emptyset_1, \emptyset_2) &= P_\psi(U_1, U_2 | \emptyset_1, \emptyset_2) \cdot 1 + P_\psi(U_1, L_2 | \emptyset_1, \emptyset_2) \cdot (-1) + \\ &\quad P_\psi(L_1, U_2 | \emptyset_1, \emptyset_2) \cdot (-1) + P_\psi(L_1, L_2 | \emptyset_1, \emptyset_2) \cdot 1 = \\ &\quad \tfrac{1}{4}[1 + \cos(\emptyset_1 - \emptyset_2 + w)] \cdot 1 + \tfrac{1}{4}[1 - \cos(\emptyset_1 - \emptyset_2 + w)] \cdot (-1) \\ &\quad + \tfrac{1}{4}[1 - \cos(\emptyset_1 - \emptyset_2 + w)] \cdot (-1) + \tfrac{1}{4}[1 + \cos(\emptyset_1 - \emptyset_2 + w)] \cdot 1 \\ &= \cos(\emptyset_1 - \emptyset_2 + w). \end{aligned} \tag{16}$$

Now choose the variable phase shifts as follows:

$$\emptyset_1' = \tfrac{1}{2}\pi, \ \emptyset_2' = \tfrac{1}{4}\pi + w, \ \emptyset_1'' = 0, \ \emptyset_2'' = (3\pi/4) + w. \tag{16}$$

Then,

$$\cos(\emptyset_1' - \emptyset_2' + w) = \cos(\emptyset_1' - \emptyset_2'' + w) - \cos(\emptyset_1'' - \emptyset_2' + w) = -\cos(\emptyset_1'' - \emptyset_2'' + w)$$

$$= 0.707, \tag{17}$$

and

$$E_\psi(\emptyset_1', \emptyset_2') + E_\psi(\emptyset_1', \emptyset_2'') + E_\psi(\emptyset_1'', \emptyset_2') - E_\psi(\emptyset_1'', \emptyset_2'') = 2.828, \tag{18}$$

in disaccord with Inequality (15). The quantity w which enters into the choice of the variable phase shifts in Eq. (16) is determinable experimentally, by varying one or the other of $\emptyset_1$ and $\emptyset_2$ until the joint probability for photon 1 to enter U_1 and photon 2 to enter U_2 becomes 0, and then using Eq. (7).

As discussed in "An Exposition of Bell's Theorem," the detection loophole can be blocked if sufficiently efficient photodetectors are developed. It may be easier to block this loophole in the experimental arrangement of Fig. 1, which is based upon the linear momentum correlation of the two photons, than in a polarization correlation experiment, because in the latter there are two competing demands on the efficiency of the apparatus: both the polarization analyzers and the photodetectors must be sufficiently efficient, and these demands are best fulfilled in different energy ranges of the photons.

In order to achieve a test of Bell's Inequality as decisive as that of Aspect et al. (1982), it would be necessary to vary the phase shifts $\emptyset_1$ and $\emptyset_2$ very rapidly, in time intervals of the order of 10 nanoseconds.[1] It is, of course, very difficult to satisfy this desideratum experimentally, but in principle it is possible, either by using acousto-optical switches, like those of Aspect et al., or by electro-optical devices.

Quite apart from the potentiality of our proposal for achieving improvements over previous tests of Bell's Inequalities, it may be pedagogically valuable. The proposed arrangement is simpler than that of the polarization correlation experiments, and opens the possibility of performing a test of Bell's Inequality as a demonstration in an undergraduate class. Furthermore, the demonstration of two-photon interference fringes in the absence of one-photon fringes would be a vivid illustration of quantum mechanical nonlocality.

FOOTNOTES

[1]Two-particle interferometry using pairs of photons produced by parametric down-conversion was reported by Ghosh and Mandel (1987), Hong, Ou, and Mandel (1987), Ou and Mandel (1988a) and (1988b), Alley and Shih (1986), and Shih and Alley (1988). The last three of these references report tests of Bell's Inequality, but in these tests quarter wave plates are introduced into the beams for the purpose of transforming momentum correlation into polarization correlation. In the proposal of the present lecture, which was briefly mentioned in Horne, Shimony, and Zeilinger (1989) and will be developed in more detail in a later paper by us, polarization correlation is completely avoided. Two-particle interferometry using pairs of photons produced in positronium annihilation was proposed by Horne and Zeilinger (1985), (1986), and (1988), but there are great obstacles in the way of realizing their proposal. Rarity and Tapster (1989) have also proposed a test of Bell's Inequality without polarization analysis, using the momentum correlation of photon pairs produced by parametric down-conversion, and had already obtained preliminary results by July, 1989.

REFERENCES

Alley, C. and Shih, Y.H., 1986, in "Proceedings of the Second International Symposium on Foundations of Quantum Mechanics in the Light of New Technology," M. Namiki et al., eds., Physical Society of Japan, Tokyo, 47.
Aspect, A., Dalibard, J., and Roger, G., 1982, Physical Review Letters, 49: 1804.

Beltrametti, E. and Cassinelli, G., 1981, "The Logic of Quantum Mechanics," Addison-Wesley, Reading, MA.

Bohm, D. and Aharonov, Y., 1957, Physical Review, 108:1070.

Einstein, A., Podolsky, B., and Rosen, N., 1935, Physical Review, 47:777.

Ghosh, R. and Mandel, L., 1987, Physical Review Letters, 59:1903.

Hong, C.K., Ou, Z.Y., and Mandel, L., 1987, Physical Review Letters, 59:2044.

Horne, M.A., Shimony, A., and Zeilinger, A., 1989, Physical Review Letters, 62:2209.

Horne, M.A. and Zeilinger, A., 1985, in "Proceedings of the Symposium on the Foundations of Modern Physics," P. Lahti and P. Mittelstaedt, eds., World Scientific, Singapore, 435.

Horne, M.A. and Zeilinger, A., 1986, "New Techniques and Ideas in Quantum Measurement Theory," D. Greenberger, ed., New York Academy of Sciences, New York, 469.

Horne, M.A. and Zeilinger, A., 1988, in "Microphysical Reality and Quantum Formalism," A. van der Merwe, et al., eds., Kluwer, Dordrecht, vol. 2, 401.

Ou, Z.Y. and Mandel, L., 1988a, Physical Review Letters, 61:50.

Ou, Z.Y. and Mandel, L., 1988b, Physical Review Letters, 61:54.

Rarity, J.C. and Tapster, P.R., 1989, "Spatially separated two color photons in fourth order interference," unpublished preprint.

Shih, Y.H. and Alley, C.O., 1988, Physical Review Letters, 61:2921.

Wu, C.S. and Shknov, J., 1950, Physical Review, 77:136.

Zeilinger, A., 1981, American Journal of Physics, 49:882.

ACKNOWLEDGEMENT

This work was partially supported by the National Science Foundation, Grants No. DIR-8810713 and No. DMR-8713559, and by Fonds zur Förderung der wissenschaftlichen Forschung (Austria), No. 6635.

A NEW VIEW ON THE UNCERTAINTY PRINCIPLE

J. Hilgevoord and J. Uffink

Department of History and Foundations of Mathematics and Science
University of Utrecht
Utrecht, The Netherlands

Summary

Upon close examination Heisenberg's microscope argument is found to depend on a relation between two quite distinct concepts of uncertainty. The first is an uncertainty in what can be predicted. The second is related to the notion of resolving power and is an uncertainty in what can be inferred (retrodiction). Quantitative measures of both kinds of uncertainties are introduced and discussed. The standard deviation is criticized as a measure of uncertainty. The usual uncertainty relations connect two uncertainties of the first kind. Uncertainties of the first and second kinds are also related by an uncertainty relation; this relation provides a general basis for the microscope argument. This new kind of uncertainty relation also allows for an adequate formulation of the uncertainty principle for line width and lifetime.

Bohr's argument with respect to the double slit problem, which is based on the uncertainty principle, is analysed and is found to depend on two uncertainties of the second kind. No corresponding uncertainty relation is known to exist; nevertheless, the validity of Bohr's conclusion can be established in a direct way.

Introduction

The uncertainty principle (UP) occupies a peculiar position in physics. On the one hand, it is often regarded as the hallmark of quantum mechanics. On the other hand, there is still a great deal of discussion about what it actually says. A physicist will have much more difficulty in giving a precise formulation of the UP then in stating e.g. the principle of relativity. Moreover, the formulations given by various physicists will differ greatly not only in their wording but also in their meaning. This peculiar state of affairs concerning the most famous principle of quantum mechanics reflects the general ambiguity of the interpretation of quantum mechanics, and one's favourite formulation of the UP will be closely related to one's favourite interpretation of quantum theory. This close link between the UP and the interpretation of quantum mechanics makes it difficult to discuss the UP in its own right. Nevertheless, one can say a number of interesting things about the UP without going deeply into the interpretational problems of quantum mechanics. In the following, we shall need only the barest minimum of interpretational postulates. In fact the only interpretational rule that will be used is the following: If ψ and ϕ are normalized states, then the number

$$| < \phi | \psi > |^2 \tag{1}$$

is the probability of finding the system in the state ϕ if it was prepared in the state ψ. By this rule alone a very substantial part of our subject can be discussed. The rule, of course, presupposes that it is possible to prepare a system in a state ψ and find it in a state ϕ. We shall just assume that these possibilities exist. We also assume that in practice it is known what is involved in preparing or finding a system in a certain state. For example, if a photon or an electron produces a black spot on a photographic plate we take this to mean that the particle has been found in a narrowly localized state. Of course, we are fully aware of the problems that are connected with such simple statements in quantum mechanics, but we will separate these problems from the discussion of the UP proper. By so doing we are thinking along the same lines, we believe, as Heisenberg in 1927, the year when he published his first paper on the UP. Indeed, firstly, Heisenberg considered the matrix elements as the essential ingredients of quantum theory. And, secondly, in trying to understand the physical meaning of the new theory, Heisenberg took as his starting point Einstein's remark that it is the *theory* which decides what can be observed. Thus, his famous discussion of thought experiments presupposes the validity of quantum mechanics. The thought experiments serve to illustrate the workings of the theory and to show its internal consistency. They do not provide a basis for the theory or lend support to its validity. Likewise, we shall base our discussion on the theory, and in particular on the interpretational rule, and investigate what follows from simply that.

In the following we shall trace some of the early history of the UP. But, at the same time, we shall develop a completely new view on the UP which, however, is already implicit in Heisenberg's original argument. The new view and the corresponding uncertainty relation provide a stronger formulation of the UP than the usual one. In addition, some of the persistent problems associated with the usual formulation are avoided.

The microscope argument

The UP was put forward by Heisenberg in 1927 in an article entitled "Über den anschaulichen Inhalt der quantentheoretischen Kinematik und Mechanik" [1]. The title of the English translation of this article reads "The physical content of quantum kinematics and mechanics" [2]. "Anschaulich" has been translated by "physical" which loses the visual element contained in the word "anschaulich". In his article Heisenberg did indeed want to make 'visible' the meaning of the quantum mechanical commutation relation

$$qp - pq = i\hbar. \tag{2}$$

Dirac, in a paper published in 1927 [3], had already remarked that this formula implies that "One cannot answer any question on the quantum theory which refers to numerical values for both q and p". This is a consequence of the fact that (2) precludes the simultaneous diagonalization of the matrices q and p. Thus in quantum mechanics it is somehow impossible to assign numerical values to both p and q. Heisenberg wanted to show how this comes about by analysing experiments designed to produce such values.

The starting point of his analysis was the demand that in order to give meaning to expressions like "the position of an object" one should indicate an experiment in which this position can be measured. For example, one may measure the position of an electron by a light microscope. Heisenberg observed that the precision with which the position of the electron can be determined in this way, depends on the wavelength of the light. In order to enhance the precision γ-rays should be used. But then the Compton effect becomes important. The photon that is scattered from the electron will change the momentum of the electron, and this change will be greater the shorter the wavelength of the photon. Thus, the more accurately the position of the electron

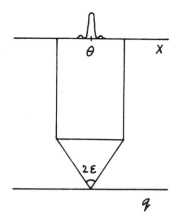

Figure 1. Heisenberg's microscope with point spread function

is determined, the less accurately is its momentum known afterwards. If q_1 and p_1 are the said inaccuracies, then, using the de Broglie relations, one easily finds

$$q_1 p_1 \sim h. \tag{3}$$

Concerning this argument of Heisenberg's the paper has an interesting 'Addition in Proof' that mentions critical remarks of Bohr, who saw the paper only after it had been sent to the publisher. Bohr remarks that in the microscope experiment it is not the change of the momentum of the electron itself that is important. The essential point is that this change cannot be precisely determined in the same experiment. In fact one may also note that Heisenberg's conclusion is somewhat rash, in view of his starting point, since he did not indicate what meaning should be given to the notion of momentum in this context.

An improved version of the microscope experiment was presented by Heisenberg in his Chicago lectures of 1929 [4]. Here Heisenberg assumes that the momentum of the electron is well determined, for example by a previous precise measurement of momentum. Next the electron is illuminated by light of wavelength λ. The light enters a microscope (fig.1). Suppose a photon that is scattered from the electron is detected at a point x of a photographic plate in the image plane of the microscope. What can be inferred from this about the position q of the electron? Because of the wave character of the light, and because of the finite aperture of the microscope, the image of a point source is a small blob which is a diffraction pattern. As a consequence, the microscope has a limited resolving power. The accuracy of the determination of the position of the electron is of the order of this resolving power, which, according to optics, is given by $\lambda/\sin \epsilon$. Hence,

$$\delta q \sim \frac{\lambda}{\sin \epsilon}. \tag{4}$$

On the other hand, the direction of the scattered photon is unknown within the angle ϵ; hence, its momentum is uncertain by an amount $\frac{h}{\lambda} \sin \epsilon$. The electron then experiences a recoil which is likewise uncertain by the same amount. Hence, the momentum after the observation has become uncertain by $\delta p \sim \frac{h}{\lambda} \sin \epsilon$, and we have

$$\delta p \delta q \sim h. \tag{5}$$

Heisenberg, in his lectures of 1929, called relation (5) an "Indeterminacy" relation (*Unbestimmtheitsrelation*). This contrasts with the term "inaccuracy" which he had used in his first paper. We are probably seeing here the influence of Bohr's remarks.

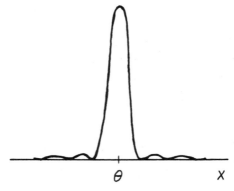

Figure 2. The point spread function $|\psi_\theta(x)|^2$

(It is interesting to note that in the English translation of Heisenberg's 1927 paper[4] the word "Ungenauigkeit", in the abstract, has been translated by "indeterminacy" which, of course, has quite a different meaning. It seems that the translator has not been able to suppress his own views on the matter.) It would be interesting to trace the first occurrence of the various terms that are used in the literature to refer to the UP. The term "uncertainty" was in use already by 1929 (e.g. Condon & Morse [5]). In the sequel we shall use the term "uncertainty" throughout as a neutral term the meaning of which will be specified in each particular case.

Let us see what the uncertainties in the case of Heisenberg's microscope actually mean. Heisenberg identified the uncertainty in the position measurement of the object, the electron, with the resolving power of the microscope. The resolving power is connected with the width of the image of a point source, the so-called point spread function. The shape of this diffraction pattern is shown in fig. 2. The position of the pattern in the image plane depends on the position of the object in a linear way, i.e. if q is the position of the object in the object plane and θ is the coordinate of the diffraction maximum, then θ is a linear function of q. (For simplicity we shall consider one dimension only, i.e. the object "plane" is a line.) If a photon is detected on the photographic plate in the image plane, i.e. if a black spot is formed, the question that arises is to which diffraction pattern the photon belongs. Our best guess, of course, is to assign a photon detected at a point with coordinate x to the pattern that has its maximum at this same point, i.e. our best guess is $\theta(q) = x$. From this the position q of the electron follows. However, we are not certain about the assignment. The photon could also belong to a slightly shifted diffraction pattern. This induces an uncertainty in the position q of the electron from which the photon was scattered. This uncertainty is called the resolving power of the instrument. It is usually quantified by Rayleigh's well-known criterion. According to this criterion two diffraction patterns which are shifted with respect to each other, become distinguishable when the maximum of the one coincides with the first minimum of the other. In our case this shift is of the order of the width of the central peak of the pattern, and this was taken by Heisenberg as the inaccuracy in the position determination of the electron by the microscope. For obvious reasons we shall call a width of the above kind a *translation width*. As another example, consider the diffraction pattern of a grating (fig. 3); it consists of a number of widely separated narrow peaks. The resolving power, in this case, is the width of the narrow peaks: if we translate the pattern by an amount of this order of magnitude it becomes distinguishable from the original pattern.

The problem with which we are confronted here can be stated as follows. We are

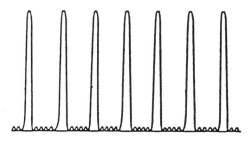

Figure 3. The diffraction pattern of a grating

given a set of probability distributions; in our case this is the set of shifted diffraction patterns characterized by the parameter θ. These probability distributions assign probabilities to the outcomes of certain experiments; in our case these outcomes are the positions of black spots on the photographic plate. Next, we are given an outcome. The problem then is to infer the probability distribution to which this outcome belongs or, equivalently, to estimate the parameter θ. Stated in quantum mechanical terms the problem is: given the outcome of an experiment, what can we say about the state of the system before the measurement?

The problem of inferring a probability distribution from a set of outcomes is a central problem of classical statistics, and we shall return to it presently. But let us first continue with Heisenberg's microscope. According to quantum theory the spread in the diffraction pattern is connected with a spread in the momentum distribution of the photon. (The momentum distribution is the absolute square of the Fourier transform of the wave function in x-space.) Because momentum is conserved in the scattering process, the momentum distribution of the electrons, after the experiment, has this same spread. Thus, by doing an experiment from which one may *infer* the position of the electron with an uncertainty δq, one can *predict* the momentum with which the electron will be found in a subsequent momentum determination, with an uncertainty δp, where δq and δp are related by relation (5). Here we encounter a second, and in physics more common, aspect of probability theory. This time we are given a probability distribution, the momentum distribution of the electrons, and we ask what can be predicted from this about the outcome of an experiment.

The above formulation of the microscope experiment conforms closely to the original formulation of Heisenberg. Note that only *one* measurement is actually performed: the determination of the photon's position. From the result of this measurement a prediction can be made about the outcome of a *subsequent* measurement of the photon's (or electron's) momentum. No *simultaneous* measurements are involved! Neither are joint probabilities, nor is the projection postulate.

Two kinds of uncertainty

The structure of the microscope argument, then, is as follows. From the detection of a spot on the photographic plate one draws an inference about the previous state of the system. This can be done with an uncertainty that is related to the translation width of the system. Let us call this an uncertainty of the *second* kind. Next, from this state, one deduces what can be predicted about the outcome of a subsequent momentum measurement. Let us call the uncertainty in this prediction an uncertainty of the *first* kind. The UP now says that the two uncertainties are related and cannot both be arbitrarily small.

From the point of view of general quantum mechanics the uncertainties of the first and second kind correspond to two ways in which one can look at the quantity

$$| < \phi | \psi > |^2,$$

namely, (1) suppose the system is prepared in a given state ψ; to what extent can we *predict* the state ϕ in which it will be found? And, (2) suppose the system is found in some given state ϕ; to what extent can we *infer* the state in which it was prepared?

Let us elaborate somewhat on this distinction.

1. Assume that a system is prepared in a state ψ. Consider a measurement performed on the system and let $\{\phi_i\}$ denote the orthonormal set of eigenstates of this measurement. To what extent can we predict in which of the states ϕ_i the system will be found? The answer is determined by the probability distribution $p_i = | < \phi_i | \psi > |^2$. Intuitively one would say that the uncertainty in this prediction depends on the shape of this probability distribution. If the probability is largely concentrated in a small subset of the possible values i the uncertainty is small; if the probability is uniformly distributed the uncertainty in this prediction is large. Thus, as a measure of uncertainty of this kind one is led to adopt a mathematical expression that measures the width or spread of the probability distribution p_i over the values of i.

2. Suppose the system is found in the state ϕ_i. What can one then infer about the state in which the system has been prepared if that state is known to belong to a given set of candidate states $\{\psi_\theta\}$? Let us for simplicity assume that this set consists of two states $\psi^{(1)}$ and $\psi^{(2)}$. The answer to our question is determined by the probabilities $p_i^{(1)} = | < \phi_i | \psi^{(1)} > |^2$ and $p_i^{(2)} = | < \phi_i | \psi^{(2)} > |^2$. For example, if only one of these probabilities is non–zero, then the measurement result allows us to rule out one possibility. On the other hand, if both probabilities are equal the measurement result gives equal support to both candidate states. In general, for arbitrary i, one would say that the measurement allows a certain inference if $p_i^{(1)} p_i^{(2)} = 0$, for all i, whereas the inference is completely uncertain if $p_i^{(1)} = p_i^{(2)}$, for all i. Thus one is led to consider as a measure of the uncertainty in inference which is connected with the considered measurement and set of candidate states, some mathematical expression that measures the overlap of the two probability distributions $p_i^{(1)}$ and $p_i^{(2)}$

This idea can be extended to the case where the set of candidate states depends on some parameter θ. Then we may say that the measurement allows an accurate inference about the value of θ if the overlap between p_θ and $p_{\theta+\delta\theta}$ already vanishes for small displacements $\delta\theta$. In fact we may adopt a typical value of $\delta\theta$, for which the overlap between p_θ and $p_{\theta+\delta\theta}$ has diminished appreciably, as a measure of the uncertainty in θ by obvious analogy with the notion of resolving power. The problem of defining a quantitative measure of uncertainty of the second kind will be considered again later on.

The key question now is whether uncertainty relations of a general kind exist between the two kinds of uncertainty. It is very gratifying that the answer turns out to be affirmative. We shall return to this subject later on, but we would like to emphasize already now that, since the two uncertainties are conceptually quite different, the quantitative measures of these uncertainties must be defined quite differently. Heisenberg, not being aware of the distinction, took the resolving power (which, as we have seen, is a translation width) as a width of the first kind, i.e. as the length of an interval on which a large portion of the total probability is situated (viz. the central diffraction peak). But generally a translation width does not have this character, as is shown by the diffraction pattern of a grating (fig. 3). For the simple pattern of fig. 2 the two kinds of width are *numerically* of the same order of magnitude and can be easily confused. Thus, Heisenberg, in the sequel of his paper, formulated the UP as a mathematical relation between two uncertainties of the first kind. And the same held, until quite recently, for all subsequent discussions of the UP. This historical fact has had a considerable influence on the further development of the subject.

The standard uncertainty relation and its problems

Let us continue with the chronological development. As a first step to a quantitative formulation of the UP Heisenberg, in his 1927 paper, considered a Gaussian wave packet in position–space. The wave packet in momentum–space, being the Fourier transform of the wave function in position–space, is itself a Gaussian also. The widths of the corresponding probability distributions are inversely proportional. This, according to Heisenberg, is a simple mathematical expression of the UP. Note that the resolving power aspect of uncertainty, which was implicit in the microscope argument, has disappeared completely.

In the same year, this result was generalized by Kennard [6] who proved the well-known relation

$$\Delta p \Delta q \geq \frac{1}{2}\hbar \tag{6}$$

where Δ denotes the standard deviation:

$$(\Delta p)^2 \;=\; <p^2> - <p>^2 = \int p^2 |\phi(p)|^2 dp - \left(\int p|\phi(p)|^2 dp\right)^2$$

$$(\Delta q)^2 \;=\; <q^2> - <q>^2 = \int q^2 |\psi(q)|^2 dq - \left(\int q|\psi(q)|^2 dq\right)^2 \tag{7}$$

and

$$\phi(p) = \frac{1}{\sqrt{2\pi}} \int e^{-ipq} \psi(q) dq \tag{8}$$

Relation (6) holds for any normalized wave function; the equality sign corresponds to Gaussians.

The Kennard relation uses the standard deviation as a measure of quantum uncertainty. At the time, this may have seemed a natural thing to do. The uncertainties of quantum mechanics seemed to be connected with measurements. In his paper of 1927, Heisenberg himself referred to q_1 as the "mean error" (mittlerer Fehler). The term "error" also appears in many later formulations of the UP. Condon & Morse in 1929 write[1]

- All measurements of position are affected by an error Δx and the simultaneous measurement of momentum is affected by an error Δp.

And Bohm in his book[2] of 1951 says:

- If a measurement of position is made with accuracy Δq, and if a measurement of momentum is made simultaneously with accuracy Δp, then the product of the two errors can never be smaller that $\sim h$.

Note, that besides using the word "error" for quantum uncertainties, these authors also refer quite explicitly to simultaneous measurements.

The use of the term "measurement error" to indicate the quantum uncertainties wrongly suggests that the latter are related to ordinary classical measurement errors. Classical measurement errors are usually distributed according to a Gaussian distribution ("law of errors") and the standard deviation is, of course, an appropriate measure of the width of a Gaussian. But quantum uncertainties refer to the outcomes of *ideal* measurements, they are not related to classical measurement errors. This is borne out by the fact that they are described by arbitrary probability distributions, not just Gaussians. Therefore, the use of the standard deviation as a measure of quantum uncertainty is not as natural as it might have seemed; actually it is not very appropriate

---

[1]ref 5, p 21
[2]ref 7, p 99

at all. As an example, consider the diffraction pattern of fig. 2. It is given by (putting $\theta = 0$)

$$|\psi(x)|^2 = \frac{a}{\pi} \frac{\sin^2(ax)}{(ax)^2} \tag{9}$$

Here π/a is the width of the central diffraction peak which, in this case, Heisenberg took as a measure of uncertainty. The standard deviation Δx in (9) diverges. Hence, relation (6) is useless already for Heisenberg's first example! Exactly the same happens in Heisenberg's second example, in which he uses a narrow slit to fix the position of the particles [4]. The diffraction pattern of the slit is again given by (9). The standard deviation diverges in many other physical applications as well; another well-known example is the Breit-Wigner form of spectral lines. The divergence is caused by the heavy weight the standard deviation places on the far away parts of the probability distribution. As a consequence, it is possible to concentrate almost all of the probability on a very narrow interval and still have a very big standard deviation. This is illustrated, again, by the distribution (9). If the parameter a increases, the distribution (9) approaches a δ-function. Nevertheless, the standard deviation is infinite for all values of a. The standard deviation, therefore, is not a good measure of the extent to which the bulk of a probability distribution is concentrated: even if 99% or more of the total probability is concentrated on an arbitrarily small interval the standard deviation may be arbitrarily large. The remarkable consequence is that from relation (6) it does not follow that the p and q distributions cannot be simultaneously arbitrarily narrow. In fact, they could both approach a δ-function without violating relation (6).

Remark

Our criticism of the standard deviation as a measure of quantum uncertainty does not rule out its usefulness for special purposes. The main example is provided by the harmonic oscillator the Hamiltonian of which is also basic to quantum optics. In the energy eigenstates the quantities Δq and Δp are directly related to the potential and kinetic energy of the oscillator, respectively. In the ground state $\Delta q = \Delta p = \sqrt{\frac{1}{2}\hbar}$, corresponding to the minimum in (6). In quantum optics q and p become related to the photon creation and annihilation operators and acquire a more abstract meaning. An important class of special states are the coherent states characterized by $\Delta q = \Delta p = \sqrt{\frac{1}{2}\hbar}$. Quite recently, still another class of states, the squeezed states, characterized by $\Delta q \Delta p = \frac{1}{2}\hbar$; $\Delta q \neq \Delta p$ has attracted attention. In a squeezed state Δq (or Δp) may be much smaller than its ground state value. The physical meaning of this fact can only be appreciated, however, by expressing it in terms of photon numbers.

Thus, the quantities $\Delta q, \Delta p$ and relation (6) may be relevant and interesting in special states of special Hamiltonians, but this does not alter their shortcomings as general expressions of the UP.

The Landau-Pollak relation

Clearly, then, the standard deviation, for a general probability distribution, does not express what we intuitively mean when we say that a probability distribution is narrow. By saying this we mean that the main part of the probability is concentrated on only a small subset of the set of possible outcomes. In that case we can predict the outcome with only a small uncertainty. A direct measure of this kind of uncertainty is given by the magnitude of this set. So, in our case, we may define the uncertainty connected with the probability distribution $|\psi(q)|^2$, as the length W_q of the smallest interval on which a large fraction α (90% say) of the total probability is situated. W_p may be defined in a similar manner for the distribution $|\phi(p)|^2$. In 1961 Landau and Pollak [8] of the Bell Telephone Company, proved that these uncertainties satisfy the relation

$$W_p W_q \geq c(\alpha) \text{ if } \alpha \geq \frac{1}{2} \tag{10}$$

$(c(\alpha)$ is of order 1). It is this relation, rather than (6), which expresses the fact that it is not possible to concentrate a large part of the probability distribution in both q- and p-space on an arbitrarily small interval. This intuitive content of the UP for uncertainties of the first kind, therefore, was not proven until 1961!

The inequality (10) provides a conceptually stronger formulation of the UP than (6) because it connects more stringent measures of uncertainty. Indeed, with the help of the Bienaymé–Chebyshev inequality, one can show that

$$W_q \le \frac{\Delta q}{\sqrt{1 - \alpha}} \, .$$

Hence, the Landau–Pollak relation already implies the existence of a lower bound to the product $\Delta p \Delta q$.

Relation (10) no longer suggests a connection with measurement errors. It has a much clearer interpretation than relation (6). The width W may be taken as an uncertainty of the first kind. Then, according to our simple minimal interpretation, relation (10) says that in a given state of the system it is not possible to predict the outcomes of both q- and p-measurements with arbitrary precision. These measurements need not be performed simultaneously. Uncertainty relations between uncertainties of the first kind express what can be simultaneously *predicted*.

This interpretation of the UP has been put forward and strongly defended by Karl Popper (1934) [9]. We have seen, however, that uncertainties of the first kind cannot tell the whole story about the UP.

In recent times other relations between uncertainties of the first kind have been derived, e.g. the so-called entropic uncertainty relation (cf. [10]), but we shall not discuss them here.

The Robertson relation

Before going on we will comment briefly on a generalization of relation (6) that was derived by Robertson in 1929 [11]:

$$\Delta A \Delta B \ge \frac{1}{2}| < AB - BA > | \tag{11}$$

Here, A and B are arbitrary Hermitian operators. This relation has played a very important, though not very fortunate, role in the further development of the subject, for it has been considered by most physicists as the supreme expression of the UP, on which all uncertainty relations should be modelled. However, it also uses the standard deviation as a measure of uncertainty and, in this respect, it fares no better than relation (6). But as compared to relation (6) it has an additional defect, namely that its right-hand side depends on the state of the system. If the state is an eigenstate of A the right hand side of (11) vanishes, as does ΔA, and no restriction at all follows for ΔB. However, what's really bad about (11) is that it directs attention to the wrong direction. The relation suggests that one must have Hermitian operators for every pair of quantities that obey the uncertainty principle. This demand has led to a fruitless search for an operator of time in quantum mechanics with the help of which one could express the uncertainty principle for energy and time [12]. In the early literature, notably already in Heisenberg's 1927 paper, the uncertainty relation between energy and time was seen as closely analogous to the one between position and momentum. By contrast, the Robertson relation (11) introduces a sharp distinction between the position–momentum and energy–time uncertainty relations. Actually, relation (11) also leads to difficulty with regard to the position–momentum relation. In relativistic quantum mechanics the status of position as an operator is problematical, and a position operator for the photon does not exist at all. This renders the Robertson relation incapable of dealing with Heisenberg's microscope experiment!

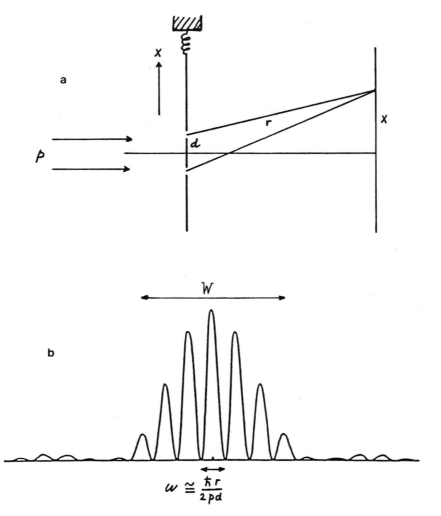

a

b

$$w \cong \frac{\hbar r}{2pd}$$

Figure 4. *(a)*. The double slit experiment, *(b)*. The double slit interference pattern

The double slit argument

The double slit experiment is another favourite illustration of the workings of the UP. The problem differs in an essential way from the microscope problem. The question now is: Is it possible to determine through which slit the particles pass and still observe interference? Bohr denied that this possibility exists on account of the UP. He argued as follows[3]. Suppose a particle passes through the screen and is detected on a photographic plate behind the screen (fig. 4). In order to locate the slit through which the particle passed, the screen is suspended from a spring so that the recoil of the screen caused by the passage of the particle can be detected. If the initial momentum of the particle is p, the recoil difference due to the passage through the one slit or the other is $\delta p_x = \frac{2pd}{r}$. For this difference to be distinguishable the initial momentum P of the screen along the x-direction must be known with an uncertainty $\delta P < \frac{2pd}{r}$. Then, by the UP, the position of the screen in this direction is uncertain by an amount $\delta Q > \frac{\hbar r}{2pd}$. But $\frac{\hbar r}{2pd}$ is precisely the width of the interference bands. Hence, Bohr concludes, the interference pattern does not appear.

It is clear from the above that the relevant width in this argument is numerically equal to the width of the interference bands. This implies that the argument cannot be based on the uncertainty relations (6) or (10) because the measures Δ and W are insensitive to the fine structure of the interference pattern (the standard deviation even diverges). On the other hand, the translation width of the pattern clearly *is* sensitive to the width of the interference bands and has the same order of magnitude. This suggests that it is, once again, an uncertainty of the *second* kind which is relevant to this problem. That this is indeed the case becomes clear if one notes that in the above argument it is the *distinguishability* of two different momentum states of the screen that is crucial. We shall return to the double slit argument in the final section.

A measure of uncertainty of the second kind

The distinction between two kinds of questions that can be asked in probability theory, which led us to distinguish between two kinds of uncertainty, is not at all typical for quantum mechanics. In particular, the problem of deciding between possible probability distributions on the basis of outcomes is the central problem of classical statistics and it was studied quite intensively at about the same time as Heisenberg invented quantum mechanics. Recently, the physicist W.K. Wootters[13] developed a concept of "statistical distance" between two probability distributions. This concept is meant to characterize the ease with which two probability distributions can be distinguished on the basis of outcomes. The larger the statistical distance, the easier it is to distinguish the two distributions. According to Wootters' criterion, the statistical distance between two probability distributions $p_i^{(1)}$ and $p_i^{(2)}$ is found to be

$$d(p^{(1)}, p^{(2)}) = \arccos \sum_i \sqrt{p_i^{(1)} p_i^{(2)}}.$$ (12)

We note the occurrence of the overlap of the two probability distributions, in accordance with our heuristic considerations, but even more striking is the occurrence in this expression of the square root of the probabilities, reminiscent of quantum mechanics, although it has nothing to do with that subject. In quantum mechanics the probability distributions arise as follows. Let $\psi^{(1)}$ and $\psi^{(2)}$ be the states between which one must decide on the basis of a measurement whose possible outcomes are characterized by the complete orthonormal set of states ϕ_i. The corresponding probabilities are $|< \phi_i|\psi^{(1)} >|^2$ and $|< \phi_i|\psi^{(2)} >|^2, (i = 1, ..., n)$, and the corresponding statistical distance is

$$\arccos \sum_i | < \phi_i|\psi^{(1)} > || < \phi_i|\psi^{(1)} > |.$$

---

[3] ref 2, p 25

Now, Wootters remarks, the quantum mechanical case differs from the classical case in the following respect. In order to distinguish between $\psi^{(1)}$ and $\psi^{(2)}$ many different experiments are available some of which suit the purpose better than others. For example, the two spin $\frac{1}{2}$ states parallel to the z-axis are more easily distinguished by measuring the spin in the z-direction than by measuring it in any other direction. Thus, we are led to define the "absolute" statistical distance between two quantum states as the distance with respect to the most discriminating experiment. As is easily seen, the most discriminating experiments are the ones that have either $\psi^{(1)}$ or $\psi^{(2)}$ among the ϕ_i, and we find

$$d(\psi^{(1)}, \psi^{(2)}) = \arccos |<\psi^{(1)}|\psi^{(2)}>|. \tag{13}$$

We may then adopt as a criterion for the distinguishability of two states, $\psi^{(1)}$ and $\psi^{(2)}$, the condition

$$d(\psi^{(1)}, \psi^{(2)}) > \arccos \beta, \tag{14}$$

or

$$|<\psi^{(1)}|\psi^{(2)}>| < \beta, \tag{15}$$

where β is some conveniently chosen positive number smaller than 1.

Condition (15) may be seen as a natural generalization of Rayleigh's criterion to arbitrary states. If the states belong to a set of states which are translated copies of each other, like the states ψ_θ in the microscope problem, we shall define these states as having translation width w if w is the smallest number for which

$$|<\psi_{\theta+w}|\psi_\theta>| = \beta. \tag{16}$$

(By this definition the states become distinguishable for the first time if they are translated with respect to each other by the translation width. By a further translation they may become indistinguishable again.) In the case of the microscope, w is of the order of the width of the central peak of the diffraction pattern and coincides, numerically, with the width W of this pattern. But in the case of the double slit pattern, the two widths are numerically very different. Whereas W is inversely proportional to the slit width, w is inversely proportional to the distance between the slits. In fact one can show [15] that for a general pattern $w \leq W$ if $\alpha^2 + \beta^2 \geq 1$.

A new uncertainty relation

We now come to the question of the existence of uncertainty relations for uncertainties of the second kind. To this end we first give a more general definition of the concept of translation width [14]. Suppose the system is symmetric with respect to translations in space. We shall consider only one spatial dimension and denote the coordinate of the spatial reference frame by x. Then in the Hilbert space of the system there exist a one-parameter group of unitary operators, representing translations of the system in space:

$$U(a) = e^{-iaP}, \quad a \in \mathbb{R}. \tag{17}$$

The Hermitian operator P, the generator of the translation group, is by definition the operator of the total momentum of the system. We assume that P has as (improper) eigenvalues the real numbers. The eigenvectors $|p>$ form a complete orthonormal set:

$$<p|p'> = \delta(p - p'); \quad \int |p><p|dp = \mathbb{1} \tag{18}$$

The set of translated states generated by a state Ψ is given by $U(a)|\Psi>$ for all values of a. By analogy with (13), the translation width of $|\Psi>$ is defined as the smallest number w_x satisfying

$$|<\Psi|U(w_x)|\Psi>| = \beta \tag{19}$$

Inserting the complete set of states $|p>$ in $<\Psi|U(a)|\Psi>$ we have

$$<\Psi|U(a)|\Psi> = \int e^{-iap}|<p|\Psi>|^2 dp = \int e^{-iap}|\phi(p)|^2 dp, \tag{20}$$

where $\phi(p) \equiv < p|\Psi >$ is the wave function of the state in p-space. Using this relation one can prove [15] the following relation between the translation width of Ψ and the width of its momentum distribution:

$$w_x W_p \geq C(\alpha, \beta), \quad \text{if } \beta \leq 2\alpha - 1. \tag{21}$$

Here w_x is defined by (19), W_p is the width of $|\phi(p)|^2$ as defined above (10) and,

$$C(\alpha, \beta) = 2 \arccos \frac{1 + \beta - \alpha}{\alpha}. \tag{22}$$

In practice C is a number of order 1. Relation (21) connects the translation width of Ψ with respect to spatial translations (a width of the second kind) with the width W_p of the momentum spectrum of Ψ (a width of the first kind). This is exactly the kind of uncertainty relation that is needed for Heisenberg's microscope argument. It connects the distinguishability of translated states ($\rightarrow$ the resolving power of the instrument) with the predictability of the momentum of these states ($\rightarrow$ the predictability of the momentum of the object (electron) after the measurement).

Note that it has not been necessary to introduce a position operator for the system. This makes the relation applicable to photons! On the other hand, the momentum appears as the fundamental operator generating translations in Hilbert space.

The uncertainty relation between lifetime and line width

The foregoing immediately suggests that the same formalism can be applied to translations in time. The corresponding operator in Hilbert space is

$$U(\tau) = e^{i\tau H} \tag{23}$$

where H is the Hamilton operator representing the total energy of the system. In contrast with the spectrum of P, the spectrum of H is assumed to be bounded from below; also, it may contain discrete eigenvalues. The matrix element $< \Psi|U(\tau)|\Psi >$ is called the "survival amplitude" of Ψ. The temporal translation width w_t, defined by

$$| < \Psi|U(w_t)|\Psi > | = \beta, \tag{24}$$

is related to the lifetime of the state: e.g. taking $\beta^2 = \frac{1}{2}$, w_t is the half life of the state. Using the completeness of the eigenstates $|E >$ of H, one has

$$< \Psi|U(\tau)|\Psi > = \int e^{i\tau E} | < E|\Psi > |^2 dE. \tag{25}$$

From this relation the uncertainty relation

$$w_t W_E \geq C(\alpha, \beta) \ , \quad \text{if } \beta \leq 2\alpha - 1 \tag{26}$$

can be derived analogously to the derivation of (21). The width W_E is a measure of the "line width" of the state. Thus, relation (26) is a general uncertainty relation between the lifetime and the line width of a state. Note that in this derivation we needed neither the existence of an operator of time, nor the exponential approximation to the survival amplitude that is necessary for the usual derivation of an uncertainty relation between lifetime and line width of an unstable state. (Note also that the standard deviation of the resulting Breit-Wigner line diverges.)

Thus, the Heisenberg microscope, and the closely related single slit experiment, as well as the relation between lifetime and line width, are all essentially illustrations of uncertainty relations of type (21) and (26) which relate uncertainties of the first and second kind. Also, an uncertainty relation between angular momentum J and angle ϕ can be derived along the same lines [14]. P, H and J appear as operators, while,

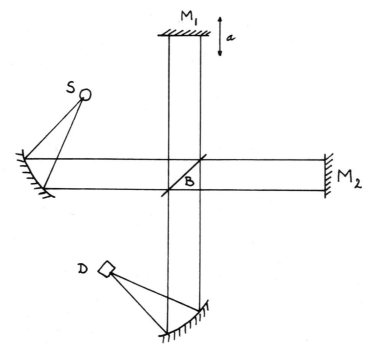

Figure 5. The spectral interferometer

x, t, and ϕ are simply parameters parametrizing the set of "translated" states. As a consequence, the corresponding uncertainty relations, unlike relations (6) and (11), are not symmetric in the variables. The reason, of course, is that the new uncertainty relations relate conceptually distinct kinds of uncertainty. From our point of view, relations (6) and (11) relate uncertainties of the first kind. One may ask whether relations also exist relating two uncertainties of the second kind. As far as we know, this is not the case.

Uncertainty and interference

Finally, we must discuss the double slit experiment. The problem here is slightly different from the preceding problems because the phenomenon of interference itself is explicitly addressed. The Michelson spectral interferometer provides a more simple situation than the double slit experiment, so let us discuss this first. In its modern form, due to Twynman and Green, it works as follows (fig. 5). Light from a source is collimated by a mirror. The resulting beam is divided by a beamsplitter B. The partial beams are reflected by mirrors M_1 and M_2 and again meet the beamsplitter. The light from the resulting overlapping beams is collimated again and detected at a detector D. The intensity of the light at D depends on the relative phase of the overlapping beams. By moving one of the mirrors this phase may be changed. As a result, the intensity at the detector changes rapidly with a period of the order of the average wave-length of the light. The "visibility" of this interference phenomenon is defined as

$$\mathcal{V} = \frac{I_{\max} - I_{\min}}{I_{\max} + I_{\min}}, \tag{27}$$

where I_{max} and I_{min} are the intensities of adjacent maxima and minima. One studies the variation of $\mathcal{V}$ as a function of the displacement a of the mirror. If the light consists of coherent pulses of length l, it is clear that $\mathcal{V}(a)$ drops to zero if the mirror is displaced over this distance because the coherent wave packets in the two beams no longer overlap. In general, $\mathcal{V}(a)$ clearly depends on the overlap of these coherent wave packets. In fact, $\mathcal{V}(a)$ turns out to be proportional to the overlap integral

$$\mathcal{V}(a) \quad \propto \quad | < \Psi | U(a) | \Psi > |, \tag{28}$$

where Ψ denotes the wave packet. If the phase of the matrix element in (28) can also be determined, then the wave number spectrum of Ψ can be obtained from (20). That's why this instrument is called a "spectral" interferometer (the wave number is $\sigma = p/h$).

It is immediately clear that this experiment measures the translation width of the wave packets. Indeed, $a = w_x$ is the distance by which the mirror should be displaced so that $\mathcal{V}(a)$ drops to the value β for the first time. Also, by (21), a lower bound W_σ on the width of the wave number spectrum of the wave packet can be obtained.

Essentially the same situation as is realized in the Michelson interferometer for light has been realized for neutrons in a beautiful instrument called a neutron interferometer[16]. It has been claimed in the literature that, by measuring the visibility $\mathcal{V}(a)$ as a function of a, the standard deviation of the neutron wave packet in position space has been obtained and the validity of relation (6) has been established. On closer inspection, however, it is found that what has actually been measured in this experiment is the translation width of the neutron wave packet [17].

From the above discussion it will be clear that the translation width is closely connected with the interference phenomenon and that it is directly measured in well-known experiments.

The double slit revisited

In a number of ways the situation here is more involved. We have seen that Bohr applied the UP to the screen with the slits, and that he considered this screen to be in such a state Ψ that a change of momentum of order $\delta P \sim \frac{2pd}{r}$ can be discerned. Hence, we are dealing with the discrimination between states which are translated in momentum space rather than in position space. Up to now we did not need to assume the existence of a position operator for our system, but if we want to apply the same formalism as before we need to assume this now. Let Q be the position operator of the screen. (Compare also [14].) Its eigenfunctions are denoted by $|q>$, where $q \in$ IR are the eigenvalues of Q, and

$$< q | q' > = \delta(q - q') ; \quad \int |q> < q| dq = 1\!1. \tag{29}$$

From

$$QP - PQ = i\hbar \tag{30}$$

and (17) follows

$$U(a)|q> = |q + a> \tag{31}$$

Because of the complete symmetry between Q and P the unitary operator

$$V(b) = e^{ibQ} , \quad b \in \text{IR} \tag{32}$$

is the translation operator in momentum space:

$$V(b)|p> = |p + b> . \tag{33}$$

Also,

$$< \Psi|V(b)|\Psi > \; = \; \int e^{ibq}|<q|\Psi>|^2 dq$$

$$= \; \int e^{ibq}|\psi(q)|^2 dq \qquad (34)$$

where $\psi(q) \equiv <q|\Psi>$ is the wave function of the screen in position–space. Again, it follows that

$$w_p W_q \geq C(\alpha, \beta) \qquad (35)$$

with w_p defined by

$$|<\Psi|V(w_p)|\Psi>| = \beta \qquad (36)$$

For the two momentum states of the screen which occur in Bohr's argument to be distinguishable, it is necessary that $w_p < \frac{2pd}{r}$. It follows from (35) that $W_q \geq \frac{r\hbar}{2pd}$. This means that the position of the screen can only be predicted with an uncertainty which is larger than the width of the interference bands of the interference pattern. According to Bohr, the interference pattern will not occur under these circumstances.

Bohr's argument has an intuitive appeal: it suggests a fuzziness of the position of the screen which causes a blurring of the interference pattern. Let us see if we can substantiate this argument with the help of the uncertainty relation (35). Suppose the interference pattern did show up. We would then be able to infer the position of the screen with an uncertainty of the order of $\frac{\hbar r}{2pd}$, the resolving power. This would mean that the translation width w_q of the state of the screen in q-space cannot be greater than $\frac{\hbar r}{2pd}$. But this does not contradict the condition $W_q \geq \frac{\hbar r}{2pd}$! The width W_q is defined as the length of the interval in q-space on which a large fraction α of the total probability is situated. Inside this interval the probability distribution is arbitrary: it could be concentrated, for example, in two narrow peaks at the boundaries of the interval. These peaks may themselves be much narrower than $\frac{\hbar r}{2pd}$, and we would have $w_q < \frac{\hbar r}{2pd}$. Hence, the condition $W_q \geq \frac{\hbar r}{2pd}$ does not rule out the simultaneous validity of the condition $w_q < \frac{\hbar r}{2pd}$ for the occurrence of the interference pattern. (In this example *two* interference patterns, separated by a distance W_q with respect to each other, would occur.)

Thus, on closer inspection, Bohr's argument is found to be inconclusive. It cannot be based on any of the known uncertainty relations. For the relations (6) and (10) this is obvious because the widths appearing in these relations have incorrect orders of magnitude. With regard to relation (35) the inconclusiveness of the argument follows from a closer scrutiny of the meaning of the uncertainties. Nevertheless, though Bohr's reasoning is not correct, his conclusion still is! A direct calculation of the interference pattern of a movable screen shows that it is the visibility (27) of the interference pattern, rather than the width of the interference bands, which is related to the distinguishability of the two momentum states. In fact, this visibility turns out to be proportional to the matrix element

$$|<\Psi|V(b)|\Psi>| \qquad (37)$$

where Ψ is the state of the screen and $b = \frac{2pd}{r}$. The same matrix element (37) is also a direct measure of the distinguishability of the two momentum states. Thus, whenever these states can be completely distinguished, that is, if the slit through which the particles pass can be determined with certainty, the visibility of the interference pattern also vanishes [18].

Acknowledgement

The authors wish to thank Sheila McNab for her help with the English.

References

[1] Heisenberg, W.: *Z. Physik* **43** (1927) 172.

[2] Wheeler, J.A. and Zurek, W.H. eds.: *Quantum Theory and Measurement*, Princeton University Press 1983, p. 62.

[3] Dirac, P.A.M.: *Proc. Roy. Soc.* **113A** (1927), 621.

[4] Heisenberg, W: *Die physikalischen Prinzipien der Quantentheorie*; S. Hirzel Verlag, Stuttgart 1930.

[5] Condon, E.U. and Morse, P.M.: Quantum Mechanics, McGraw-Hill 1929.

[6] Kennard, E.H.: *Z. Physik* **44** (1927) 326.

[7] Bohm, D.: *Quantum Theory*, Prentice Hall 1951.

[8] Landau, H.J. and Pollak, H.O.: *Bell. Syst. Techn. Journal* **40** (1961) 65.

[9] Popper, K.: *Logik der Forschung*, Springer, Wien 1935.

[10] Maassen, H. and Uffink, J.: *Physical Review Letters* **60** (1988) 1103.

[11] Robertson, H.P.: *Physical Review* **34** (1929) 163.

[12] Allcock, G.R.: *Annals of Physics* (New York) **53** (1969) 253.

[13] Wootters, W.K.: *Physical Review* **D23** (1981) 357.

[14] Hilgevoord, J. and Uffink, J.: in: *Microphysical Reality and Quantum Formalism*, eds. A. van der Merwe et al. (Kluwer 1988) p. 91.

[15] Uffink, J. and Hilgevoord, J.: *Foundations of Physics* **15** (1985) 925.

[16] Greenberger, D.M.: *Reviews of Modern Physics* **55** (1983) 875.

[17] Uffink, J.: *Physics Letters* **108A** (1985) 59.

[18] Uffink, J. and Hilgevoord, J.: *Physica* **B151** (1988) 309.

MEASUREMENT PROBLEMS IN QUANTUM FIELD THEORY IN THE 1930'S

Arthur I. Miller

Department of Physics
Harvard University
Cambridge, MA 02138
USA

and

Department of Philosophy
University of Lowell
Lowell, MA 01854
USA

On 18 September 1931 Werner Heisenberg wrote to Niels Bohr that "my work seems to be somewhat gray on gray." The bright future for quantum physics that seemed to be just over the horizon in Fall of 1927 darkened with the formulation of relativistic quantum mechanics. The principal problems were: P.A. M. Dirac's 1928 electron theory predicted negative energy states that resisted interpretation; in 1928, Heisenberg and Wolfgang Pauli found that in quantum electrodynamics the electron's self energy is divergent; and the continuous energy spectrum of β-particles in the supposedly two-body final state of nuclear β-decay implied that energy was not conserved in nuclear reactions and that perhaps quantum mechanics was not valid within the nucleus.

At the 20-25 October 1930 Solvay Conference Bohr, Dirac, Heisenberg and Pauli concurred that fundamental difficulties in quantum electrodynamics might be clarified through investigating measurability of electromagnetic field quantities. Upon return to Copenhagen from Brussels, Bohr continued discussing field measurements with Lev Landau who happened to be visiting. In December 1930 Landau went on to Zürich where he interested Pauli's assistant Rudolf Peierls in field measurements. Their deliberations led to a joint 1931 publication entitled, "Extension of the Uncertainty Principle to Relativistic Quantum Theory" (Landau and Peierls, 1931). They concluded that electromagnetic field quantities could not be measured in the quantum domain and assumed the root of the difficulty to be the negative energy states in Dirac's theory of the electron. Consequently, they wrote, "it would be surprising if the formulation of quantum electrodynamics bore any resemblance to reality." Landau and Peierls' results bore not only on the validity of quantum electrodynamics, but more generally on how a theory of submicroscopic phenomena ought to be structured. Fundamental problems that everyone assumed to have been settled in 1927 surfaced again.

The situation became severe enough that, in 1932, Bohr reanalyzed the field measurement problem with his assistant Léon Rosenfeld. Their results are in the 1933 paper "On the Question of the Measurability of Electromagnetic Field Quantities" (Bohr and Rosenfeld, 1933). Besides demonstrating inadequacies and errors in Landau and Peierls' conclusions, they deduced new complementarities in quantum field theory that further elucidated the measurement situation in nonrelativistic quantum mechanics. The Bohr-Rosenfeld opus can be rightly considered to be the sequel to Bohr's (1928) complementarity paper. It is typically cited without any further discussion as the place where field measurements are carefully examined. The style is vintage Bohr. Every long sentence is a gem in its construction: omit a single word and the sentence loses its meaning. The mechanical contrivances for field measurements rival James Clerk Maxwell's models of the ether.

I will use published papers and correspondence to explore this fascinating and often forgotten episode in the history of physics which bears on physics today. I will proceed as follows:

(1) To set the stage I begin with Heisenberg's discussion of field measurability in 1929 which was the first attempt at this problem;
(2) The paper of Landau and Peierls;
(3) Reactions by Bohr, Heisenberg and Pauli;
(4) The Bohr-Rosenfeld paper;
(5) I will conclude with some affects of their analysis on physics in the 1930's and the relevance of their work to physics today.

1. HEISENBERG'S DISCUSSION OF FIELD MEASURABILITY IN 1929

In the chapter entitled "Critique of Wave Theory" in his 1929 University of Chicago lectures Heisenberg argues that since there are restrictions on the particle view of matter, and since in quantum physics there is a wave-particle duality, then there must be restrictions on the wave picture (Heisenberg, 1930).

Even in classical electromagnetic theory there are limits to the measurement accuracy of field quantities because only the averaged value of wave amplitudes over finite regions of space and time can be measured. But whereas in classical physics one can consider ideally shrinking a measurement volume to zero, what is the situation in quantum electrodynamics?

Consider a region of space of side δl and volume $\delta v = (\delta l)^3$ filled with light of wavelength $\lambda > \delta l$. Light of wavelength $\lambda < \delta l$ is not interesting because its wave effects average out to zero. The energy and momentum of light in this spatial region is

$$E = \frac{\delta v}{8\pi} (E^2 + H^2) \tag{1}$$

$$G = \frac{\delta v}{4\pi c} (E \times H) \tag{2}$$

If E and H were known exactly there would be an inconsistency with particle theory because δv could be reduced enough so that the usual inequalities for light quanta

$$E > h\nu \tag{3}$$

$$G > \frac{h\nu}{c} \tag{4}$$

would be violated. Eqs.(1) - (4) lead to uncertainties in E and H of

$$(\Delta E_x)(\Delta H_y) > \frac{hc}{(\delta l)^4} \tag{5}$$

which refers to simultaneous knowledge of E_x and H_y in the same spatial volume.

Without proof Heisenberg states that in different volume elements E_x and H_y can be measured to any degree of accuracy, which is rather bold because he has not yet even described a method to measure fields. Heisenberg next moves to correct this deficiency because as had been de rigueur in fundamental investigations for Bohr and Heisenberg since 1927, "it must be possible to trace the origin of the uncertainty in a measurement of the electromagnetic field to its experimental source."

Heisenberg suggests a measurement of E_x and H_y with two collimated beams of electrons passing through a volume element δv. The uncertainty relation in Eq.(5) can be obtained by taking into account two inaccuracies: (1) the angular deflection of each beam must be greater than its dispersion at the collimating slit; and (2) the field produced by one electron (as a particle) influences the field measured by another electron in an intrinsically uncontrollable manner owing to the uncertainty of each electron's position.

To complete the analysis Heisenberg derives the uncertainty relation in Eq.(5) from the equal time commutation relation for E_x and H_y, that is, from the formalism of quantum electrodynamics.

In summary thus far: Heisenberg's notion that there are uncertainty relations for the electromagnetic field is correct and so is his idea to take into account effects of the test charge on the field to be measured. But his detailed analysis turned out to be incorrect. This was made forcefully clear by Bohr and Rosenfeld in 1933, in their response to results of Landau and Peierls, to which we turn next.

2. THE PAPER OF LANDAU AND PEIERLS

Landau and Peierls begin by claiming the necessity to extend the uncertainty relations for energy and time, and momentum and position, into relativistic situations. Consider a system with energy (momentum) E (P) before measurement and E' (P') after measurement. Assume that the energy and momentum of the measuring apparatus are known. Since conservation of momentum applies exactly, the momentum uncertainties in the initial and final states are equal

$$\Delta P = \Delta P'. \tag{6}$$

According to time-dependent perturbation theory the most favorable situation for conservation of energy is

$$\Delta |E - E'| > \frac{\hbar}{\Delta t}. \tag{7}$$

Over a long time Δt those final states are given preference for which energy is conserved. Eq.(7) is also the uncertainty relation for energy and time which is rooted in the unknown interaction energy between measuring instrument and system being measured. Since $\Delta(E - E') = (v - v')\Delta P$, where v (v') is the system's initial (final) velocity, then

$$|v - v'|\Delta P > \frac{\hbar}{\Delta t}. \tag{8}$$

The upper limit on $|v' - v|$ from relativity is c. Therefore,

$$\Delta P \, \Delta t > \frac{\hbar}{c} \tag{9}$$

and so, except for a long time measurement, "introduction of the concept of momentum [into relativistic quantum mechanics] is meaningless."

In measuring the change of momentum of a charged body an additional inaccuracy enters due to the radiation emitted. Assuming now that velocities before and after measurement are much less than c, the emitted radiation energy is

$$E_{radiation} = \frac{e^2}{c^3} \frac{(v' - v)^2}{\Delta t} \tag{10}$$

which has to be taken into account in the energy balance and so yields another inaccuracy

$$\Delta P \, \Delta t > \frac{e^2}{c^3} |v' - v|. \tag{11}$$

Their key point is that whereas Eq.(11) gives no new information for electrons, it yields a new result for macroscopic charged bodies where the charge e is a very large integral multiple of the elementary charge. The new result is obtained by multiplication of Eqs. (8) and (11) which gives

$$\Delta P \, \Delta t > \frac{\hbar}{c} \sqrt{\frac{e^2}{\hbar c}} . \tag{12}$$

Since here $e^2 \gg \hbar c$ then this uncertainty relation is more stringent than the one for momentum and position from nonrelativistic quantum mechanics

$$\Delta x \, \Delta p_x \sim \hbar . \tag{13}$$

Eq.(12) is central to their derivation of uncertainties in field measurements, to which they turn next.

Landau and Peierls propose to measure an electric field by measuring the change in momentum of a heavy charged test particle placed in the field. They use a heavy test body in order to exclude interference by the body's own magnetic field. After a time Δt the body's momentum is measured to within an accuracy ΔP. Then the magnitude of the electric field strength can be known with an accuracy

$$e \, \Delta E \, \Delta t > \Delta P. \tag{14}$$

Multiplication of Eqs. (12) and (14) gives

$$\Delta E > \frac{\sqrt{\hbar c}}{(c \, \Delta t)^2} \tag{15}$$

with a similar result for the uncertainty of the magnetic field strength ΔH

$$\Delta H > \frac{\sqrt{\hbar c}}{(c \, \Delta t)^2} . \tag{16}$$

Eqs.(15) and (16) are for separate field measurements.

For simultaneous measurement of E and H Landau and Peierls include the effect of the magnetic field of the charged probe

$$\Delta H_{probe} = \frac{v'}{c} \frac{e}{(\Delta l)^2} \tag{17}$$

where Δl is the distance between the test charge and the magnetic needle that is set up to measure H, which yields

$$\Delta E \, \Delta H > \frac{\hbar c}{(c\Delta t)^2 (\Delta l)^2} \, . \tag{18}$$

According to Eqs.(15), (16) and (18) there are no measurement uncertainties for static fields where $\Delta t = \infty$. But in the quantum range, write Landau and Peierls, "field strengths are not measurable quantities," that is, not exactly measurable quantities and, in no small part, this is due to the uncertainty relation in Eq.(11). As further support they recall that the vacuum expectation value of the square of the electric field strength is infinite. This catastrophe they relate to the $\Delta t = 0$ limit of Eq.(15) which gives "an infinite indeterminancy of the field strength."

Glaringly absent in these calculations is any connection between the commutation relations for the electromagnetic fields and the uncertainty relation in Eq.(18), and there is no discussion of averaged values of fields. Landau and Peierls place the field measurability problem among the other woes of quantum electrodynamics that include lack of meaning of momentum in the theory. But in their view the most insidious problem in quantum electrodynamics is the unmeasurability of the electron's position which is an inherent part of the theory owing to negative energy states. Consequently, they conclude, "it would be surprising if the formulation of [quantum electrodynamics] bore any resemblance to reality. And so, in the "correct relativistic quantum theory...there will therefore be no physical quantities and no measurements in the sense of wave mechanics." As yet further support for their radical view of overthrow of foundations they offer the continuous energy spectrum of β-rays in nuclear β-decay.

3. REACTIONS BY BOHR, HEISENBERG AND PAULI

Landau and Peierls wrote their paper in December to early January of 1930-1931. Heisenberg received a preprint and reacted immediately in a letter to Bohr of 23 January 1931 (Pauli, 1985): "Landau and Peierls totally misunderstand the limits of the uncertainty relations in quantum electrodynamics." Heisenberg elaborated in a letter to Peierls of 26 January 1931:

"It is very clear to me -- and beautifully put in your work -- that in the present quantum mechanics of relativity effects there are many too many unobservables inserted into the theory....I have a criticism against your work that I have also often maintained against Bohr: I have always believed that existing concepts collectively fail. But what interests me are those few relations that do not fail, e.g., in 1924 the Burger-Dorgelo sum rules....I shall study your work closely, whether actually all concepts fail, as you maintain. And I hope to learn from it" (emphasis in original).

Heisenberg goes on to criticize their assertion that there is no uncertainty relation for static fields as trivial because the time averaged values of any operators over long time intervals always commute.

Concludes Heisenberg, "Hopefully before long you will write a paper on concepts that do not fail."

On 12 March 1931 Heisenberg wrote to Pauli that Bohr agreed with Landau and Peierls' uncertainty relations, but found their derivations sloppy. Bohr's principal criticism was that their uncertainty relations "in no way mean that the relativistic quantum mechanics is too narrow and must give way to a more general formalism. [Also] in nonrelativistic wave mechanics only a small portion of all operators are measurable." Heisenberg agreed with Bohr and emphasized that "The nonmeasurability of operators is not a good criterium for the non-closedness of a theory." Heisenberg has clearly come a long way from his apparently positivistic stance in the 1925 matrix mechanics paper.

Pauli was pungently direct in a letter to Peierls of 3 July 1931 written from Ann Arbor Michigan, where he had lectured on the Landau-Peierls paper. Like Bohr, Pauli believed that:

"[The] foundation for the inequalities of the field strengths is not correct....It is obviously incorrect that the radiated energy $(e^2 (v - v')^2)/(c^3 \Delta t)$ is an undetermined energy variation. (On account of that result you regard that all considerations are in question only for macroscopic probe bodies with charge e, for which $e^2/\hbar c > 1$.) It may be that the radiated energy also contains an indeterminancy on account of the uncertainty in its time duration, but in first approximation the radiated energy is certainly a <u>determined</u> variation. Also the equation

$$(v - v') \Delta P > (e^2 (v - v')^2)/(c^3 \Delta t)$$

is surely not correct as an uncertainty relation. This, indeed, follows from the fact that it does not contain h. If it is correct then an indeterminancy principle for the momentum of charged particles would have to be postulated in classical theory"(emphasis in original).

With the following caveat Pauli encouraged Peierls to publish his results: "You must publish (with Landau) either an improved derivation of the inequalities for the field strengths, or you must publically own up to it that these inequalities (for charged macroscopic probe bodies) are presently not substantiated."

In a letter of 18 January 1933 to Heisenberg, Pauli expresses doubts over Landau and Peierls' uncertainty relations because he had thought of a possible way to measure with arbitrary accuracy the momentum change or recoil of the probe body due to its own field. Consider placing the probe body inside of a very large chest that is permeable to light. The recoil of the chest resulting from absorption of the energy radiated from the momentum change of the probe body can be measured accurately at least for highly-charged macroscopic bodies, thereby avoiding the uncertainty relation Eq.(11). But Pauli had a "feeling" that this was not the case.

Unknown to everyone, Bohr had already acted to correct the situation, as he writes to Pauli on 25 January 1933:

"[W]e surely all agree that we stand before a new developmental phase, which requires new methods. What to me stands at the heart of the situation is that we should not anticipate this development through misuse of apparently logical arguments. The necessary caution to the danger of the Landau and Peierls mentality has indeed led to a closer investigation of the limits of the quantum electrodynamical formalism. Even if I could not find an error in Landau and Peierls' arguments on the measurement of electromagnetic field quantities, their criticism of the formalism's foundation always made me uncomfortable [since] the particle problem does not enter explicitly. Together with Rosenfeld I have in the autumn undertaken the task and we have arrived at the result that full agreement stands between the principal limits of the measurements of the electromagnetic force and the exchange relations of the field components in the formalism. We show that the measurement disturbance caused, according to Landau and Peierls, by the emitted radiation of the probe particle can be totally eliminated....It is demonstrated, e.g., that the average value of all electric and magnetic field components, taken on the same space-time region, are fully commutable, and that consequently the ones cited in the example in Heisenberg's book on the exactness proof of the formalism are not at all suitable."

Pauli's letter to Peierls of 22 May 1933 relays Bohr's opinions and results with Rosenfeld on field measurements which Pauli had read and found no errors. Pauli goes right to the heart of the matter:

"If a measurement inaccuracy-inequality exists that does not emerge directly from the theoretical formalism as an exchange relation, then one must have assignable physical grounds for one such relation. For the relations in your work these grounds are the factual errors due to the negative energy states....But for the field measurement the negative energy states play no role [because the exchange relations] do not contain the charge, mass, or dimensions of the probe body."

4. THE BOHR-ROSENFELD PAPER

This being the case, Bohr and Rosenfeld (1933) use the unequal time commutation relations for free fields:

$$[\overline{E_x^{(I,)}}, \overline{E_x^{(II)}}] = [\overline{H_x^{(I)}}, \overline{H_x^{(II)}}] = i\sqrt{\hbar}c(\overline{A_{xx}^{(I, II)}} - \overline{A_{xx}^{(II, I)}}) \quad (19)$$

$$[\overline{E_x^{(I,)}}, \overline{E_y^{(II)}}] = [\overline{H_x^{(I)}}, \overline{H_y^{(II)}}] = i\sqrt{\hbar}c(\overline{A_{xy}^{(I, II)}} - \overline{A_{yx}^{(II, I)}}) \quad (20)$$

$$[\overline{E_x^{(I,)}}, \overline{H_x^{(II)}}] = 0 \quad (21)$$

$$[\overline{E_x^{(I,)}}, \overline{H_y^{(II)}}] = [\overline{E_y^{(I,)}}, \overline{H_x^{(II)}}] = i\sqrt{\hbar}c(\overline{B_{xy}^{(I, II)}} - \overline{B_{yx}^{(II, I)}}) \quad (22)$$

where I and II refer to different space-time regions, bars over field quantities designate averaged values, e.g.,

$$\overline{E_x^{(I)}} = \frac{1}{V_I T_I} \int dv \int E_x dt \quad (23)$$

where the volume I is of extent L_I and of duration T_I and

$$\overline{A_{xx}^{(I, II)}} = -\frac{1}{V_I V_{II} T_I T_{II}} \int_{T_I} dt_1 \int_{T_{II}} dt_2 \int_{V_I} dv_1 \int_{V_{II}} dv_2 (\frac{\partial^2}{\partial x_1 \partial x_2} - \frac{1}{c^2} \frac{\partial^2}{\partial t_1 \partial t_2})(\frac{1}{r}\delta(t_2 - t_1 -$$

$$\frac{r}{c}) \quad (24)$$

$$\overline{A_{xy}^{(I, II)}} = -\frac{1}{V_I V_{II} T_I T_{II}} \int_{T_I} dt_1 \int_{T_{II}} dt_2 \int_{V_I} dv_1 \int_{V_{II}} dv_2 \frac{\partial^2}{\partial x_1 \partial y_2}(\frac{1}{r}\delta(t_2 - t_1 - \frac{r}{c}) \quad (25)$$

$$\overline{B_{xy}^{(I, II)}} = -\frac{1}{V_I V_{II} T_I T_{II}} \int_{T_I} dt_1 \int_{T_{II}} dt_2 \int_{V_I} dv_1 \int_{V_{II}} dv_2 \frac{1}{c} \frac{\partial^2}{\partial t_1 \partial z_2}(\frac{1}{r}\delta(t_2 - t_1 - \frac{r}{c}). \quad (26)$$

They emphasize that the appearance of δ-functions in the nonaveraged commutation relations mean that the "quantum field theoretical quantities are not true pont functions [and] unambiguous meaning can be given only to" averaged values.

Bohr and Rosenfeld go on to remind everyone that a consequence of noncommutability of operators are uncertainty relations:

$$\Delta \overline{E_x}^{(I)} \Delta \overline{E_x}^{(II)} \sim \hbar |\overline{A_{xx}^{(I, II)}} - \overline{A_{xx}^{(II, I)}}| \quad (27)$$

$$\Delta \overline{E_x}^{(I)} \Delta \overline{E_y}^{(II)} \sim \hbar |\overline{A_{xy}^{(I, II)}} - \overline{A_{yx}^{(II, I)}}| \quad (28)$$

$$\Delta \overline{E_x}(I)\,\Delta \overline{H_x}(II) = 0 \tag{29}$$

$$\Delta \overline{E_x}(I)\,\Delta \overline{H_y}(II) \sim \hbar |\,\overline{B_{xy}(I,\,II)} - \overline{B_{yx}(II,\,I)}\,|. \tag{30}$$

Several results follow immediately from symmetry properties of the A's and B's, for example, when $T_I = T_{II}$ and $V_I = V_{II}$, then all commutators vanish. So the averages of all field quantities are exactly measurable when two space-time regions coincide. This contradicts Heisenberg's 1929 result in Eq.(5).

Field fluctuations must be considered:

"The fluctuations in question are intimately related to the impossibility, which is characteristic of the quantum theory of fields, of visualizing the concept of light quanta in terms of classical concepts."

In fact, it will emerge from the analysis to follow that the more exact is the measurement of the average value of the electric field strength, the less exact can we know the photon composition of the field being measured. A new complementarity emerges in quantum field theory.

In order to assess the affect of fluctuations on field measurements, Bohr and Rosenfeld define a critical field strength **S** to be the square root of the order of magnitude of the averaged value of the vacuum fluctuation. For field strengths greater than **S** fluctuations can be neglected. They define the order of magnitude of a second critical field **U** to be the square root of the product of uncertainties in field strengths, for example, the square root of the terms in Eq. (27).

The space-time region that Bohr and Rosenfeld investigate is the one where L > cT. The region where L < cT is uninteresting because wave properties are averaged out in the time integration. The quantities **S** and **U** are: for L < cT

$$\mathbf{S} \sim \mathbf{U} \sim \frac{\sqrt{\hbar c}}{LcT} \tag{31}$$

and for L > cT

$$\mathbf{S} \sim \frac{\sqrt{\hbar c}}{L^2} \text{ and } \mathbf{U} \sim \sqrt{\frac{\hbar}{L^3 T}}. \tag{32}$$

Consequently, for L > cT

$$\frac{\mathbf{U}}{\mathbf{S}} \sim \sqrt{\frac{L}{cT}} \tag{33}$$

and so for L ≫ cT fluctuations can be neglected. Having estimated these critical field strengths, Bohr and Rosenfeld turn to the assumptions required for physical field measurements.

Like Landau and Peierls, Bohr and Rosenfeld define measurement of the electric field strength through the change in momentum of a suitable test body

$$p_x' - p_x = \rho\,\overline{E_x}\,VT. \tag{34}$$

But unlike Landau and Peierls, Bohr and Rosenfeld define all terms in the operational manner that, in Bohr's opinion, is necessary for quantum mechanics: ρ is the uniform charge density distribution over the test body that fills the volume V (Bohr and Rosenfeld go on to show that optimum measurement conditions, that is, minimum $\Delta \overline{E_x}$, for $L > cT$ require a large number of elementary charges on the test probe, thereby permitting the idealization of a continuous volume charge density. This is opposite to classical electromagnetic theory in which the idealization is that field measurements can be made with point charges at each space-time point.); the time interval Δt over which the change of momentum occurs is much less than the temporal extent T of the volume V; and the average field is being measured. Although increasing the test body's mass reduces its acceleration during the momentum measurement, quantum mechanics gives an intrinsic uncertainty in position of amount

$$\Delta x \sim \frac{\hbar}{\Delta p_x} \tag{35}$$

where Δp_x is the accuracy of the momentum measurement. However, continued Bohr and Rosenfeld, this state of affairs does not restrict the field measurement because another quantity is available, namely, the charge density ρ. From Eqs.(34) and (35) they obtain for an order of magnitude accuracy

$$\Delta \overline{E_x} = \frac{\hbar}{\rho \Delta x VT}. \tag{36}$$

So even a small measurement error Δx can be offset by a sufficiently large ρ which can reduce the measurement error $\Delta \overline{E_x}$.

What about the field generated by acceleration of the test body which is superposed on the field to be measured? Landau and Peierls had concluded that it was just this additional field that prevented any accurate measurement of electromagnetic field quantities because it brought in restrictions that went beyond the uncertainty relation $\Delta x \Delta p_x \sim \hbar$. But they were wrong ab initio for the following reasons: they used formulae for essentially point charges as test probes; they did not use averaged field values; the only way that Bohr and Rosenfeld could reproduce their results was not to distinguish between unaveraged and averaged field quantities, or between the time interval Δt during which the test body undergoes a change in momentum and the characteristic duration T of the space-time measurement region.

Rather than Landau and Peierls' method for calculating effects of the electromagnetic reaction force on a point probe, Bohr and Rosenfeld use a test body with a uniform volume charge distribution. They estimate the momentum transfer to be

$$p_x{}' - p_x \sim \frac{\Delta p_x \ (\Delta x)^2 \ \rho^2}{\mathbf{U}^2} \frac{\Delta t}{T} \tag{37}$$

and not the one obtainable from Landau and Peierls' Eq.(10). So no matter what the values are of ρ, Δp_x, or $\mathbf{U}$, for $\Delta t \ll T$, effects of the electromagnetic reaction force can be lessened. But, needless to say, Δt can never be zero and, at first sight, effects of the radiation reaction force cannot be separated from the field to be measured. Bohr and Rosenfeld will demonstrate that this is not the case for a single field measurement in one spatial region.

But first they deal with Pauli's suggestion to exactly measure the momentum change in Eq.(11), thereby avoiding any field uncertainties. Since from Eq.(37) the radiation reaction force is always mixed with the field to be measured, then any attempt to disentangle the radiation fields produced in the momentum measurements by investigating the composition of the test probe's field would impair the measurement in question.

Next Bohr and Rosenfeld deal with an assumption implicit throughout their analysis: the test body behaves like a rigid body. The momentum transfer measurement can be performed either by collision of the test body with another suitable heavy body, or by means of the Doppler effect. It will turn out that in either case the test body must be brought to rest and then returned to its original position. They prove that in both cases the measurement can be analyzed classically and without use of relativity. Since to optimize measurement accuracy a macroscopic test body must be chosen, then atomic considerations against a rigid body can also be neglected

With these preliminaries settled, Bohr and Rosenfeld turn to calculating the electromagnetic reaction force on the test body, for which they use classical electrodynamics. Consider two space-time regions I and II with volumes V_I and V_{II} and time extent T_I and T_{II}. What is the electromagnetic field produced in II due to a field measurement in I? Assume that V_I is filled with two charge distributions ρ_I and $-\rho_I$. They are attached to a fixed frame that serves also as a coordinate system. During an interval $\Delta t_1 \ll T$ the test body with charge density ρ_I is removed from the frame and experiences a displacement through a distance $D_x^{(I)}$ and is then brought to rest and remains so for a time $\Delta t_2 \ll T$; during a time interval $\Delta t_3 \ll T$ the test body is moved back to its initial position where it coincides with the neutralizing body and is reattached to the rigid frame. For infinitesimal Δt's the test body's field is produced by an electric dipole of extent $D_x^{(I)}$, whose average value in region II is

$$\overline{E_x^{(I,\ II)}} = D_x^{(I)} \rho_I V_I T_I \overline{A_{xx}^{(I,\ II)}} \tag{38}$$

$$\overline{E_y^{(I,\ II)}} = D_x^{(I)} \rho_I V_I T_I \overline{A_{xy}^{(I,\ II)}} \tag{39}$$

$$\overline{H_x^{(I,\ II)}} = 0 \tag{40}$$

$$\overline{H_y^{(I,\ II)}} = D_x^{(I)} \rho_I V_I T_I \overline{B_{xy}^{(I,\ II)}}. \tag{41}$$

These equations are from classical electrodynamics. What about quantum-mechanical restrictions? The field intensity produced by the momentum measurement of the test probe is $\rho \Delta x$. So the field energy in the measurement volume V is $\rho^2 (\Delta x)^2 V$. Consequently, the number of light quanta n in the test body's field is

$$n \sim \frac{\rho^2 (\Delta x)^2 VL}{hc} \sim \frac{\rho^2 (\Delta x)^2 L}{cT\mathbf{U}^2}. \tag{42}$$

The maximum accuracy for the quantum domain is obtained by making ρ as large as possible and $\mathbf{U}$ as small as possible. Since they consider $L > cT$, then the number of light quanta in the field produced by the probe body is huge and so the more accurate are the classically calculated fields.

There are now two cases to be considered: measurement of the average of a single field in one spatial region and measurement of the average of a single field component over two spatial regions.

Consider measurement of $\overline{E_x}$ in a space-time region I. Then, by definition,

$$p_x^{(I)'} - p_x^{(I)} = \rho_I V_I T_I (\ \overline{E_x^{(I)}} + \overline{E_x^{(I,\ I)}}\ ), \tag{43}$$

$\overline{E_x^{(I)}}$ is the averaged value of $\overline{E_x}$ if no field measurement were made. $\overline{E_x^{(I,I)}}$ is the field in I produced by the test body in I, which is calculated from classical electrodynamics. As before

$$\Delta \overline{E_x^{(I)}} = \frac{\hbar}{\rho_I \Delta x_I \ V_I T_I} + \rho_I \Delta x_I \ V_I T_I \mid \overline{A_{xx}^{(I,I)}} \mid \tag{44}$$

where the displacement is predictable within Δx_I. The larger the value of ρ_I, the greater will be the contribution from $\overline{E_x^{(I,I)}}$. For minimal value of $\Delta \overline{E_x}$ (with respect to Δx), Bohr and Rosenfeld obtain

$$\Delta \overline{E_x \ min}^{(I)} \sim \sqrt{\hbar \mid A_{xx}^{(I,I)} \mid}, \tag{45}$$

which is the critical field value **U** in Eq. (32). Although this value can be less than the one of Landau and Peierls, nevertheless it could exclude any measurement of fields in the quantum domain.

Can we get rid of the probe's field? Yes, claim Bohr and Rosenfeld. Field effects of the test body can be completely compensated because the electromagnetic field quantities of the test probe are mathematically analogous to spring forces. For example, setting regions I and II equal in Eq. (38) yields that $E_x^{(I,I)}$ is proportional to the charge's displacement $D_x^{(I)}$ from equilibrium. Bohr and Rosenfeld consider a contrivance where the test body is attached to a fixed frame with a spring whose force constant K is

$$K = \rho_I V_I T_I \ \overline{A_{xx}^{(I,I)}} \tag{46}$$

and which provides a force counter to the forces arising from the test body's own fields in Eq. (38). Similar considerations hold for the fields in Eqs. (39) - (41).

Consequently, the reaction force of the probe body can be eliminated and the accuracy of a single field measurement "is restricted solely by the limit set for the classical description of the field effects of the test body."

Here Bohr and Rosenfeld draw a far-reaching comparison between measurements in nonrelativistic quantum mechanics and quantum field theory, which clarifies Bohr's often repeated statement that ultimately all measurements are accomplished with classical concepts. In nonrelativistic quantum mechanics and quantum field theory there is an essential nonseparability between object and measurement system. But there is an important difference in the role of classical physics in these two theories. In nonrelativistic quantum mechanics the accuracy of classical mechanics in the measurement process is impaired owing to the uncontrollable interaction between object and measuring system that gives rise to statistical expectations. In quantum field theory optimal accuracy in measurement requires ever more well defined classical field concepts. This result is rooted in the complementarity between the photon constitution of the field to be measured and the field effects of the macroscopic test body.

What about measurement of a single field, for example, $\overline{E_x}$, in two different space-time regions?

The relevant equations are:

$$p_x^{(I)'} - p_x^{(I)} = \rho_I \ V_I T_I \ (\ \overline{E_x^{(I)}} + \overline{E_x^{(I, \ I)}} + \overline{E_x^{(II, \ I)}} \ ) \tag{47}$$

$$p_x(II)' - p_x(II) = \rho_{II} V_{II} T_{II} (\overline{E_x^{(II)}} + \overline{E_x^{(II,II)}} + \overline{E_x^{(I,II)}}) \qquad (48)$$

where $\overline{E_x^{(I)}}$ ($\overline{E_x^{(II)}}$) is the field to be measured when it is averaged over region I (II); $\overline{E_x^{(I,I)}}$ ($\overline{E_x^{(II,II)}}$) is the field due to the test body's acceleration in I (II); and $\overline{E_x^{(I,II)}}$ ($\overline{E_x^{(II,I)}}$) is the field produced by the test body's acceleration in II (I) that affects the field measurement in I (II). The contributions $\overline{E_x^{(I,I)}}$ and $\overline{E_x^{(II,II)}}$ can eliminated with springs attached to the test bodies. What about the contributions from $E_x^{(I,II)}$ and $E_x^{(II,I)}$?

Analogous to the case of a single field measurement Bohr and Rosenfeld obtain

$$\Delta \overline{E_x^{(I)}} = \frac{\hbar}{\rho_I \Delta x_I V_I T_I} + \rho_{II} \Delta x\, V_{II} T_{II} | \overline{A_{xx}^{(II,I)}} | \qquad (49)$$

$$\Delta \overline{E_x^{(II)}} = \frac{\hbar}{\rho_{II} \Delta x_{II} V_{II} T_{II}} + \rho_I \Delta x\, V_I T_I | \overline{A_{xx}^{(I,II)}} | . \qquad (50)$$

Increasing either of the charge densities increases the error in the other field measurement.

The minimal product of the uncertainties is clearly

$$\Delta \overline{E_x^{(I)}} \, \Delta \overline{E_x^{(II)}} \sim \hbar [| \overline{A_{xx}^{(I,II)}} | + | \overline{A_{xx}^{(II,I)}} |] \qquad (51)$$

which differs from the uncertainty relation in Eq.(27) owing to the plus sign. In general, Eq.(51) agrees with Eq.(27) when one of the δ-functions in Eq. (27) vanishes. But, more importantly, Eq.(51) contradicts what one expects for the measurement of two field averages over the same space-time domain.

Consequently, "demonstration of the agreement between measurability and quantum electromagnetic formalism requires a more refined measuring arrangement." They accomplish agreement through further mechanical "contrivances" that include insertion of a third test body that is neutral and connected to test bodies with springs and which can communicate with test body II with light signals. These "contrivances" yield the minimum product of the uncertainties in Eq.(27). Therefore the quantum electrodynamical formalism agrees with the concepts of classical electrodynamics and with thought experiments that take account of quantum mechanical restrictions. Contrary to Landau and Peierls, the field of the probe body does not impair field measurement, but is "an essential feature of the ultimate adaptation of quantum field theory to the measurability problem." (For discussion of sequel investigations by Bohr and Rosenfeld see Corinaldisi (1953).)

CONCLUSION

What was the affect of the Bohr-Rosenfeld paper?

The paper made an enormous impact on Bohr. As he wrote to Pauli (1985) on 15 February 1934, "eliminating electrons with their divergent self energy from measurements of the electromagnetic field...was for me a great liberation." That the "field concept" must be applied with caution could well reside in the fact that "all field actions can be observed only through their effects on matter." Bohr began to look at fundamental problems differently. For example, continued Bohr, since we can consider electron theory to be an idealization valid for actions greater than h, then why not consider quantum electrodynamics to be an idealization whose concepts are valid for charges much larger than the fundamental charge. Then there ought to be a correspondence principle argument for taking quantum electrodynamics down into smaller space-time regions. Evidence for such a correspondence principle is "Einstein's 1916 derivation from essentially correspondence principle arguments [from the old Bohr atomic theory] of the radiation law that has as one of its consequences fluctuation phenomena." Bohr suggested that further investigations of field quantization and measurement theory could be the route to a consistent theory of matter free of divergences. And this path "would make research purely illusory toward further construction of hole theory." As a start Bohr suggested (with no details given) applying a version of quantum electrodynamics and measurement theory to atomic physics problems.

On 23 February 1934 Pauli wrote to Heisenberg that Bohr's statements on the application of field theory and measurement theory to atomic physics are "ganz konfus."

Pauli's comments notwithstanding, in 1935 Heisenberg proposed a reformulation of quantum electrodynamics somewhat analogous to Bohr's. In 1935 the situation in quantum electrodynamics was desperate (see Miller, 1991). Attempts by Heisenberg, Pauli, and Victor Weisskopf at a quantized density matrix formalism in which the electron's self energy is finite had failed. Heisenberg wrote to Pauli on 25 April 1935 that the situation was similar to the one in 1922, "Wir wissen dass alles falsch ist." In a subsequent letter Heisenberg wrote to Pauli that with regard "to the discussion of these difficulties we should remember Bohr and Rosenfeld's analysis from which emerged that only for the average values of field strengths over a definite space-time region stand simple indeterminancy relations." Heisenberg proposed a formulation of the quantized density matrix theory where all δ-functions that give rise to light cone singularities are replaced with a nonsingular function Δ that has a finite width, thereby smearing out singularities. Heisenberg emphasized that in this new quantum electrodynamics analysis of measurement interferences will be as important as in quantum mechanics. Suffice it to say that Heisenberg abandoned this line of research.

This paper explores an episode in the history of quantum measurement theory in which confusions of the late 1920's surfaced again, in particular, the relationship between commutativity of operators and their uncertainty relations, and vice versa for Landau and Peierls. We learn from the Bohr-Rosenfeld paper that in certain cases this relationship is a deep-rooted complementary one. A spin off from Bohr and Rosenfeld's analysis were attempts by Bohr and then Heisenberg to reformulate quantum field theory with methods of measurement theory, a unique turn that may be worth bearing in mind today.

ACKNOWLEDGEMENT

This research is supported in part by a grant from the National Science Foundation.

REFERENCES

Bohr, N., 1928, The Quantum Postulate and the Recent Development of Atomic Theory, Nature (Supplement): 580.
--- and L. Rosenfeld, 1933, Zur Frage der Messbarkeit der elektromagnetischen Feldgrössen, Mat.-fys. Medd. Dan. Vid. Selsk., 12; translated by A. Peterson as On the Question of the Measurability of Electromagnetic Field Quantities, in: Selected Papers of Lèon Rosenfeld, R.S. Cohen and J. Stachel, eds., Reidel, Dordrecht.
Corinaldisi, E., 1953, Some Aspects of Measurability in Quantum Electrodynamics, Nuovo Cimento, 10:83-100.

Heisenberg, W., 1930, Die physikalischen Prinzipien der Quantentheorie, Herzl, Leipzig; 1930, translated by C. Eckart and F.C. Hoyt as The Physical Principles of the Quantum Theory, Dover, New York.

Landau, L. and R. Peierls, 1931, Erweiterung des Unbestimmtheitsprinzips für die relativistische Quantentheorie, ZsP, 69: 56; 1965, translated in: Collected Papers of Landau, D. ter Haar, ed., Gordon and Breach, New York.

Miller, A.I., 1991, Source Book in Quantum Field Theory, Harvard University Press, Cambridge, MA.

Pauli, W., 1985, Wissenschaftlicher Briefwechsel mit Bohr, Einstein, Heisenberg, U.A., Band II: 1930-1939, K.v. Meyenn, ed., Springer-Verlag, Berlin.

ON THE COLLAPSE OF THE WAVE-FUNCTION

David Z. Albert

Dept. of Philosophy
Columbia University
New York City

I-INTRODUCTION

There is a conventional wisdom about what a workable theory of the collapse of the wave-function ought to be able to do, which runs like this:

(i) It ought to guarantee that <u>measurements always have outcomes</u>[1] (that is: it ought to guarantee that there can never by any such thing in the world as a superposition of 'measuring that A is true' and 'measuring that B is true').

(ii) It ought to preserve the familiar statistical connections between the outcomes of those measurements and the wave-functions of the measured systems just prior to those measurements (that is: it ought to guarantee that a measurement of non-degenerate observable 0 on a system in the state $|\psi\rangle$ yields the result o' with probability $|\langle\psi|\phi\rangle|^2$, where $0|\phi\rangle=o'|\phi\rangle$).

(iii) It ought to be consistent with everything which is experimentally known to be true of the dynamics of physical systems (for example: it ought to be consistent with the fact that isolated microscopic physical systems have never yet been observed <u>not</u> to behave in accordance with linear quantum-mechanical equations of motion; that such systems, in other words, have never yet been observed to undergo <u>collapses</u>).

Bell[2] has recently suggested that an interesting theory of the collapse of the wave-function due to Ghirardi, Rimini and Weber[3] looks as if it may be able to do all that; but the present note will show how, on closer examination, it begins to look much less so.

Sixty-Two Years of Uncertainty
Edited by A. I. Miller
Plenum Press, New York, 1990

II-THE PROPOSAL OF GHIRARDI, RIMINI AND WEBER

Ghirardi, Rimini and Weber's idea (which is formulated for non-relativistic Schrodinger Quantum mechanics) goes like this: The wavefunction of an N-particle system

$$\psi(\bar{r}_1 \ldots \bar{r}_N, t) \tag{1}$$

usually envolves in accordance with the Schrodinger equation; but every now and then (once in something like $\frac{1}{N} \cdot 10^{15}$ sec.), at random, but with fixed probability per unit time, the wave-function is suddenly multiplied by a normalized Gaussian (and the product of those two separately normalized functions is multiplied, at that same instant, by an overall normalizing constant). The form of the multiplying Gaussian is

$$K \exp(-[\bar{x}-\bar{r}_k]^2/2\Delta^2) \tag{2}$$

where $\bar{r}_k$ is chosen at random from the arguments $\bar{r}_n$, and the width of the Gaussian, Δ, is of the order of 10^{-5} cm. The probability of this Gaussian's being centered at any particular point $\bar{x}$ is stipulated to be proportional to the absolute square of the inner product of (1) (evaluated at the instant just prior to this 'jump') with (2). Then, until the next such 'multiplication' or 'jump' or 'collapse' (as these sudden events have variously been called), everything proceeds, as before, in accordance with the Schrodinger equation. The probability of such jumps per particle per second (which is taken to be something like 10^{-15}, as we mentioned above), and the width of the multiplying Gaussians (which is taken to be something like 10^{-5} cm.) are new constants of nature.

That's the whole theory. No attempt is made, and no attempt need be made, to 'explain' the occurrence of these 'jumps'; that such jumps occur, and occur in precisely the ways stipulated above, can be thought of as a new fundamental law; a beautifully straightforward and absolutely physicalist law of collapse, wherein (at last!) there is no talk at a fundamental level of 'measurements' or 'amplifications' or 'recordings' or 'observers' or 'minds'.

Given what is experimentally known to be true at present, this theory can very probably do (iii). Here's why: for isolated microscopic systems (i.e., systems consisting of small numbers of particles) 'jumps' will be so rare as to be completely unobservable in practice; and Δ has been chosen large enough so that the violations of conservation of energy which those jumps must necessarily produce will be very very small (over reasonable time-intervals), even in macroscopic systems.[4]

Ghirardi, Rimini, and Weber and Bell think that this theory can very probably do (i) and (ii) too. Here is what they seem to have in mind: they suppose (if we read them correctly) that every measuring instrument must necessarily include some sort of <u>pointer</u>, which indicates the outcome of the measurement, and that the pointer (if this instrument really deserves to be called a <u>measuring</u> instrument) must necessarily be a macroscopic physical object, <u>and</u> (this is what will turn out to be problematic) that the pointer must necessarily assume macroscopically different <u>spatial</u> <u>positions</u> in order to indicate different such outcomes; and it turns out that if all of that is the case, then the GRW theory can do (i) and (ii).

It works like this: suppose that the GRW theory is true. Then, for measuring instruments such as were just described, superpositions like

$\alpha|A\rangle|$Measuring Instrument Indicates That 'A'$\rangle$

$+\beta|B\rangle|$Measuring Instrument Indicated That 'B'$\rangle$ (3)

(which will invariably be superpositions of macroscopically different localized states of some macroscopic physical object) are just the sorts of superpositions that don't last long. In a very short time, in only as long as it takes for the pointer wave-function to get multiplied by one of the GRW Gaussians (which will be something on the order of $\frac{1}{N} \cdot 10^{15}$ seconds, where N is the number of elementary particles in the pointer) one of the terms in (3) will disappear, and only the other will propagate, and the measurement will have an outcome. Moreover, in accordance with (ii), 'the probability that one term rather than another survives is proportional to the fraction of the norm which it carries'. the details are spelled out quite nicely in Ref. 2.

The question, of course, is whether all measuring instruments (or, rather, whether all reasonably <u>imaginable</u> measuring instruments) really <u>do</u> work like the ones described above.

III-STERN-GERLACH EXPERIMENTS

Here is a standard sort of Stern-Gerlach arrangement for measuring the z-spin of a spin-$\frac{1}{2}$ particle: the measured particle, to begin with, is passed through a magnetic field which is non-uniform in z direction. That field splits the wave-function of the particle into spatially separate $\sigma_z = +\frac{1}{2}$ and $\sigma_z = -\frac{1}{2}$ components[5]. Those two components move (freely, perhaps, or perhaps under the influence of additional fields) towards two different points (call one A and the other B) on a florescent screen. The screen works like this: a particle striking the screen at, say, point B, knocks atomic electrons in the screen in the vicinity of B into excited orbitals. A short time later, those electrons return to their ground states, and (in the process) emit photons, and thus the vicinity of B becomes a luminous dot, which can be observed directly by an experimenter.

We want to inquire whether or not the GRW theory entails that a measurement such as this has an outcome. That will depend on whether or not there ever necessarily comes a time, in the course of such a measurement, when the position of a macroscopic object, or the positions of some gigantic collection of microscopic objects, is <u>correlated</u> to the measured z-spin. With all this in mind, let's rehearse the stages of the measuring-process again:

First the wave-function of the particle is magnetically separated into $\sigma_z = +\frac{1}{2}$ and $\sigma_z = -\frac{1}{2}$ components. No outcome of the z-spin measurement (no collapse, that is) will be precipitated by that, since, as yet, nothing in the world save the position of that particle[6] (nothing, that is, save a single microscopic degree of freedom) is correlated to the z-spin. Let's keep looking.

Next, the particle hits the screen, and at that stage the florescent electrons get involved. Consider however, whether those florescent electrons get involved in such a way as to precipitate (via GRW) an outcome of the z-spin measurement. Here is the crucial point: the GRW 'collapses' are invariably collapses onto eigenstates of position (or, more precisely, onto narrow Gaussians in position-space); but it is the <u>energies</u> of those florescent electrons, and <u>not</u> their positions, that get correlated, here, to the z-spin to be measured! The GRW collapses aren't the right <u>sorts</u> of collapses to precipitate an outome of the measurement here.

Let's make this point somewhat more precise. Suppose that the initial state of the measured particle is an eigenstate of x-spin. Then, just after the impact of the particle on the screen, the state of the particle and of the various florescent electrons in the vicinities of A and B will look (approximately; ideally) like this:

$$\frac{1}{\sqrt{2}}|\sigma_z = +\frac{1}{2}, \ \bar{x}=A\rangle_{MP} \cdot |\uparrow\rangle_{e_1} \cdots |\uparrow\rangle_{e_N} \cdot |\downarrow\rangle_{e_{N+1}} \cdots |\downarrow\rangle_{e_{2N}}$$

$$+ \frac{1}{\sqrt{2}}|\sigma_z = -\frac{1}{2}, \ \bar{x}=B\rangle_{MP} \cdot |\downarrow\rangle_{e_1} \cdots |\downarrow\rangle_{e_N} \cdot |\uparrow\rangle_{e_{N+1}} \cdots |\uparrow\rangle_{e_{2N}} \qquad (4)$$

where 'MP' is the measured particle, $e_1 \ldots e_N$ are florescent electrons in the vicinity of A, $e_{N+1} \ldots e_{2N}$ are florescent electrons in the vicinity of B, $|\uparrow\rangle$ represents an excited electronic state, and $|\downarrow\rangle$ is a ground state. Suppose, now that a GRW 'collapse' (that is: a multiplication of (4) by a Gaussian of the form (2), where $\bar{r}_n$ is the position-coordinate of one of the florescent electrons) occurs. Consider whether this sort of collapse will make one of the terms in (4) go away, and allow only the other to propagate. The problem, once again, is that these aren't the right sorts of collapses for that job; because $|\uparrow\rangle$ can't be distinguished from $|\downarrow\rangle$ in terms of the position of anything. (Here's a somewhat more precise way to put it: the position differences between $|\uparrow\rangle$ and $|\downarrow\rangle$, which do, in fact, exist, are far smaller than the 10^5cm widths of the multiplying Gaussians). Indeed, such a collapse will leave (4) almost entirely unchanged (except, perhaps, in the wave-function of some single one of the many many fluorescent electrons).

We have left aside the whole question of the probability of such a collapse here, but it ought to be noted in passing that that probability might well be extremely low. It's well known, after all, that the unaided human eye is capable of detecting very small numbers of photons; so perhaps only very small numbers of florescent electrons need, in principle, be involved here! It would be interesting to calculate those numbers; but however that calculation comes out, it appears (for the reasons described in the previous paragraph) that the GRW theory won't entail that an outcome of the z-spin measurement emerges at this stage, either.

We shall have to look still elsewhere. The next stage of the measuring-process involves the decay of the excited electronic orbitals, and (in the process) the emission of photons. If the first term in (4) obtained, the photons would be emitted at A; if the second term obtained, the photons would be emitted at B. Those two states, then, can be distinguished, at least at the moment of emission, in terms of the positions of the photons. Now, so far, GRW's theory has been applied by them only to nonrelativistic systems of particles. Photons, on the other hand, are purely relativistic particles, and it isn't completely clear how GRW might treat them. If photons can't experience GRW collapses, then of course no outcome can possibly emerge at this stage. But let's suppose that photons can experience GRW collapses. The problem at this stage of the measurement will be that that distinguishability in terms of positions will be extremely short-lived. In almost no time, in too little a time for a GRW collapse to be likely to occur (supposing that A and B are, say, a few centimeters apart, on a flat screen) the two photon wave-functions described above will almost entirely overlap in position-space, and the distinguishability in terms of positions will go away, and we shall be in just such a predicament as we found ourselves at the previous stage of the measurement. No outcome, it seems, will emerge here, either.

But now we're running out of stages. The measurement (according to the conventional wisdom above measurements) is already _over_! By now, after all, we have a recording; by now genuinely macroscopic changes (that is: changes which are thermodynamically irreversible, changes which are directly visible to the unaided human eye) have already taken place in the measuring apparatus. The technical details of real Stern-Gerlach experiments have of course been oversimplified or idealized or just left out of the present account, but those details are beside the point (any number of _other_ experimental arrangements, which, like this one, are free of macroscopic moving parts, would have served our purposes here equally well); the _point_ is simply that genuine recordings need _not_ entail macroscopic changes in the _position_ of anything. Changes in the _internal_ states of large numbers of microsystems (changes, say, in atomic energy levels) can be recordings too.

That's what's overlooked in the GRW proposal. What the GRW theory requires in order to produce a collapse isn't merely that the recording in the measuring apparatus be macroscopic (in any or all of the senses of 'macroscopic' just described), but rather that the recording-process involve macroscopic changes in the _position_ of something. The problem is that _no_ changes of that latter sort are involved in the kinds of measurements we have considered here.

IV- INSIDE HIS HEAD

Suppose, after all this, that we wanted to stick with the GRW theory anyway. What would that entail?

Well, we would have to deny that the measurement described above is over even once a recording exists. We would have to insist (and certainly this _is_ an ineluctable _fact_, when you come right down to it) that no measurement is _absolutely_ _over_, no measurement _absolutely_ _requires an outcome_, until there is a _sentiment observer_ who is actually _aware_ of that outcome.

So, if we wanted to try to stick with this theory in spite of everything, the thing to do would be to insist that as a matter of fact we _haven't_ run completely out of stages yet, and to _go on_ looking, in those latter stages, for an outcome of this experiment (even though we've already looked right up to the retina of the observer and not found one); and of course the only place _left_ to look at this point is going to be on the _inside_ of the _nervous system_ of the _observer_.

This is going to be an uncomfortable position to be in; it's going to feel as queer as hell to tell oneself that the possibility of entertaining some particular fundamental theory of the whole physical world now hinges on the answers to certain detailed questions about the physiology of human beings; but; let's try to press on and see how it might work. Here's one idea: Consider the two different physical states of the observer's retina (call them RA and RB) which arise in consequence of it's being struck by light from one or the other of the two luminous spots on the fluorescent screen. Perhaps the retinal states RA and RB macroscopically differ from one another in (among other things) the <u>positions</u> of some gigantic collection of microscopic physical objects (the positions of some collection of ions, say). Whether or not that turns out to be so is of course a question for neurophysiology (and I presume it is a question for the <u>future</u> of that subject), but the idea would be that if it <u>does</u> turn out like that, then the <u>observer's retina itself</u> can play the role of GRW's macroscopic pointer in this experiment: <u>it</u> (the retina) can bring about the collapse, <u>it</u> can suffice to finally precipitate an <u>outcome</u>.

Now of course it might turn out that RA and RB do <u>not</u> differ from one another in the positions of any sufficiently gigantic collection of physical objects, it might turn out, say, that the retina works more or less like a florescent screen, and (consequently) that no outcome of this experiment can emerge at the stage of the retina, either; and in that case we would presumably turn our attention next to the optic nerve; and <u>then</u>, finally (if things go badly with the optic nerve too) to the observer's brain itself.

If things were ever to get to <u>that</u> point, <u>then</u> the possibility of continuing to entertain the GRW theory would hinge directly on whether or not the brain state associated with seeing a luminous dot on the florescent screen at point A (call that BA) differs from the brain state associated with seeing a luminous dot on the florescent screen at point B (call that BB) in terms of the positions of any gigantic collection of physical components of the observer's visual cortex. If those two brain-states <u>do</u> differ in that way (and that, once again, will be a question for neurophysiology), then the GRW theory <u>could</u> continue to be entertained.

But consider how things would stand if it were to emerge, after all this, that BA and BB do <u>not</u> differ in precisely that way, if it were to emerge that they differ only in <u>other</u> physical ways than that. Then I think the game would finally genuinely be over; then we would <u>really</u> be out of stages.

This needs to be said with some care. Suppose that a wiseguy were to suggest that the spin-measurement is not yet completely finished even at the stage we're at now, that it isn't _really_ finished until, say, the observer _writes the outcome down on a piece of paper_. It's easy to see that at _that_ point, when the observer actually _writes that down_, the _right_ sort of a collapse, the sort of a collapse we've been _waiting_ for, _would_, finally, certainly, occur; because the state of the ink in the written words "luminous dot at A" clearly differs from the state of the ink in the written words "luminous dot at B" in the terms of the _positions_ of a _gigantic_ collection of ink-molecules. The problem is going to be that this collapse has come really absolutely _too late_ to do it's job.

Here's the point: what we know by pure introspection, what it's hard even to _imagine_ what it would be _like_ to be in _doubt_ about, is that all the Stern-Gerlach experiments we ever do have outcomes, if nowhere else, at least in our _minds_; and that they have outcomes there _before_ we ever come to write those outcomes down; and if the collapse has any job to do at all, it's job is to underwrite at least precisely _that_. If this wiseguy really supposes that that somehow fails to be so, if he supposes that we can somehow be _mistaken_ even in thinking that we _think_ we see a spot at some definite position on the screen, then it gets hard to see what he takes to be the _evidence_ that there _are_ such things as collapses in the first place, and it gets hard to imagine precisely what _job_ he wants the collapse to _do_!.

Suppose, on the other hand that the human neurophysiology works out O.K.; suppose it turns out that the human brain states BA and BB which I described above (the states which correspond to seeing a luminous dot at point A and seeing a luminous dot at point B) differ by, among other things, the _positions_ of some gigantic number of ions. How would things stand then?

Well, that would certainly be a big relief. That would mean that the GRW theory entails that at the point when the visual cortex of the experimenter gets into the game, then (and _only_ then) an outcome of this experiment _does_ finally emerge. Of course this outcome comes _astonishingly_ late, but (if we're willing to entertain the possibility that our senses often radically deceive us even about what's right out there in the macroscopic external world, before we actually _look_ at it) it would have to be admitted that _this_ outcome _does_ come _technically_ on time. But what if we were to begin to worry at that point that there may be _other_ sorts of sentient observers in the universe (dolphins, say, or androids, or martians) whose neurophysiologies might, for all we know, be very different? If there really _are_ other genuinely sentient observers who do such experiments somewhere, or if there _could_ be any in the future, then the possibility of entertaining the GRW theory is presumably going to hinge on the answers to detailed questions about their physiologies too; and of course worries like that could never finally be put to rest unless it be by means of some sort of general principle.

Here's the sort of question we're going to have to ask ourselves: is there anything about what it means to be a sentient observer, or anything about what it means for such an observer to have a belief, which entails, as a matter of principle, that the physical brain states of such an observer which correspond to different such beliefs (beliefs about the outcome of some experiment, say) differ from one another in terms of the positions of any gigantic collection of physical objects?

But it seems to me that when you put the question this way (and here, at last, is the single absolutely fatal fundamental trouble with this particular way of attempting to cook up a collapse), it gets pretty clear that the answer is going to have to be no.

V-GENERAL CONSIDERATIONS ON THE POSSIBILITY OF ANY THEORY OF THE COLLAPSE SATISFACTORILLY DOING IT'S JOB

Let's see if there's a possibility of doing better. Suppose we were able to improve on this theory. Suppose that we were able to cook up some absolutely ultimate and unsurpassable sort of a collapse-theory, some theory which is patently as good as any theory of the collapse can ever possibly be. Suppose, that is, that we were able to cook up a theory which is consistent with everything we know to be true of the behaviors of isolated microscopic systems, and which entails (somehow) that superpositions of states which differ from one another in terms of anything macroscopic whatever[7] (not just in terms of the positions of any gigantic number of particles) don't last long. And suppose that it were clear that this theory can indeed guarantee that an experiment carried out by an ordinary human observer invariably has an outcome. And suppose that we were to begin to wonder whether even this ultimate theory (whether, that is, any theory of the collapse) can guarantee that any experiment carried out by any possible sentient observer invariably has an outome too.

Here's the sort of question we'd have to ask ourselves then: is there anything about what it means to be a sentient observer, or anything about what it means for a sentient observer to measure something, to ascertain something, which entails, as a matter of principle, that the states of the world which correspond to different outcomes of a measurement need to differ from one another, physically, in any macroscopic way at all?

Well, I can't see how. I think not.

Let's work through an example. Let me tell you a science fiction story about a man with something like a <u>microscope eye</u>.

First we'll need to set things up. The story involves a device for producing a correlation between the position of a certain microscopic particle P and the z-spin of an electron; a sort of <u>measuring-device</u> for z-spin in which the "pointer" is this particle called P. Here's how the device works: (See fig. 1) if P starts out in it's middle position, and if an electron whose z-spin is up is fed through the device, then the z-spin of that electron is unaffected by it's passage through the device, and the device is unaffected too, except that P ends up, once the electron has passed all the way through, in its upper position; and if P starts in its middle position, and if an electron whose z-spin is down is fed through the device, then the z-spin of that electron is unaffected by it's passage through the device, and the device is unaffected too, except that P ends up in its <u>lower</u> position.

Here's how the story goes: Suppose that sometime in the distant future somebody named John undergoes a technically astonishing neurosurgical procedure which leaves him looking like this (see fig. 2): John has a little door on either side of his head, and a device like the one I just described is now sitting in the middle of his brain, and (here comes the crucial point) the particular way in which that device is now hooked up to the rest of John's nervous system makes John behave as if his <u>occurrent beliefs</u> about the spins of electrons which happen to pass through that device are determined <u>directly</u> by the position of P.

Here's what I mean. Suppose that John is presented with an electron in an eigenstate of z-spin, with z-spin up, and that John is requested to ascertain what the value of the z-spin of that electron <u>is</u>. What he does is to take the electron into his head through his right door, and pass it through his surgically implanted device (with P initially in its middle position), and then expel it from his head through his left door. And when that's all done (when P is in it's upper position, but when, as yet, the value of the z-spin of the electron isn't recorded anywhere in John's brain <u>other</u> than in that position of P), John announces that he is, at present, <u>consciously aware</u> (as vividly and as completely as he is now aware of anything, or has ever been aware of anything in his life, he swears) of what the value of the z-spin of the electron is, and that he would be delighted to tell <u>us</u> that value is, if we would like to know.

Let's think about that for a minute. There's going to be a temptation here to suppose that John must somehow be mistaken. There will be a temptation to say: "Look, how can it <u>be</u> that John is now <u>consciously aware</u> of what the value of the z-spin of the electron is?

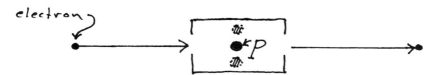

Figure 1. The z-spin measuring-device.

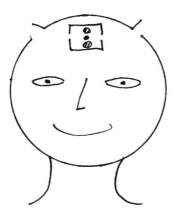

Figure 2. John.

Nothing in John's brain, other than the position of P, is <u>correlated</u> to the spin of the electron now; and P isn't a <u>natural</u> part of John's brain <u>at all</u>; P is just a <u>surgical implant</u>, P is (after all) just a <u>particle</u>!" And imagine that John will say "Look; whether or not P happens to be one of the components of my brain that I was <u>born</u> with is completely <u>beside</u> the point; I tell you now, from my own introspective experience, that if (as you tell me) nothing is now correlated to the z-spin of that electron other that the position of P, then things must now be wired up in such a way that the position of P itself directly determines my present occurrent conscious beliefs about the value of the z-spin of that electron!"

And John is now indeed in a position to announce, correctly, what the value of the z-spin of that electron is, if he's asked to; and it's pretty clear that John can in principle be wired up so as to reproduce, in his present state, <u>any</u> of the behaviors of a genuine "knower" of that z-spin <u>whatsoever</u>; and John will claim (and who will we be to argue with him?) that, after all, noone is better qualified than he himself to judge whether his own psychological experiences are really "genuine" ones or not!

Anyway, here's what's certain: the possibility of entertaining <u>any</u> theory of the collapse <u>whatsoever</u> is now going to hinge on whether or nor we decide to <u>believe</u> what John says. The point is that nothing <u>macroscopic</u> has <u>happened</u> in this story yet; and so if John now actually <u>has</u> an ocurrent belief about the z-spin of the electron, as he says he does, and as <u>all</u> of his behavior testifies he does, then John can indeed come to have such beliefs <u>without</u> anything macroscopic <u>happening</u>; and so there cannot possibly <u>be</u> a theory of the collapse of the wave-function which <u>precludes</u> the development of superpositions of states corresponding to <u>different</u> such beliefs (which, once again, is what a theory of the collapse of the wave-function is supposed to be <u>for</u>), and the <u>reason</u> that there can't be any such theory is that any such theory would now (in the light of John's experience) be required to insert that collapse at a level (the level of isolated microscopic systems) at which we know, by experiment, that no collapses ever occur.

Any theory <u>whatsoever</u> of the collapse of the wave-function, then, is going to have to entail that everybody like John is <u>radically</u> mistaken even about what <u>his own psychological experiences</u> are. That's the fundamental trouble, the <u>necessary</u> trouble, with them all.

REFERENCES

1) Of course, measurements need not have outcomes until they're <u>over</u>, until a <u>recording</u> exists in the measuring-device! So, if (i) is

to be a meaningful physical requirement of a satisfactory theory of the collapse, then something is going to have to be said about <u>what</u> a recording <u>is</u>. It will be best (it will make our argument as strong and as general as possible, as the reader will presently see) to be very <u>conservative</u> about that here; so no change in the physical state of a measuring-device will be called a recording here unless that change is macroscopic, irreversible, and visible to the unaided eye of a human experimenter.

2)   J.S. Bell, "Are There Quantum Jumps?" in "Speakable and Unspeakable in Quantum mechanics Cambridge University Press (1988).

3)   G.C. Ghirardi, A. Rimini and Weber, Phys. Rev. D34, 470 (1988).

4)   See also P. Pearle "Combining Stochastic Dynamical State-Vector Reduction with Spontaneous Localization" to be published in Phys. Rev. A.

5)   Unless, of course, the initial wave-function of the particle is an eigenfunction of σ_z. We shall be interested in cases where it isn't.

6)   Actually, the <u>first</u> thing that gets correlated to the z-spin in an arrangement like this is the momentum, or something approxima-ting the momentum, of the measured particle; but that momentum (since the initial wave-function of the particle is taken to be reasonably well localized) quickly (<u>before</u> the particle hits the screen) get translated into a position, which can be 'read-off' from the screen.

7)   Whatever it is that that (macroscopic) turns out to mean.   Let it mean anything you like.

Acknowledgement

Sections I-III of this paper describe work done in collaboration with Lev Viadman of the University of South Carolina.

OLD AND NEW IDEAS IN THE THEORY OF QUANTUM MEASUREMENT

Gian Carlo Ghirardi° and Alberto Rimini°

°Dipartimento di Fisica Teorica
Università di Trieste
Trieste, Italia

°Dipartimento di Fisica Nucleare e Teorica
Università di Pavia
Pavia, Italia

1. Introduction

1.1 Textbook interpretation of the wave function

The conceptually simplest interpretation of the quantum-mechanical wave function is that adopted by most textbooks. In the celebrated book by Dirac[1] one reads — *each state of a dynamical system at a particular time corresponds to a ket vector . . . if the ket vector corresponding to a state is multiplied by any complex number, not zero, the resulting ket vector will correspond to the same state* (pages 16, 17). And later — *a measurement always causes the system to jump into an eigenstate of the dynamical variable that is being measured* (page 36). Similar sentences can be found, e. g., in the book by Messiah[2] (pages 249 and 251). In the above statements the term state refers to a single system, not to a statistical ensemble of systems. This emerges clearly from the second statement, concerning reduction, which is quite incomprehensible in this form if the ket vector is not referred to a single system. In most textbooks the wave function up to a factor is interpreted just in this way — as the state of the single considered system.

A few remarks are appropriate about the above interpretation. First, it is strictly related to the acceptance of the existence of stochastic transitions in correspondence with measurements. Second, it does not imply that any system always is in a definite state (i.e. has a wave function). In fact, the contrary may very well happen, even for an isolated system. Finally, such an interpretation allows to regard the wave function as an objective property of the system.

The textbook interpretation of the wave function, as it is common experience of all physicists, works perfectly well in ordinary conditions, even though it presents some problematic aspects with systems consisting of distant microscopic parts. But, as we shall see, it breaks down when one tries to make the theory of quantum measurement.

1.2. Macroscopic objects and quantum mechanics

Macroscopic objects obey classical mechanics to an extremely high degree of accuracy. What does it happen when we apply quantum mechanics to them? When this question is posed, one first notes that Ehrenfest's theorem assures us that quantum mean values evolve according to classical laws. This fact is comforting. One then notes that the principles of quantum mechanics consider the existence of macroscopically extended wave functions, of the type

$$(1.1) \qquad |\psi_{strange}\rangle = |\psi_{here}\rangle + |\psi_{there}\rangle,$$

and that, on the other hand, our experience of the behaviour of macroscopic particles never reveals anything corresponding to such wave functions. To reconcile, by means of a suitable interpretation of the quantum-mechanical wave function, the existence of $|\psi_{strange}\rangle$ with our rooted conviction that macroscopic objects really are somewhere is a difficult task. One can observe that quantum effects like diffraction, tunnelling, spreading, which could give rise to extended wave functions, are found to be negligible for macroscopic particles. Again, this is a comforting fact. But this is a practical, dynamical result. No principle of quantum mechanics forbids $|\psi_{strange}\rangle$, and in fact wave functions of a similar type are met with in the theory of quantum measurement where peculiar dynamical conditions take place.

The above considerations take the attribution of some local character to macroscopic objects for granted. Is this fully legitimate? Somewhat joking, we shall explain our point of view by discussing some statements about macroscopic objects. Committing ourselves completely to classical ideas we are tempted to make the following assertion: *macroscopic objects have real properties including a definite position.* One can object that predicating real properties to anything is hazardous. Then we retire to a weaker assertion: *the behaviour of macroscopic objects allows to think that they have real properties including a definite position.* However, one can still object that what we are claiming is not a merit of the behaviour of macroscopic objects, but of the way we describe them. So we retire again to the third form of the statement: *the description of macroscopic objects allows to think that they have real properties including a definite position.* A further possible objection is that the description to which the latter statement makes reference is not necessarily unique. In this way we arrive at the last and weakest form of the statement: *there exists a description of macroscopic objects which allows to think that they have real properties including a definite position.* This form should be accepted by almost everybody and still expresses, we think, a significant nontrivial feature of macroscopic objects. Following the ideas of Bell,[3] we designate such a feature as the definite and local character of the macroscopic world.

Unless we decide that quantum mechanics shall not be applied to macroscopic objects (but also this choice would be problematic, in the absence of a precise criterion for recognizing the macroscopic character), we would like that their definite and local character should follow from the principles of the theory.

This work deals with an attempt at (*i*) incorporating the definite and local character of macroscopic objects in the principles of quantum mechanics and (*ii*) describing reduction as a real physical process thereby recovering the textbook interpretation of the wave function.

1.3. Some principles of quantum mechanics

We recall here some principles of quantum mechanics which will be used in the following discussions. The first principle is of course the Schrödinger equation

$$(1.2) \qquad i\hbar\frac{d}{dt}|\psi\rangle = H|\psi\rangle,$$

which rules the time evolution of the wave function when the system is left undisturbed. The second principle is the reduction postulate with the associated probability rule

(1.3)
$$|\psi\rangle \rightarrow |\psi_a\rangle = |\phi_a\rangle/\|\phi_a\|, \qquad |\phi_a\rangle = P_a|\psi\rangle,$$
$$\Pr(a) = \|\phi_a\|^2.$$

The reduction postulate has to be applied when a measurement is performed. Finally, we recall the tensor product rule, according to which the space of state vectors of a composite system is the tensor product of the spaces of state vectors of the constituent systems (with some inessential complications when identical particles are there), i. e.

(1.4)
$$\psi(1,2) = \sum_i \phi_i(1)\chi_i(2) \neq \phi(1)\chi(2).$$

We point out that two radically different principles of evolution are there — the Schrödinger equation which is deterministic and linear and the reduction postulate which is stochastic and nonlinear. The decision whether to use one or the other rests upon the distinction between proper quantum systems and measuring apparatuses. Such a distinction appears to be easy in practice, but it is not clear in principle.

The tensor product rule is responsible for the fact that systems having interacted with other systems usually are not associated to a wave function. We meet here that characteristic trait of quantum mechanics called entanglement of the state vectors or nonseparability. Quantum entanglement raises problems when the entangled systems are distant and when one of the entangled systems is macroscopic. Also the suspicion that the whole Universe might be entangled is cause of worry.

2. Theory of quantum measurement

2.1. The program

The reduction postulate and the probability rule deal with the measured system S alone. The measuring apparatus A is supposed to be there, of course, but it remains outside the dynamical description. The theory of measurement consists in applying the quantum rules to the composite system $S + A$ and in deducing from such a description the usual postulate dealing with S alone or, less ambitiously, in showing that the two descriptions are not contradictory.

The theory of measurement, besides being a natural exigence within quantum theory, owes its interest to the fact that it has heavy implications on the concept of state and on the interpretation of the wave function. A characteristic fact is that there is not one theory of measurement, there are several. Moreover, there are conceptual frames according to which no theory of measurement should be made.

We shall make reference to an ideal description of the dynamics of measurement. It can be shown[4] that resorting to more realistic descriptions does not change the conclusions.

The elements of the description of measurement are very simple. Let v, with eigenvalue equation

(2.1)
$$v|\psi_n\rangle = v_n|\psi_n\rangle,$$

be the dynamical variable of the system S which is being measured. We exclude the inessential complications related to the possible degeneracy of v. Let G, with eigenvalue equation

(2.2)
$$G|A_{n,r}\rangle = G_n|A_{n,r}\rangle,$$

be the dynamical variable of the apparatus A which serves as a pointer. Then, if $|\psi_m\rangle$ is the state vector of S at the beginning of the measurement, it must happen that

$$(2.3) \qquad\qquad |\psi_m\rangle|A_0\rangle \xrightarrow{\text{S.E.}} |\psi_m\rangle|A_{m,s_m}\rangle,$$

where $|A_0\rangle$ is a suitable initial state vector of A and the arrow indicates the time evolution during the measurement as given by the Schrödinger equation applied to the system $S + A$. Eq. (2.3) expresses simply the conditions that the apparatus ascribe to S the state vector $|\psi_m\rangle$ and that the state vector of S be left unchanged. The time evolution is requested to be ruled by the Schrödinger equation because $S + A$ is left undisturbed during the measurement.

2.2. The difficulty

If $|\psi\rangle = \sum_m a_m|\psi_m\rangle$ is the state vector of S at the beginning of the measurement, then, according to the reduction postulate and the probability rule, the system $S + A$ undergoes during the measurement the process

$$(2.4) \qquad\qquad |\psi\rangle|A_0\rangle \quad \begin{array}{l} \xrightarrow{|a_1|^2} \quad |\psi_1\rangle|A_{1,s_1}\rangle \\[1em] \xrightarrow{|a_2|^2} \quad |\psi_2\rangle|A_{2,s_2}\rangle, \end{array}$$

$$\dotfill$$

with obvious meaning of symbols. Since the output is not uniquely determined one says that the final situation is a mixture, represented by the statistical ensemble made up with the various outputs with their proper weights.

On the other hand, the same state vector of $S + A$ at the beginning of the measurement evolves, by the Schrödinger equation, according to

$$(2.5) \qquad\qquad |\psi\rangle|A_0\rangle \quad \longrightarrow \quad \sum_m a_m|\psi_m\rangle|A_{m,s_m}\rangle.$$

This result follows unescapably from eq. (2.3) and the linear character of the Schrödinger equation. The final state vector is now unique, one says it is a pure state, and it can be attributed either to a single system or to an ensemble according to one's choice about the interpretation of state vectors.

The problem of the theory of quantum measurement is that the result (2.5) implied by the Schrödinger equation seems to contradict both the result (2.4) given by the reduction postulate and common sense, which is unable to give a meaning to a superposition of macroscopically distinguishable states.

2.3. A model for the dynamics of measurement

Models of the system $S + A$ having the characteristic dynamical property (2.3) can easily be constructed. If one is not particularly interested in describing the amplification process which is always present in a quantum measurement, it is an acceptable schematization to consider the measuring apparatus A as consisting of the only pointer, taken to be a single macroscopic object moving in one dimension. The following model belongs to a well known type.[5]

Let v be the variable of S to be measured and Q, P be the position and momentum variables of the pointer. The Hamiltonian of $S + A$ is assumed to be

$$(2.6) \qquad\qquad \begin{aligned} H &= H_S + H_A + H_{\text{int}}, \\ H_S &= 0, \\ H_A &= P^2/2M, \\ H_{\text{int}} &= (d\beta/dt)\, l(v)\, P, \end{aligned}$$

where $\beta(t)$ is zero for $t \leq t_0$, is one for $t \geq t_0$ and is a rounded step in between, so that $d\beta/dt$ is a bell–shaped function contained in the interval (t_0, t_1). The length $l(v)$ is a function of v such that the distances $|l(v_n) - l(v_m)|$ are all macroscopic. The form of H_S is irrelevant for our purposes. Then, if $|A_0\rangle$ is a state vector of A such that $|\langle Q|A_0\rangle|^2$ is centered around $Q = 0$ with spread much smaller than the distances $|l(v_n) - l(v_m)|$ and $|\langle P|A_0\rangle|^2$ is centerd around $P = 0$ with reasonable spread, one finds[5] in the time interval (t_0, t_1) the evolution (2.3) where

(2.7) $$|A_{m,s_m}\rangle \simeq \exp\left(-\tfrac{i}{\hbar}l(v_m)P\right)|A_0\rangle$$

are well separated state vectors lying around the positions $l(v_m)$. The model works, but of course it induces the evolution (2.5), not the reduction (2.4).

3. Review of possible solutions

We expound here concisely the most significant, in our opinion, solutions proposed to the problem raised in sect. 2.2. Of course the review is far from being complete.

Before going through the various points of view, it is necessary to consider and give an answer to a crucial question — is it possible, by means of suitable experiments on the system $S + A$, to distinguish between the statistical ensemble corresponding to the mixture and the one corresponding to the pure state? The answer is yes, it is,[4] provided no limitation is accepted on the measurements to be performed on $S + A$ or, if $S + A$ interacts with some other system E, on $S + A + E$.

3.1. No theory of measurement

Logically classical instruments. According to Bohr, one must include among the principles of quantum mechanics the following proposition: the working of measuring instruments must be accounted for in purely classical terms.[4] According to this principle, the pretension of applying quantum mechanics to A is logically wrong, so that the problem is solved radically.

If taken as a criticism to quantum mechanics the position of Bohr can be shared. It is true that the classical character of instruments is necessary for formulating certain principles of quantum mechanics. It is also true that such a classical character cannot be deduced from quantum mechanics as it stands. What is difficult to understand is how one can be satisfied with such a situation. Furthermore, to be consistent, this point of view would require an exact criterion to distinguish between systems to be described classically and systems to be described quantistically.

Hidden variables. Hidden variables transform quantum mechanics into a theory of a classical type. As a consequence, no theory of measurement is necessary or, if it is done, it is trivial.

We note however that, if we refer to theories which reproduce exactly quantum mechanics when the supplementary variables are properly averaged, the problem of making compatible the description (2.4) obtained by applying the reduction postulate with the description (2.5) obtained by applying dynamics to the system $S + A$ is still there and probably it cannot be solved without invoking the practical impossibility of revealing interference when macroscopic objects are involved.[6]

3.2. Limitations on measurability

Limitations on measurability of $S + A$. Several authors, like Daneri, Loinger and Prosperi[7] or Jauch,[8] use the argument that measuring apparatuses are classical in the sense that only a set of commuting observables can be measured on them. It is then possible to show that the ensemble corresponding to the mixture and the one corresponding to the pure state cannot be distinguished by further experiments on $S + A$. A different limitation is considered by Hepp,[9] according to whom the system $S + A$ is infinite or nearly, while actual observables are local in a suitable sense. Again the two ensembles cannot be distinguished by further experiments on $S + A$.

An objection raised against theories based on limitations on measurability of $S + A$ is that such a limitation has a practical nature and should not be used to solve problems of principle.

Limitations on measurability of $S + A + E$. A deeply different type of limitation has been considered by Joos and Zeh.[10] These authors observe that $S + A$ necessarily interacts with its environment E. Then eq. (2.3) must be replaced by

$$
\begin{aligned}
|\psi_m\rangle|A_0\rangle|E_0\rangle &\longrightarrow \\
\longrightarrow |\psi_m\rangle|A_m\rangle|E_0\rangle &\longrightarrow \\
\longrightarrow |\psi_m\rangle|A_m\rangle|E_m\rangle.
\end{aligned}
$$

(3.1)

A convincing discussion shows that the states $|E_m\rangle$ are practically orthogonal. Then the ensemble corresponding to the mixture and the one corresponding to the pure state cannot be distinguished by measurements which do not involve also E, a system which could be as large as the rest of the world.

It is difficult to maintain the objection that the limitation considered by Joos and Zeh is only of a practical nature. For this reason, we regard the point of view based on the consideration of environment as conceptually superior to other approaches resorting to limitations on measurability.

We note however that, in all theories considered in this subsection, the solution of the problem is achieved at the level of the statistical ensemble. For all these theories, therefore, the interpretation of the wave function as the state of the single physical system is lost.

3.3. Wave function of the Universe

This point of view, originally due to Everett,[11] is most commonly indicated by the misleading name many–world interpretation or, more rarely, by the name many–mind interpretation. The theory says that the Universe has a wave function which describes its state (in some sense) and which evolves according to a Schrödinger equation. When a measurement takes place, no reduction occurs. The time evolution is of the type (2.5) and is conveniently written

$$
\begin{aligned}
&|\psi\rangle|A_0\rangle|O_0^1\rangle\cdots|O_0^k\rangle|R_0\rangle \\
&\longrightarrow \sum_m a_m|\psi_m\rangle|A_m\rangle|O_m^1\rangle\cdots|O_m^k\rangle|R_m\rangle,
\end{aligned}
$$

(3.2)

where $O^1,\ldots,O^k$ are observers and R indicates the rest of the world. Let, e. g., the observer O^1 be ME. Then — *I am only aware of one outcome of a measurement because the ME that makes this statement is the ME associated with one particular outcome. There are other MEs, which are associated with different terms in the wave function, and which are aware of different outcomes.*[6]

We agree that no convincing *logical* argument allows to reject this theory. However we prefer to refuse it, because it seems to us that it destroys the separation between physical systems and observers. As for the question in which we are interested, whether or not the theory allows to interpret the wave function as the state of the single considered physical system, the answer is that no physical system, except the Universe, has a wave function.

3.4. Actual reduction

Special role of the observer. One can observe that the measurement process never terminates as indicated in eq. (2.3). In reality, the apparatus A is always measured by another apparatus B, which is measured ..., etc. Therefore eq. (2.3) must be replaced the so called von Neumann chain

$$
\begin{aligned}
|\psi_m\rangle|A_0\rangle|B_0\rangle \cdots |K_0\rangle &\longrightarrow \\
\longrightarrow |\psi_m\rangle|A_m\rangle|B_0\rangle \cdots |K_0\rangle &\longrightarrow \\
\longrightarrow |\psi_m\rangle|A_m\rangle|B_m\rangle \cdots |K_0\rangle &\longrightarrow \\
\cdots \quad \cdots \cdots \cdots \cdots \cdots \cdots \cdots \quad &\cdots \\
\longrightarrow |\psi_m\rangle|A_m\rangle|B_m\rangle \cdots |K_m\rangle, &
\end{aligned}
$$

(3.3)

where $A, B, \ldots$ are apparatuses and K is the observer's mind or consciousness. If we decide to apply the reduction postulate, thereby violating the Schrödinger equation, it is completely irrelevant at which stage this is done. According to Wigner,[12] to ascribe the violation to the last and most peculiar element of the chain appears as the least unreasonable choice. Then the evolution (2.5) is replaced by the transition

$$
\begin{aligned}
\sum_m |\psi_m\rangle|A_0\rangle|B_0\rangle \cdots |K_0\rangle &\longrightarrow \\
\longrightarrow \sum_m |\psi_m\rangle|A_m\rangle|B_0\rangle \cdots |K_0\rangle &\longrightarrow \\
\longrightarrow \sum_m |\psi_m\rangle|A_m\rangle|B_m\rangle \cdots |K_0\rangle &\longrightarrow \\
\cdots \quad \cdots \cdots \cdots \cdots \cdots \cdots \cdots \cdots \cdots \quad &\cdots \\
\longrightarrow |\psi_{\overline{m}}\rangle|A_{\overline{m}}\rangle|B_{\overline{m}}\rangle \cdots |K_{\overline{m}}\rangle &
\end{aligned}
$$

(3.4)

which contains reduction.

Like many people, we refuse this type of solution of the problem. However, it must be admitted that it is simple and effective. Reduction actually takes place — the interpretation of the wave function as the state of the single physical system is recovered.

Real physical process. One can try to describe reduction as a real physical process. Unlike the theory ascribing reduction to observer's consciousness, this point of view requires a definite modification of the Schrödinger equation. Such a modification must have practically unobservable consequences in all ordinary situations and must become effective only in those special conditions in which superpositions of macroscopically distinguishable states emerge.

Again, reduction actually takes place and the interpretation of the wave function as the state of the single physical system is recovered.

4. Spontaneous localization

4.1. Preliminary remarks

The reduction process (1.3) is stochastic and nonlinear. This suggests that any modification of the Schrödinger equation aiming at incorporating reduction shall have the same features. Spontaneous localization is a working proposal for such a modification. The idea of a stochastic modification of the Schrödinger equation is not new. Previous attempts, however, have left unsolved two problems:

i) the preferred basis problem — which are the states to which the stochastic process leads?
ii) the system dependence problem — how can the process become ineffective in going from macroscopic to microscopic systems?

Quantum mechanics with spontaneous localization gives a definite answer to the first question and shows that the answer to the second one follows.

There are two main frameworks in which the idea of spontaneous localization can be accomodated. The earliest and most elementary framework (QMSL) is characterized by the taking place of instantaneous jumps of the state vector. In the more refined continuous approach (CSL) jumps are replaced by a diffusion process in the Hilbert space. The CSL theory is more elegant and powerful. Still we first expound QMSL because of its much more intuitive character. For the same reason we limit ourselves, within QMSL, to the case of systems of distinguishable particles.

4.2. Assumptions

Quantum mechanics with spontaneous localization makes the following assumptions. [13,14,15]

1) Each particle of a system of n distinguishable particles labelled by index i experiences, with mean frequency λ^i, a sudden spontaneous process.

2) In the time intervals between two successive sudden spontaneous processes the system evolves according to the Schrödinger equation.

3) The sudden spontaneous process is a localization described by

$$(4.1) \qquad \begin{aligned} |\psi\rangle &\to |\psi^i_x\rangle = |\phi^i_x\rangle / \||\phi^i_x\|\,, \\ |\phi^i_x\rangle &= L^i_x |\psi\rangle, \end{aligned}$$

where L^i_x is a norm–reducing, positive, selfadjoint, linear operator in the n–particle Hilbert space, representing the localization of particle i around the point x. The probability density for the occurrence of x is assumed to be

$$(4.2) \qquad \mathcal{P}_i(x) = \||\phi^i_x\|^2.$$

This requires that

$$(4.3) \qquad \int d^3x \left(L^i_x\right)^2 = 1.$$

4) The localization operator L^i_x is chosen to be

$$(4.4) \qquad L^i_x = (\alpha/\pi)^{3/4} \exp\left(-\tfrac{1}{2}\alpha \left(q^i - x\right)^2\right).$$

The length $1/\sqrt{\alpha}$ which characterizes the process is small on a macroscopic scale but large with respect to atomic distances, say

$$(4.5) \qquad 1/\sqrt{\alpha} \approx 10^{-5}\,\mathrm{cm}.$$

It measures the accuracy of the localization process.

Note the similarity between eqs. (4.1) and (4.2) on the one side and the reduction rule (1.3) on the other.

4.3. Localization as a whole

We discuss here the characteristic property which makes spontaneous localization an acceptable candidate for being responsible for reduction.[14] Let us consider a macroscopic object, consisting of N distinguishable particles; in general, it will have many macroscopic degrees of freedom. If it can be considered as macroscopically rigid, it will be described by the centre–of–mass position and the space orientation. If it is deformable, it will possess a field of position variables. If the considered system is actually a macroscopic object, it has a definite macroscopic state and a definite microscopic internal structure. Its wave function is the product of a macroscopic wave function and a structural (i. e. referring to the microscopic internal degrees of freedom) wave function. We assume that the structural wave function is sharply peaked (with respect to $1/\sqrt{\alpha}$).

Let us write

$$(4.6) \qquad q_i = Q + \tilde{q}_i(r),$$

where Q is the centre–of–mass position and r represents a set of $3N - 3$ independent variables. The set r, as defined here, includes orientation and possible other macroscopic variables, besides the structural degrees of freedom. For the sake of simplicity, we feign that the set r doesn't contain macroscopic variables. Then the wave function has the structure

$$(4.7) \qquad \psi(q, s) = \Psi(Q)\chi(r, s),$$

where the structural wave function χ is peaked around the value r_0 of r. We have indicated by s the spin variables. The action of the localization operator for the i-th particle is

$$
\begin{aligned}
(4.8) \qquad & L_x^i\big[\Psi(Q)\chi(r,s)\big] \\
& = (\alpha/\pi)^{3/4} \exp\left(-\alpha\,(q^i - x)^2/2\right)\Psi(Q)\chi(r,s) \\
& = (\alpha/\pi)^{3/4} \exp\left(-\alpha\,(Q + \tilde{q}^i(r) - x)^2/2\right)\Psi(Q)\chi(r,s) \\
& \simeq (\alpha/\pi)^{3/4} \exp\left(-\alpha\,(Q + \tilde{q}^i(r_0) - x)^2/2\right)\Psi(Q)\chi(r,s) \\
& = \big[L_{y^i}^{\text{c.m.}}\Psi(Q)\big]\chi(r,s),
\end{aligned}
$$

where the localization operator for the centre of mass $L_{y^i}^{\text{c.m.}}$ is defined by

$$(4.9) \qquad L_X^{\text{c.m.}} = (\alpha/\pi)^{3/4}\exp\left(-\alpha(Q - X)^2/2\right)$$

and localizes around the position

$$(4.10) \qquad X = y^i = x - \tilde{q}^i(r_0).$$

According to eq. (4.8), in the considered conditions, the localization process does not affect the internal structure of the system and each localization of a single component particle is equivalent to the localization of the centre of mass, so that

$$(4.11) \qquad \lambda^{\text{c.m.}} = \sum_i \lambda^i.$$

In suitable conditions a similar conclusion can be drawn also in cases in which the wave function has not the form (4.6). Let us consider a product wave function of the type

$$(4.12) \qquad \psi_1(q_1)\,\psi_2(q_2)\ldots\psi_N(q_N),$$

where the spin variables have been omitted for a quick notation. The wave function (4.12), with a suitable time dependence in each factor, could describe, e. g., a beam of particles. In correspondence with each single particle wave function ψ_i, consider the same wave function displaced by D as defined by

(4.13)
$$\psi_i'(q_i) = \psi_i(q_i - D)$$

and construct the N-particle wave functions

(4.14)
$$\left(\tfrac{1}{\sqrt{2}}\right)^N (\psi_1(q_1) + \psi_1'(q_1))(\psi_2(q_2) + \psi_2'(q_2)) \ldots (\psi_N(q_N) + \psi_N'(q_N))$$

and

(4.15)
$$\left(\tfrac{1}{\sqrt{2}}\right)(\psi_1(q_1)\psi_2(q_2) \ldots \psi_N(q_N) + \psi_1'(q_1)\psi_2'(q_2) \ldots \psi_N'(q_N)).$$

A wave function like (4.14), with a suitable time dependence in each factor, could be produced, e. g., in a neutron interference experiment. A wave function of the type (4.15) is not easily produced in ordinary conditions. However, consider a particle detector and suppose that the second term in (4.15) represents the electrons and ions involved in the discharge when this has taken place and the first term represents the same particles when the discharge has not taken place. Then a wave function of the type (4.15), suitably completed in each term with factors referring to other parts of the system $S + A$, could just be the realization of the superposition appearing in eq. (2.5). A little reflection shows that nothing special happens to the independent particle wave function (4.14). Simply, a fraction of particles given by the single particle rate λ^{micro} times the time of flight along separated paths is affected by the process so that, in the example of neutrons, it does not contribute to the interference pattern. On the contrary, the wave function (4.15), similarly to (4.7), is reduced to one of its terms with frequency $\sum_i \lambda^i$ (provided $D \gg 1/\sqrt{\alpha}$, of course).

With the aim of getting a very small single particle frequency and an appreciable rate for macroscopic N, we tentatively assume

(4.16)
$$\lambda^i \approx 10^{-16} \sec^{-1} \approx 1 / (10^7 \div 10^8 \, \text{years}) \qquad \text{roughly equivalent to}$$
$$\lambda(1\text{g}) \approx 10^7 \sec^{-1}.$$

Eq. (4.5) for α and eq. (4.16) define the orders of magnitude of the constants appearing in the model.

5. Application of spontaneous localization

5.1. Microscopic systems

According to our tentative choice for the parameters of the localization process, a microscopic particle is practically never localized, so that nothing changes in its dynamics even in the case in which it has an extended wave function. Nothing is expected to change in experiments like, e.g., neutron interferometry.

5.2. Structure of macroscopic objects

The structure of systems having a sharply localized internal wave function is not changed.[14] It seems very unlikely that some effect can be revealed in the cases in which some constituents have an extended wave function, like in superconducting devices.[16]

5.3. Macroscopic particles

If an extended wave function of a macroscopic particle is created by some kind of Schrödinger dynamics, the spontaneous localization process transforms it into one of its localized components in times of the order of $1/\lambda$ in the average. If the creation of an extended wave function is slow, the spontaneous localization process, without otherwise interfering with the Schrödinger dynamics, constrains the system from the outset into a way leading to one of the localized components.[5]

The question is whether the squeeze of the wave packet, caused by a localization followed by a Schrödinger evolution and a further localization and so on, gives rise to a relevant stochastic behaviour contradicting classical determinism. A quantitative answer can be given.[13,14] For a dynamical variable V let us define

$$\langle V \rangle = \overline{\langle \psi | V | \psi \rangle},$$

(5.1)

$$\langle\langle V \rangle\rangle = \left\langle \left(V - \langle V \rangle \right)^2 \right\rangle,$$

where the line represents the average over the statistical ensemble generated by the localization process. For a free particle, one can prove that

(5.2)

$$\langle q_l \rangle = \langle q_l \rangle_S,$$

$$\langle p_l \rangle = \langle p_l \rangle_S,$$

$$\langle\langle q_l \rangle\rangle = \langle\langle q_l \rangle\rangle_S + \left(\alpha \lambda \, \hbar^2 / 6 \, m^2 \right) t^3,$$

$$\langle\langle p_l \rangle\rangle = \langle\langle p_l \rangle\rangle_S + \left(\alpha \lambda \, \hbar^2 / 2 \right) t,$$

where the suffix S indicates the value corresponding to the pure Schrödinger evolution from the same initial condition. For a macroscopic particle and for any reasonable choice of the initial wave packet, $\langle\langle q_l \rangle\rangle_S$ is practically constant for enormous times. The time T for which the two terms in $\langle\langle q_l \rangle\rangle$ are equal is given by

(5.3)

$$T = \left(6 \langle\langle q_l \rangle\rangle_S \, m^2 / \alpha \lambda \, \hbar^2 \right)^{1/3}$$

For $m = 1\mathrm{g}$ and $\langle\langle q_l \rangle\rangle_S^{1/2} = 10^{-5} \, \mathrm{cm}$, using the orders of magnitude (4.5), (4.16) for the parameters, one finds $T \approx 100 \, \mathrm{years}$. This is a long time for keeping something isolated from uncontrollable influences. Some effect of the type discussed here could perhaps be detectable for a mesoscopic particle.

5.4. Measurement

The model. Let us apply QMSL to the system $S + A$, where A is the model apparatus presented in section 2.3.[5] We recall that, in the absence of localizations, the evolution

(5.4)

$$|\psi_m\rangle|A_0\rangle \quad \longrightarrow \quad |\psi_m\rangle|A_{m,s_m}\rangle$$

takes place in the time interval (t_0, t_1). The state vectors $|A_{m,s_m}\rangle$ are well separated, i. e. their position spreads are much smaller than the distances between two pointer positions $|l(v_n) - l(v_m)|$. Note that, since S is microscopic, its localizations can be disregarded and that, since A has a well localized internal wave function, its localizations as a whole, only, have to be considered.

Suppose first that the time $1/\lambda$ is larger than $t_1 - t_0$, so that we can think that, first, the above evolution is completed and, then, a localization of A takes place. The result of the localization is

$$(5.5) \qquad L_X^A \sum_m a_m |\psi_m\rangle |A_{m,s_m}\rangle = \sum_m a_m |\psi_m\rangle L_X^A |A_{m,s_m}\rangle.$$

Each $L_X^A |A_{m,s_m}\rangle$ is nonzero only for X belonging to one interval I_m of a set of nonoverlapping intervals. For $X \in I_{\overline{m}}$ the surviving term is

$$(5.6) \qquad a_{\overline{m}} |\psi_{\overline{m}}\rangle L_X^A |A_{\overline{m},s_{\overline{m}}}\rangle.$$

The probability of $X \in I_{\overline{m}}$ is

$$(5.7)
\begin{aligned}
\Pr(I_{\overline{m}}) &= |a_{\overline{m}}|^2 \int_{I_{\overline{m}}} dX \langle A_{\overline{m},s_{\overline{m}}}| \left(L_X^A\right)^2 |A_{\overline{m},s_{\overline{m}}}\rangle \\
&= |a_{\overline{m}}|^2 \int dX \langle A_{\overline{m},s_{\overline{m}}}| \left(L_X^A\right)^2 |A_{\overline{m},s_{\overline{m}}}\rangle = |a_{\overline{m}}|^2.
\end{aligned}$$

The model can be solved exactly.[5] It is found that the above conclusions hold whatever the relation between $1/\lambda$ and $t_1 - t_0$ is.

The general case. In any measurement process

$$(5.8)
\begin{aligned}
|\psi_m\rangle |A_0\rangle &\longrightarrow |\Phi_m\rangle, \\
\sum_m a_m |\psi_m\rangle |A_0\rangle &\longrightarrow \sum_m a_m |\Phi_m\rangle = |\Phi\rangle,
\end{aligned}$$

where the state vectors $|\Phi_m\rangle$ are macroscopically distinguishable. We suggest that this implies that a macroscopic number N of microscopic "pointer" constituents of A are confined in different macroscopically distant spatial regions in the different state vctors $|\Phi_m\rangle$. Then $|\Phi\rangle$ has a structure of the type (4.15) and a single localization of a single "pointer" constituent is sufficient to reduce $|\Phi\rangle$ to one of its terms $|\Phi_m\rangle$. The reduction rate is $N\lambda^{\mathrm{micro}}$.

The question arises whether N is in all cases large enough.

EPR situation. Let us consider an EPR–Bohm set–up where the system wave function is

$$(5.9)
\begin{aligned}
\psi &= \chi \, \psi_L(1)\, \psi_R(2), \\
\chi &= \tfrac{1}{\sqrt{2}}\left(u_+(1)u_-(2) - u_-(1)u_+(2)\right),
\end{aligned}$$

with obvious meaning of symbols. Describing the apparatuses according to the previously discussed model, the wave function of $S + A$ before any measurement is

$$(5.10) \qquad \psi A_0(Q_L) A_0(Q_R),$$

where Q_L and Q_R are the pointer coordinates of the left and right apparatuses, respectively. When the first measurement, say on the left, takes place, the dynamical evolution is, in our schematization,

$$(5.11) \qquad u_\pm(1) A_0(Q_L) \xrightarrow{\text{S.E.}} u_\pm(1) A_\pm(Q_L),$$

where A_+ and A_- lie in the macroscopically separated intervals I_+ and I_-, respectively. Defining Ψ such that

$$(5.12) \qquad \psi A_0(Q_L) A_0(Q_R) \xrightarrow{\text{S.E.}} \Psi,$$

a trivial repetition of the argument given at the beginning of the present subsection shows that for $X \in I_+$

$$(5.13) \qquad L_X^{A_L} \Psi = \tfrac{1}{\sqrt{2}} u_-(2) \psi_R(2) A_0(Q_R) u_+(1) \psi_L(1) L_X^{A_L} A_+(Q_L)$$

and similarly for $X \in I_-$. It is seen that nothing changes with respect to the application of the reduction postulate.

6. Diffusion processes in Hilbert space

As discussed in sections 1–3, it is interesting to try to describe reduction as a real physical process. In sections 4 and 5 a conceptual framework of this type, QMSL, has been discussed. Among the inconveniences of QMSL there is the fact that the stochastic part of its principle of evolution, though perfectly definite, is not expressed through a compact mathematical equation. In the present section we shall present a general framework, based on the consideration of continuous Markov processes in the Hilbert space, which overcomes this difficulty and, furthermore, will allow to give a simple treatment of the case of systems containing identical constituents.

We consider[17,18] the Ito stochastic equation for the state vector

$$(6.1) \qquad d|\phi_B(t)\rangle = (C\,dt + A \cdot dB)\,|\phi_B(t)\rangle,$$

where $C, A \equiv \{A_i\}$ are operators on the Hilbert space of the system and $B \equiv \{B_i\}$ is a real Wiener process satisfying

$$(6.2) \qquad \overline{dB_i} = 0, \qquad \overline{dB_i\,dB_j} = \delta_{ij}\gamma\,dt.$$

In what follows we shall consider also the case in which i is a continuous index; correspondingly a Dirac delta function has to replace the Kronecker symbol. Equation (6.1) will be referred to as the raw equation; it has to be noted that it does not preserve, in general the norm of the state vector. In fact, using stochastic calculus one gets

$$(6.3) \qquad \begin{aligned} d\,||\phi||^2 &= \langle\phi|d\phi\rangle + \langle d\phi|\phi\rangle + \overline{\langle d\phi|d\phi\rangle} \\ &= \langle\phi|\,(A + A^\dagger)\,|\phi\rangle \cdot dB + \langle\phi|\,(C + C^\dagger)\,|\phi\rangle dt + \langle\phi|A^\dagger \cdot A\,|\phi\rangle\gamma\,dt. \end{aligned}$$

Given an initial state vector $|\phi(0)\rangle$, the evolution equation transforms it, with a given probability, into the state $|\phi_B(t)\rangle$ according to the particular realization $B(t)$ of the Wiener process. We have now to give a physical meaning to the states $|\phi_B(t)\rangle$, taking into account that they have different norms for different sample functions $B(t)$. One could simply prescribe that the state has to be normalized and that the probability of occurrence of such a state is just the probability of the specific process leading to it. We make another choice, which is the analogue of assumption (4.2) of QMSL and of the postulate of standard quantum mechanics about the probability of finding a result in a measurement. Precisely, we assume that the physical probabilities for the occurrence of the normalized vectors are obtained from the raw ones by weighting them by the squared norms of the states $|\phi_B(t)\rangle$. Thus, the states which acquire a larger norm by the raw process weigh more in the ensemble.

The proposed cooking procedure is consistent only if the average of the weight factors equals 1, which amounts to require that the raw process conserve the stochastic average of the norm:

(6.4)
$$d\overline{||\phi||^2} = \overline{d||\phi||^2} = 0.$$

This condition, as easily verified, implies

(6.5)
$$C + C^\dagger = -\gamma A^\dagger \cdot A.$$

If we denote by $-\frac{i}{\hbar} H$ the skew-Hermitian part of the operator C, the raw equation becomes

(6.6)
$$d|\phi\rangle = \left(-\frac{i}{\hbar} H \, dt + A \cdot dB - \frac{1}{2}\gamma A^\dagger \cdot A \, dt\right) |\phi\rangle.$$

One has now to perform the cooking procedure according to the prescriptions indicated above. The cooking, due to the linearity of the raw equation and to the Markov nature of the stochastic process, can be performed whenever one wants. The result is expressed concisely in the physical equation

(6.7)
$$d|\psi\rangle = \left\{ \left[-\frac{i}{\hbar} H - \frac{1}{2}\gamma \left(A^\dagger - A_\psi\right) \cdot A + \frac{1}{2}\gamma \left(A - A_\psi\right) \cdot A_\psi \right] dt + \left(A - A_\psi\right) \cdot dB \right\} |\psi\rangle,$$
$$A_\psi = \frac{1}{2}\langle\psi| \left(A + A^\dagger\right) |\psi\rangle,$$

which is the fundamental equation of the theory. It is important to remark that this equation is nonlinear and stochastic.

7. Reduction properties

The most important feature of eq. (6.7), from the point of view which interests us here, consists in the fact that, when the operators A are appropriately chosen, it induces a continuous dynamical reduction of the state vector.

We assume that the operators A are selfadjoint and commute among themselves. To discuss the dynamical reduction, let us disregard the Hamiltonian term in eq. (6.7). We have then the norm conserving physical nonlinear stochastic equation for the state vector

(7.1)
$$d|\psi\rangle = \left[-\frac{1}{2}\gamma \left(A - A_\psi\right)^2 dt + \left(A - A_\psi\right) \cdot dB \right] |\psi\rangle,$$
$$A_\psi = \langle\psi|A|\psi\rangle.$$

Since we have assumed that the operators A commute among themselves we can introduce the projection operators P_σ on their common eigenmanifolds and write

(7.2)
$$A = \sum_\sigma a_\sigma P_\sigma,$$

where $a_\sigma \neq a_\tau$ for $\sigma \neq \tau$. Defining

(7.3)
$$z_\sigma = \langle\psi|P_\sigma|\psi\rangle$$

we get for the random variables z_σ the stochastic equations

(7.4)
$$dz_\sigma = 2 z_\sigma \sum_\tau z_\tau \left(a_\sigma - a_\tau\right) \cdot dB.$$

It is immediate to prove[18] that, in the limit for large t, the variables z_σ attain either the value 0 or the value 1, and that the probability of z_σ taking the value 1 is equal to the squared norm of the σ component of $|\psi(0)\rangle$.

Concluding, the above considered dynamics is such that, in the long run, any given initial state is driven (or, equivalently, the homogeneous ensemble decomposes into pure subensembles associated to states lying) into one of the common eigenmanifolds of the operators A, with the appropriate probabilities. Obviously the specific time rate of the reduction process, as well as the competition of this process with the evolution induced by the Hamiltonian, depend on the specific details of the operators and parameters appearing in eq. (6.7).

It is important to remark that, after we have guaranteed the reduction to take place, we can resort to the use of the statistical operator formalism to investigate specific physical consequences of the theory, such as mean values of dynamical variables, etc. The derivation of the evolution equation for the statistical operator is a trivial task. We define

$$(7.5) \qquad \rho = \overline{|\psi\rangle\langle\psi|}.$$

With reference to the general form (6.7) of the dynamical equation, we get immediately

$$(7.6) \qquad \frac{d\rho}{dt} = -\frac{i}{\hbar}[H,\rho] + \gamma A\,\rho\cdot A^\dagger - \tfrac{1}{2}\gamma\{A^\dagger\cdot A,\rho\}.$$

It is worth noticing that this equation defines the infinitesimal generator of a quantum dynamical semigroup.[19,20] If one considers dynamical equations for the statistical operator derived from the assumption of the occurrence of hitting processes, one can only derive a particular case of this equation, i.e. the one in which $A^\dagger\cdot A = 1$. We refer the reader to ref. 18 for a detailed discussion of this point and of the relations between the continuous and discontinuous cases.

8. Objective and subjective reductions

Recently, H. Stapp[21] in an interesting paper has stressed the importance of distinguishing between different reduction mechanisms which have a quite different conceptual status. We call objective a reduction process when it derives from a dynamical equation which actually drives the state vector in one of a set of orthogonal subspaces whose direct sum is the whole Hilbert space. Stapp uses a somewhat richer and more sophisticated language calling *Heisenberg's objective reductions* these processes which describe in a mathematically exact way the *transition from the possible to the actual*. As we have proved in the previous section the stochastic nonlinear equation (6.7) actually exhibits such a property, when the operators A are selfadjoint and commuting. Note that when this happens the matrix elements of the statistical operator between state vectors belonging to different subspaces tend to zero.

The simplest example of a dynamics inducing objective reductions is obtained by considering the case of 1 particle and by choosing for the operators A appearing in eq. (6.7) the form

$$(8.1) \qquad A = A(r) = (\alpha/\pi)^{3/4}\exp\left[-\tfrac{1}{2}\alpha(q-r)^2\right].$$

The statistical operator equation is then

$$(8.2) \qquad \begin{aligned} \frac{d\rho}{dt} &= -\frac{i}{\hbar}[H,\rho] \\ &+ \gamma\left[\left(\frac{\alpha}{\pi}\right)^{3/2}\int d^3r\,\exp\left[-\tfrac{1}{2}\alpha(q-r)^2\right]\rho\exp\left[-\tfrac{1}{2}\alpha(q-r)^2\right] - \rho\right]. \end{aligned}$$

Taking $\lambda = \gamma$, (8.2) is the same equation for the statistical operator as given by QMSL in the case of a single particle (in spite of the fact that the two processes are different).

Let us now consider subjective reduction mechanisms. By this expression (or, to be precise, by the expression *subjective von Neumann's reductions*) Stapp denotes a mechanism leading to the suppression of the off-diagonal matrix elements of the statistical operator in a given *preferred basis*. To analyze this case and to compare it with the objective process, let us consider the case of 1 particle and assume that it is subjected to the action of a Hermitian stochastic white noise potential. If we disregard for simplicity the free Hamiltonian, we have, in the coordinate representation, the evolution equation for the state vector

$$(8.3) \qquad i\hbar \frac{\partial \psi(q,t)}{\partial t} = V(q,t)\psi(q,t),$$

with

$$(8.4) \qquad \begin{aligned} &\overline{V(q,t)} = 0, \\ &\overline{V(q,t)V(q',t')} = \lambda\hbar^2 \exp\left[-\tfrac{1}{4}\alpha(q-q')^2\right]\delta(t-t'). \end{aligned}$$

For a given potential, the solution of eq. (8.3) is

$$(8.5) \qquad \psi(q,t) = \exp\left[-\tfrac{i}{\hbar}\int_0^t d\tau\, V(q,\tau)\right]\psi(q,0).$$

The coordinate representation of the statistical operator is then correctly defined through the stochastic average of $\psi(q,t)\psi^*(q',t)$. We have

$$(8.6) \qquad \begin{aligned} \rho(q,q',t) &= \overline{\exp\left[-\tfrac{i}{\hbar}\int_0^t d\tau\,[V(q,\tau)-V(q',\tau)]\right]}\rho(q,q',0) \\ &= \exp\left[-\lambda\left(1 - \exp\left[-\tfrac{1}{4}\alpha(q-q')^2\right]\right)t\right]\rho(q,q',0). \end{aligned}$$

By taking the time derivative we get

$$(8.7) \qquad \frac{\partial\rho(q,q',t)}{\partial t} = -\lambda\left(1 - \exp\left[-\tfrac{1}{4}\alpha(q-q')^2\right]\right)\rho(q,q',t),$$

which is simply the coordinate representation of the non-Hamiltonian terms of eq. (8.2). The stochastic dynamics (8.3) leads therefore to the same equation for the statistical operator as QMSL and as the process defined by eqs. (6.7) and (8.1) and causes the damping of the off-diagonal elements in the coordinate representation. However if one considers the wave function (8.5) corresponding to a specific sample function $V(q,t)$, one has

$$(8.8) \qquad |\psi(q,t)|^2 = |\psi(q,0)|^2.$$

We note that, in particular, if the initial state vector is a linear superposition of two states which are localized in two well separated spatial regions L and R, then, for any given realization of the stochastic potential the probability that, at any time t, the particle be at L or at R remains constant. In other words no state vector is driven, contrary to what happens in QMSL and in the process (6.7), (8.1), within one of the two spatially separated regions. The suppression of the off-diagonal elements of the statistical operator represents simply the fact that the components of the state vector acquire, due to the stochastic nature of the potential, random phases in different spatial regions. It is important to stress the completely different conceptual significance of the two considered mechanisms.

With reference to the problem we are discussing in this paper, Stapp asserts that, since Heisenberg's objective reductions require fundamental changes in the laws of nature (changing Schrodinger's dynamics, introducing new constants of nature, etc) the principle of economy suggests to make resort to von Neumann's subjective reductions induced by the stochastic quantum noise associated to the 2.7°K cosmic background radiation. We do not think appropriate to enter here into a debate on this point. In our opinion, however, such a type of solution is not fully satisfactory. Actually the author feels the necessity, in order to give a firm background to his position, to make resort to a model describing how our perceptions are committed to our memory. We consider the author's arguments quite interesting, but we are inclined to stick to theories yielding objective reductions.

9. Systems containing identical constituents

The general formalism CSL introduced in section 6, as already stated, allows to obtain a simple treatment of the case of systems containing identical constituents. To this purpose let us consider the creation and annihilation operators $a^\dagger(q, s)$, $a(q, s)$ of a particle at point q with spin component s, satisfying canonical commutation or anticommutation relations. In terms of these operators we define a locally averaged number density operator

$$(9.1) \qquad N(x) = (\alpha/2\pi)^{(3/2)} \sum_s \int d^3q \, \exp\left[-\tfrac{1}{2}\alpha(q - x)^2\right] a^\dagger(q, s) a(q, s).$$

The operators $N(x)$, for different values of the parameter x, commute among themselves. Denoting by

$$(9.2) \qquad |q, s\rangle = N a^\dagger(q_1, s_1) \ldots a^\dagger(q_n, s_n) |0\rangle$$

the symmetrized (antisymmetrized) improper state containing n particles at the indicated positions, we have

$$(9.3) \qquad N(x)|q, s\rangle = n_x |q, s\rangle,$$

with

$$(9.4) \qquad n_x = (\alpha/2\pi)^{(3/2)} \sum_i^N \exp\left[-\tfrac{1}{2}\alpha(x - q_i)^2\right].$$

With reference to our general stochastic dynamical equation (7.6) we identify now the discrete index i with the continuous index x and the operators A_i with $N(x)$. In this way we get the physical stochastic nonlinear differential equation for the state vector

$$(9.5) \qquad \begin{aligned} d|\psi\rangle = \Big[&-\tfrac{i}{\hbar}H\,dt - \tfrac{1}{2}\gamma \int d^3x \, \big(N(x) - N_\psi(x)\big)^2 dt \\ &+ \int d^3x \, \big(N(x) - N_\psi(x)\big) dB(x) \Big] |\psi\rangle, \end{aligned}$$

where

$$(9.6) \qquad N_\psi(x) = \langle\psi|N(x)|\psi\rangle$$

and the Wiener process $B(x)$ satisfies

(9.7)
$$\overline{dB(x)} = 0,$$
$$\overline{dB(x)dB(y)} = \gamma \delta^3 (x - y)\, dt.$$

It is understood that in the case in which the system contains various types of identical particles (e.g. electrons and nucleons) a sum over all different constituents which enter into play has to appear in eq. (9.5).

It is evident that equation (9.5) respects the symmetry character of the state vector. Moreover, since the operators $N(x)$ are selfadjoint and commuting, the non-Hamiltonian part of the dynamical equation induces a continuous objective reduction onto the "common subspaces" of these operators. Actually, due to the fact that the operators have a continuous spectrum and to the presence of the Hamiltonian term in the evolution equation the dynamics leads to "well localized state vectors", similarly to what happens in QMSL in the case of distinguishable particles.

As usual, we can immediately derive from (9.5) the equation for the statistical operator

(9.8)
$$\frac{d\rho}{dt} = -\frac{i}{\hbar}[H,\rho] - \frac{1}{2}\gamma \int d^3x \, [N(x),[N(x),\rho]].$$

It is immediately checked that, in the case of a single particle, eq. (9.8) reduces to the corresponding equation for QMSL provided one makes the choice

(9.9)
$$\lambda = \gamma \left(\alpha/4\pi\right)^{3/2}.$$

We could study in detail the relevant physical implications of the introduced theory. This however has been done elsewhere[18] and we will limit ourselves here to discuss a simplified model which allows to derive in a straightforward way the main consequences which are of interest for the subsequent discussion. The simplifications consist in disregarding the Hamiltonian term and discretizing the space.

We divide the space into cells of volume $(\alpha/4\pi)^{-3/2}$ and denote by N_i the number operator counting the particles in the i-th cell. As follows from the discussion of the preceeding section in the considered case the dynamical evolution drives the state vector into a manifold such that the number of particles present in any cell is definite. The simplified equation for the statistical operator reads

(9.10)
$$\frac{d\rho}{dt} = -\frac{1}{2}\gamma \left(\alpha/4\pi\right)^{3/2} \sum_i [N_i,[N_i,\rho]].$$

If we denote by $|n_1, n_2, \ldots, n_i, \ldots\rangle$ the state with the indicated occupation numbers for the various cells, the solution of eq. (9.10) reads, in the considered basis

(9.11)
$$\langle n_1, n_2, \ldots |\rho(t)| n'_1, n'_2, \ldots\rangle =$$
$$\exp\left[-\frac{1}{2}\gamma \left(\alpha/4\pi\right)^{3/2} \sum_i \left(n_i - n'_i\right)^2 t\right] \langle n_1, n_2, \ldots |\rho(0)| n'_1, n'_2, \ldots\rangle.$$

Equation (9.11) shows that linear superpositions of states containing different number of particles in the various cells are dynamically reduced to one of the superposed states with an exponential time rate depending on the expression $\frac{1}{2}\gamma \left(\alpha/4\pi\right)^{3/2} \sum_i \left(n_i - n'_i\right)^2$.

10. Choice of the parameters

The theory presented in the previous section contains two parameters which, if the dynamical equation is considered as describing fundamental physical processes, acquire the status of new constants of nature. The parameter γ, whose dimensions are those of a volume divided by a time, expresses the strength of the non–Hamiltonian terms in the equation, the other parameter α, having the dimensions of 1 divided a squared length, is related to the localization accuracy of the process. We will try to get now some indications about the possible values of these parameters.

We note that stringent physical requirements determine the order of magnitude of an upper bound for α: the localization accuracy must in fact be appreciably larger than the spreads around the lattice equilibrium positions in solids. In fact, when one considers a body whose internal state is such that the spreads of the relative coordinates are of order $1/\Delta$, then, by going through an argument completely analogous to the one used for QMSL one can prove that, if and only if

$$(10.1) \qquad 1/\sqrt{\alpha} \gg 1/\Delta,$$

the centre of mass and internal motions decouple and the internal motion is governed, to a very high degree of accuracy, by the standard quantum dynamics. In accordance with this argument we shall take

$$(10.2) \qquad 1/\sqrt{\alpha} \gg 10^{-8} \, \text{cm} .$$

Note that the choice $1/\sqrt{\alpha} \approx 10^{-5}$ cm adopted within QMSL largely satisfies this constraint.

Given for granted that condition (10.2) is satisfied we discuss now various physical consequences of the theory which have to be taken into account in fixing the parameter values.

Excitations and dissociations of atomic and nuclear bound states. The evaluation of the probability of occurrence of such processes can be made by using the CSL equations but the effect can be more intuitively understood by making reference to the QMSL model. Consider a harmonic oscillator potential such that its ground state has an extension $1/\Delta$. Suppose that a system in such a state suffers a hitting process with a localization accuracy $1/\sqrt{\alpha}$. One can then easily evaluate[22] the probability P_{ND} that after the hitting the system be again found in the ground state. One gets

$$(10.3) \qquad P_{ND} = \left[1 + \alpha/(4\Delta^2)\right]^{-1/2} .$$

Since the extension of our system is of atomic or nuclear dimension, eq. (10.2) holds and therefore the probability P_{D+E} that the system be dissociated or excited by the hitting turns out to be

$$(10.4) \qquad P_{D+E} = 1 - P_{ND} = \alpha/(8\Delta^2) .$$

We recall now that the hitting frequency has the expression (9.9) in terms of the parameters γ and α. There follows that the probability of dissociation or excitation per unit time of our systems Q_{D+E} turns out to be

$$(10.5) \qquad Q_{D+E} = \gamma \left(\alpha/4\pi\right)^{3/2} \frac{\alpha}{8\Delta^2},$$

which for an atom ($1/\Delta \approx 10^{-8}$ cm) becomes

$$(10.6) \qquad Q_{D+E}^{\text{atom}} \simeq 10^{-17} \gamma \left(\alpha/4\pi\right)^{3/2} \alpha \, \text{cm}^2 .$$

Increase of the mean energy value. The occurrence of the spontaneous localization processes induces a steady increase of the mean value of the energy of the physical system. This can be easily evaluated by the dynamical equation for the statistical operator. In the case of a single particle one has

(10.7)
$$\Delta E_1 = \gamma \left(\alpha/4\pi\right)^{3/2} \alpha \hbar^2 t / 2m.$$

The total energy increase for N particles turns out to be simply N times the above energy increase for a single particle. If we take for m the nucleon mass we then have a total energy increase per unit time for a system of N nucleons

(10.8)
$$\Delta E_N / t \approx 10^{-30} N\gamma \left(\alpha/4\pi\right)^{3/2} \alpha \text{ erg cm}^2$$
$$\approx 10^{-7} \gamma \left(\alpha/4\pi\right)^{3/2} \alpha \text{ erg cm}^2,$$

where in the last expression we have assumed N of the order of the Avogadro's number.

Associated to the energy increase there is a temperature increase. With reference to an ideal monoatomic gas the increase in temperature per unit time turns out to be

(10.9)
$$\Delta T / t \approx 10^{-7} \gamma \left(\alpha/4\pi\right)^{3/2} \alpha \text{ cm}^2 \sec {}^\circ\text{K year}^{-1}.$$

Reduction Rates. It is important to evaluate the reduction rates which are characteristic of the considered theory. In fact since practical measurements, yielding definite values of macroscopic dynamical variables, are often accomplished in time intervals of the order of nanoseconds, the suppression of the linear superpositions must occur with a rate which is consistent with these times. Again one can deal with this problem with complete rigour[18] but for the present purpose of estimating the order of magnitude, the discretized model that has been considered above is sufficient. Referring for simplicity to a homogeneous macroscopic body of density D, we consider two states corresponding to a rigid displacement of the body in a certain direction. Let us denote by V the volume of the body in one position which is not covered by the body in the other position. With reference to eq. (9.11), the number of cells contributing to the sum in the exponential is now given by $2V \left(\alpha/4\pi\right)^{3/2}$, and for each of them $(n_i - n_i')^2 = \left(D \left(\alpha/4\pi\right)^{-3/2}\right)^2$. The exponential damping factor becomes then

(10.10)
$$\exp\left[-\gamma \left(\alpha/4\pi\right)^{3/2} V \left(\alpha/4\pi\right)^{3/2} \left(\alpha/4\pi\right)^{-3} D^2 t\right]$$
$$= \exp\left[-\gamma D^2 Vt\right] = \exp\left[-\gamma D n_{\text{out}} t\right],$$

where

(10.11)
$$n_{\text{out}} = DV.$$

Obviously, for low densities (smaller than 1 particle per cell) the expression $(n_i - n_i')^2$ takes only the value 0 or 1 and the damping factor is simply

(10.12)
$$\exp\left[-\tfrac{1}{2}\gamma \left(\alpha/4\pi\right)^{3/2} Nt\right],$$

N being the number of cells which are occupied in one state and not in the other.

With reference to eq. (10.10), we note that the reducing effect is more subtle in the present case than in QMSL. However the physical meaning of the effect is transparent: the appeareance of n_{out} expresses the fact that, while in QMSL all displaced particles contribute to the reduction rate, here, due to identity of particles, those which lie in the same region do not contribute.

We can now put together all the above results to identify acceptable ranges of values for the parameters of the theory. In the table we list, in the first column the quantities which are of interest for us, i.e. the probability of dissociation or excitation per unit time of an atom, the total energy increase per unit time of a system containing $N \approx 10^{23}$ particles, the corresponding temperature increase per unit time, the life time for the suppression of the coherence between states localized in spatial regions separated by distances larger than $1/\sqrt{\alpha}$ in the case of one particle and the corresponding life time in the case of a macroscopic object with density 10^{24} cm^{-3}. In the second column we give the expressions for the indicated quantities in terms of the parameters. In the third column we express the same quantities in terms of the parameter γ when the choice 10^{-5} cm is made for $1/\sqrt{\alpha}$. Finally in the last column we give the numerical values for the considered quantities when the choice 10^{-30} cm^3 sec^{-1} is made for γ. Note that the indicated choice for γ is of the order of the one resulting from the relation (9.9) when one makes the original choices (4.5) and (4.16) of QMSL for α and λ.

TABLE 1

Quantity	Expression	$1/\sqrt{\alpha} = 10^{-5}$ cm	$\gamma = 10^{-30}$ cm^3 sec^{-1}
Q_{D+E}^{atom}	$10^{-17} \gamma \left(\frac{\alpha}{4\pi}\right)^{3/2} \alpha$ cm^2	$10^7 \gamma$ cm^{-3}	10^{-23} sec^{-1}
$\Delta E_N / t$	$10^{-7} \gamma \left(\frac{\alpha}{4\pi}\right)^{3/2} \alpha$ erg cm^2	$10^{17} \gamma$ erg cm^{-3}	10^{-13} erg sec^{-1}
$\Delta T / t$	$10^{-7} \gamma \left(\frac{\alpha}{4\pi}\right)^{3/2} \alpha$ °K cm^2 sec y^{-1}	$10^{17} \gamma$ °K cm^{-3} sec y^{-1}	10^{-13} °K y^{-1}
$\tau_{1\,part}$	$2 \left(\gamma \left(\frac{\alpha}{4\pi}\right)^{3/2}\right)^{-1}$	$10^{-14} \gamma^{-1}$ cm^3	10^{16} sec
τ_{macro}	$10^{-24} \gamma^{-1} n_{out}^{-1}$ cm^3	$10^{-24} \gamma^{-1} n_{out}^{-1}$ cm^3	$10^6 n_{out}^{-1}$ sec

From the last column of the table one can remark:

a) In the case of a macroscopic object for a displacement of 10^{-5} cm one has $n_{out} \approx 10^{19}$ so that the life time for suppressing the coherence turns out to be 10^{-13} sec, a quite short value.

b) The most significant data of the table seem to be those giving the dissociation probability per unit time for an atom and the temperature increase per year. In particular with our choice of the parameters, when one takes into account that the age of the universe is 10^{10} y, one gets a total temperature increase from the beginning of the universe of 10^{-3} °K, a value to be compared with the background radiation of 2.7 °K.

We can summarize the analysis performed in this and in the preceeding sections by saying that the continuous spontaneous localization process based on eq. (9.5), i.e. CSL, gives a unified description of physical phenomena at all levels. In fact, when applied to a microscopic system it agrees with the standard quantum description of such systems, when used to study the interaction of a microscopic system with a macroscopic one in measurement–like situations it gives wave packet reduction with definite pointer positions and when applied to the description of a macroscopic object it gives classical mechanics (essentially to the same accuracy as standard quantum mechanics) but it forbids the occurrence of the disturbing superpositions of macroscopically different states.

The theory seems to yield a consistent solution to the problem of the quantum theory of measurement and of the quantum description of macroscopic objects. The agreement with quantum predictions for microscopic objects seems to be out of experimentally feasible tests. Difficulties might be encountered in the following cases:

a) If the persistence of macroscopic quantum coherence for times larger than those allowed by the theory would be proved.

b) If one would find processes in which macroscopic superpositions occur but they involve the displacement of a number of particles so small that reduction does not take place in sufficiently short times.

Obviously we are not claiming that one should necessarily adhere to the point of view we have presented here. Actually we are aware that to accept or refuse it is, to a large extent, a matter of taste. In any case we think that it is interesting to have shown that this way exists. For people willing to adopt this attitude there are still some problems which arise naturally within CLS. The first one is that of trying to get a relativistic generalization of the theory, a problem which looks highly difficult.[23] Another problem of interest could be that of ascribing the stochastic terms in the evolution equation to some other specified physical mechanism. It has been repeatedly suggested in the literature that gravity could be the cause of the phenomenon of wave packet reduction. We consider therefore useful to present in the next section some recent investigations in this direction which fit within the general framework presented in section 6.

11. A reduction model involving gravity

Recently an interesting spontaneous reduction model has been presented[24] which exhibits some nice features. In particular in it reduction is related to gravity and no parameters appear besides the gravitational constant.

The starting point consists in the introduction of a stochastic evolution equation for the state vector of a macroscopic homogeneous object which induces for the statistical operator the equation

(11.1) $$\frac{d\rho}{dt} = -\frac{i}{\hbar}[H,\rho] - \frac{1}{2}\frac{G}{\hbar}\int\int\frac{d^3r_1\,d^3r_2}{r_{12}}\,[f(r_1),[f(r_2),\rho]]\,,$$

where $f(r)$ is the mass density operator at r and G is Newton's gravitational constant. In the case of a sphere of mass M, volume V and radius R

(11.2) $$f(r) = (M/V)\theta(R - |q - r|),$$

q being the centre–of–mass position operator. Obviuosly, this is not the whole story; a dynamical reduction model is physically interesting only if it can be derived from a microscopic dinamics. In the considered case, since the reducing mechanism is related to the mass density, one cannot deal with point- like particles. Therefore the natural choice is that of making reference to an extended model for all microconstituents. In ref. 24 it is suggested that the mass density operator for a composite system is simply given by the sum over all nucleons of the mass density operators $f_i(r)$ for extended nucleons themselves:

$$(11.3) \qquad f_i(r) = (m/v)\theta(R_n - |q_i - r|),$$

where m, v, R_n are the nucleon mass, volume and radius, respectively, and q_i is the position operator of the i-th nucleon. The contribution from electrons and other light particles are disregarded.

The idea is fascinating since, as remarked above, it relates reduction to gravity and gets rid of all constants. However, one can prove[25] that the physical implications of the model when the choice (11.3) is made are very similar to those deriving from a strengthened CSL in which $1/\sqrt{\alpha} \approx 10^{-13}$ cm. The preceeding discussion should have made clear that such a choice has the following unpleasant physical implications.

i) The decoupling of the centre-of-mass and relative motions is no more correct and remarkable changes with respect to standard quantum mechanics occur in the internal dynamics of solids.

ii) The dynamical equation (11.1) for a macroscopic object cannot be derived consistently from its microscopic analogue.

iii) The dissociation probability for atoms and nuclei becomes unacceptably high.

iv) The total energy increase is extremely high:

$$(11.4) \qquad \Delta E_N / t \approx 10^3 \text{erg sec}^{-1} \qquad \text{for} \qquad N \approx 10^{23}.$$

The model cannot therefore be considered, as it stands, as an acceptable reduction model. However, the idea on which it is based is very interesting and, with a modification, it can be adapted to avoid the above mentioned troubles and to describe point like particles, keeping the connection of the reduction mechanism with gravitational effects. There is a price to pay for this, and it consists in the fact that one parameter besides G appears in the model.

The idea is quite simple:[25] the mass density operator $f(r)$ appearing in eq. (11.1) is assumed to have the expression

$$(11.5) \qquad f(r) = \sum_i m_i N_i(r),$$

where the sum is extended to all types of massive elementary particles of the theory and $N_i(r)$ is the average number density operator (9.1) of CSL, in which the choice 10^{-5} cm has been kept for the parameter $1/\sqrt{\alpha}$. The resulting model still exhibits all appealing features of CSL and has the property of involving only one new parameter and of relating reduction to gravitational effects.

12. Conclusions

According to common opinion the crucial conceptual points of quantum mechanics are

i) indeterminism,

ii) nonlocality, strictly related to nonseparability,

iii) measurement theory,

iv) the description of macroscopic objects,

v) the objectiv meaning of the wave function.

QMSL and CSL make plausible that one can find a consistent solution of a realistic type for the last three, by making resort to a nonlinear and stochastic modification of the quantum dynamics. What about the other puzzling features?

Nonlocal behaviours mediated by entangled and extended wave functions of microscopic systems, like (5.9), remain present in QMSL and CSL as in standard quantum mechanics. In a sense, the nonlocal effects become more dramatic, as it is always the case when a realistic interpretation is imposed on quantum theory. We note that the same formal feature, nonseparability, of the quantum–mechanical state space is at the origin both of the nonlocal behaviours and of the cumulative effects which allow to spontaneous localization to be effective at the macroscopic level and ineffective for few particles. In fact, cumulativity depends essentially on the fact that wave functions like (4.7) or (4.15) are, in a suitable sense, N times entangled.

As for indeterminism, theories like QMSL and CSL plainly incorporate it in the unified principle of evolution. The combined effect of localization and Schrödinger evolution gives rise to the additional spreads present in eqs. (5.2), but these terms are so small that, to our knowledge, they do not contradict experimental facts. The only effect of stochastic spontaneous localization is suppression of the embarassing, even though undetectable, linear superpositions of macroscopically distinguishable states.

References

1. P.A.M. Dirac, *The principles of quantum mechanics* (Clarendon Press, Oxford, 1958).
2. A. Messiah, *Mécanique quantique*, tome I (Dunod, Paris, 1959).
3. J.S. Bell, *Speakable and unspeakable in quantum mechanics* (Cambridge University Press, Cambridge, 1987).
4. B. d'Espagnat, *Conceptual foundations of quantum mechanics*, second edition (W. A. Benjamin, Inc., Reading (Mass.), 1976).
5. F. Benatti, G.C. Ghirardi, A. Rimini, T. Weber, Il Nuovo Cimento **100 B** (1987), p. 27.
6. E. Squires, *The mystery of the quantum world* (Adam Hilger Ltd, Bristol and Boston, 1986).
7. A. Daneri, A. Loinger and G.M. Prosperi, Nuclear Physics **33** (1962), p. 297; Il Nuovo Cimento **44 B** (1966), p. 119.
8. J.M. Jauch, Helvetica Physica Acta **37** (1964), p. 293.
9. K. Hepp, Helvetica Physica Acta **45** (1972), p. 237.
10. E. Joos and H.D. Zeh, Zeitschrift für Physik B – Condensed Matter **59** (1985), p. 223.
11. H. Everett III, Reviews of Modern Physics **29** (1957), p. 454.
12. E.P. Wigner, in *The scientist speculates,* edited by I.J. Good (Heinemann, London and Basic Books, New York, 1961), p. 284.
13. G.C. Ghirardi, A. Rimini, T. Weber, in *Quantum Probability and Applications II (Heidelberg, 1984),* edited by L. Accardi and W. von Waldenfels, Lecture Notes in Mathematics, vol. 1136 (Springer, Berlin, 1985), p. 223; in *Fundamental Aspects of Quantum Theory (Como, 1985),* edited by V. Gorini and A. Frigerio (Plenum Press, New York and London, 1986), p. 57.
14. G.C. Ghirardi, A. Rimini, T. Weber, Physical Review D **34** (1986), p. 470; Physical Review D **36** (1987), p. 3287.
15. J.S. Bell, in *Schrödinger — Centenary celebration of a polymath,* edited by C.W. Kilmister (Cambridge University Press, Cambridge, 1987), p. 41.
16. A.I.M. Rae, preprint, University of Birmingham.
17. P. Pearle, Physical Review A **39** (1989), p. 2277.
18. G.C. Ghirardi, P. Pearle, A. Rimini, Trieste preprint IC/89/44.
19. G. Lindblad, Communications in Mathematical Physics **48** (1976), p. 119.
20. E.B. Davies, *Quantum theory of open systems* (Academic Press, London–New York–San Francisco, 1976).

21. H.P. Stapp, preprint LBL – 26968.
22. F. Benatti, G.C. Ghirardi, A. Rimini, T. Weber, Il Nuovo Cimento **101 B** (1988), p. 333.
23. P. Pearle, these proceedings.
24. L. Diósi, preprint KFKI – 1988 – 55/A.
25. G.C. Ghirardi, R. Grassi, A. Rimini, Trieste preprint IC/89/105.

TOWARD A RELATIVISTIC THEORY OF STATEVECTOR REDUCTION

Philip Pearle

Hamilton College
Clinton, New York 13323

1. INTRODUCTION

"For each measurement, one is required to ascribe to the ψ-function a quite sudden change... .The abrupt change by measurement ... is the most interesting point of the entire theory. ... For *this* reason one can *not* put the ψ-function directly in place of the physical thing ... because from the realism point of view observation is a natural process like any other and cannot *per se* bring about an interruption of the orderly flow of events."

This remark of Schrodinger (from his famous "Cat Paradox" paper[1] of 1935) motivates the following research program. Take as a fundamental postulate the "realism point of view" that the ψ-function directly represents "the physical thing". Then modify Schrodinger's own equation so that it reduces the state-vector, that is to say, it describes the "orderly flow" of the ψ-function during a measurement from a superposition of possible outcomes to a single actual outcome.

It wasn't until 1966, when Bohm and Bub[2] suggested a term to add to Schrodinger's equation, that this research program was actually initiated. Their term, nonlinear in ψ, also depends upon certain fixed "hidden variables" whose values determine what the measurement outcome will actually be.

A decade later I suggested a class of terms, nonlinear in ψ, any one of which, when added to Schrodinger's equation, performs the requisite task of state-vector reduction[3]. In the initial proposal, randomly distributed phase factors in the statevector determine the measurement outcome. I soon thereafter[4] intro-duced a different determining mechanism, a randomly fluctuating source. This has remained the mechanism for subsequent work. (However, no convincing physical identification of the source has yet been made, although it has been suggested that it is related to fluctuations of the metric[5,6,7].) In section 2, I will review some high points of this program between the years 1976 and 1986.

What guidance is there in the development of such theories? In the absence of a conflict between experiment and ordinary quantum theory that might point the way, the Hope has been that only one theory would recommend itself on physical and aesthetic grounds. There are strong physical constraints to be satisfied: agreement with the predictions of quantum theory for all presently known experiments and a satisfactory evolution that reduces the statevector. At present, I believe there is a reasonably attractive theory[8,9,10,11], the

Sixty-Two Years of Uncertainty
Edited by A. I. Miller
Plenum Press, New York, 1990

Continuous Spontaneous Localization (CSL) theory. It was made possible by the introduction of important physical ideas in 1986 by Ghirardi, Rimini and Weber[12,13] (GRW). Section 3 contains a brief introduction to the CSL theory. For, one of the main points of this paper is to show (in Section 4) that the CSL theory is Galilean invariant, even though the Schrodinger equation depends explicitly upon time.

Most exciting is the possibility of constructing a Lorentz invariant generalization of the CSL theory. What the Bell inequality argument[14] shows is that a realist who also holds fast to the ideas of special relativity will have to give up some old notions and adopt some new ones. When asked what notions might have to be abandoned, Bell remarked[15]: "For me its a dilemma. I think its a deep dilemma, and the resolution of it will not be trivial: it will require a substantial change in the way we look at things." In the rest of this paper I wish to present some tentative steps that have been made toward a CSL relativistic quantum field theory. Based upon this, I will try to indicate what a few of these notions appear to me to be.

2. HIGH POINTS OF THE EARLY PROGRAM

I want to focus on five points.

2.1 Equations Can Be Found

A number of equations that reduce the statevector were introduced by Diosi[16], Gisin[17] and myself[4,18]. They perform as follows. Consider a measurement with two possible outcomes a_1 and a_2. As a result of normal hamiltonian evolution during t<0, the statevector $| \psi,t>$ evolves into the superposition

$$| \psi,0> = \alpha_1(0)| a_1> + \alpha_2(0)| a_2> \qquad (2.1)$$

where $| a_1>$ and $| a_2>$ are macroscopically distinguishable states. Stochastic differential equations (i.e. equations that depend upon white noise) were constructed which govern the further evolution of $\alpha_n(t)$. They have the property that $|\alpha_n(t)|^2 \to 1$, $|\alpha_{\neq n}(t)|^2 \to 0$ for large t, for a fraction $|\alpha_n(0)|^2$ of the sample white noise functions driving the time evolution. Thus the experimental results are predicted to occur with the frequencies given by ordinary quantum theory.

2.2 Martingale Property

How is this achieved? The crucial element is a simple principle[4]. Suppose the equations are made to preserve $\Sigma_n|\alpha_n(t)|^2=1$. Furthermore, let the diffusion equation (which can immediately be constructed from the stochastic differential equation) for each $|\alpha_n(t)|^2$ in the interval 0 to1 have absorbing boundaries at 0 and 1. These two properties ensure that reduction takes place, i.e. precisely one $|\alpha_n(t)|^2 \to 1$ for each white noise sample function. Then, agreement with quantum theory is assured if the diffusion is such that the Martingale property

$$d<|\alpha_n(t)|^2>/dt = 0 \qquad (2.2)$$

is satisfied (<...> denotes the average over all white noise functions). This is because Eq. (2.2) implies $|\alpha_n(0)|^2= <|\alpha_n(\infty)|^2>$ and reduction implies $<|\alpha_n(\infty)|^2> = 0 \cdot Prob(|\alpha_n(\infty)|^2=0) + 1 \cdot Prob(|\alpha_n(\infty)|^2=1)$, so $|\alpha_n(0)|^2 = Prob(|\alpha_n(\infty)|^2=1)$.

2.3 Experimental Tests: Interrupt and Interfere

Since a reduction theory has different statevector dynamics than quantum theory, it is to be expected that predictions of the two will differ for certain kinds of experiments. A reduction theory is constructed to agree with the predictions of quantum theory when the reduction dynamics is allowed to go to completion. But suppose the reduction initiated by one measurement is inter-rupted and another measurement is quickly interposed. Then the $|a_n(t)|^2$ are "caught out", possessing neither their quantum theory values nor their reduced values at the start of the second measurement, so it might be expected that the predictions for the experiment consisting of the two measurements would differ from quantum theory's.

However, it turns out, because of the Martingale property, that the predic-tions are different from quantum theory's only if the second measurement involves interference between the first measurement's reducing states[19]. An example of such an experiment is two-slit neutron interference, and was per-formed by Zeilinger et al[20]. The two equal-amplitude wavepackets of a neutron exiting the macroscopically separated slits are presumed to subsequently engage in reduction dynamics. When the packets overlap at the detector they do not any more have equal amplitudes, and so the interference pattern's contrast ought to be diminished compared to the case of no reduction. There was no observed dim-unition of contrast, to one percent accuracy, which was translated to mean that the presumed characteristic reduction time has to be larger than 8 sec. For com-parison, GRW[12] proposed a characteristic time of 10^{16} sec.

2.4 Violation of Conservation Laws

For experiments where the predicted results differ from quantum theory's, it is possible to have violation of conservation laws[21]. For example, consider a double Stern-Gerlach apparatus designed to take a beam of spin 1/2 particles, split it in the z-direction, and then recombine it. Suppose a single particle with spin initially pointing in the +x-direction enters the apparatus. The two macro-scopically separated wavepackets with spins ±1/2 in the z-direction are expected to undergo reduction dynamics, so that when they are recombined their ampli-tudes are no longer equal. Thus the particle no longer with certainty has the x-angular momentum +1/2 it started with: its x-angular momentum is not conserved. Where has it gone? Perhaps into the fluctuating medium that causes the reduction, but so far this back-reaction has received no formulation.

2.5 Superluminal Communication

It has been pointed out by Gisin[22,23,24] that the density matrix evolution equation has to be of the Lindblad[25] type (i.e. $d\rho/dt$ depends only upon ρ) if the reduction theory is to avoid superluminal communication. This places a severe constraint[26] upon the kinds of admissable reduction theories.

2.6 The SL Theory

In spite of the good deal of progress, outlined above, made toward construc-ting and understanding reduction theories, it was recognized[18] that there were two outstanding problems which stalled further development. The *preferred basis problem* was that the macroscopically distinguishable states | a_n> which compete in the reduction "game"[27] were put in by hand. One would hope that what is meant by macroscopically distinguishable would be completely determined by the theory. The *trigger problem* was that the turning on and off of the reducing terms also had to be done by hand. It was expected that these terms would depend upon some parameter like separation or energy density distribution or complexity in such a way as to grow large and dominate the dynamics only when the states | a_n> differ macroscopically, but no formulation of this was found. Then along came the Spontaneous Localization (SL) theory of GRW[12].

The SL theory does not follow the program outlined above entailing a modifi-cation of Schrodinger's equation. Instead it invokes an abrupt change of the statevector (which I shall call a "hit"), the multiplication of a particle's wave-function by a gaussian of width $\approx 10^{-5}$cm. However, this is not quite the "abrupt change by measurement" to which Schrodinger objected because the hitting is conceived of as a natural process that is always taking place, but infrequently (once every 10^{16} sec. per particle on average). This scarcely affects the quantum dynamics of a microscopic particle. However, when a macroscopic object, composed of many particles, is in a superposition of spatially separated states, the hit of any one particle reduces the statevector to representing the object surrounding the localized particle. This will occur at a rapid rate equal to the number of particles multiplied by the single particle rate. Thus, GRW resolve the preferred basis problem, roughly speaking, in favor of the position basis. The hitting process, which does not need to be turned on or off, triggers the reduction and resolves the trigger problem.

3. THE CSL THEORY

With the advent of the ideas of GRW it became possible to construct a modi-fied Schrodinger equation that incorporates them[8]. Remarkably, the equation is, like quantum theory, linear in the statevector. This linearity makes it possible to utilize in the CSL theory the extensive formalism already developed for quantum theory. For example, the path integral formalism can be applied to the CSL theory[9]. More importantly for this paper, the infinitesimal generators of relativ-istic transformations can be constructed. The CSL theory bears another resem-blance to ordinary quantum theory in that its rule for predicting probabilities is quadratic in the statevector.

I will first discuss the simplest example that illustrates the behavior of the CSL theory. Then I will briefly present the nonrelativistic CSL theory of a single particle moving in one dimension (see reference 11 in this volume for a more extensive discussion).

3.1 Simplest Example

Consider the equation for the statevector

$$d|\psi,t>/dt = [Aw(t)-\lambda A^2]|\psi,t> \tag{3.1}$$

with initial condition given by Eq. (2.1). In Eq. (3.1), the Hermitian operator A's two eigenvectors are $|a_1>$ and $|a_2>$, with eigenvalues a_1 and a_2. w(t) is a white noise function[28], a gaussian random process characterized by expectation values

$$<w(t)> = 0, \quad <w(t)w(t')> = \lambda\delta(t-t'), \tag{3.2}$$

which may also be written as $w(t) = dB(t)/dt$, where B(t) is a brownian motion function with probability density

$$\rho(B) = (2\pi\lambda t)^{-1/2}exp-B^2/2\lambda t \tag{3.3}$$

The first thing to notice is that there is no factor i in Eq. (3.1): the "hamiltonian" is anti-hermitian. This means that the statevector's norm will not remain constant. For this reason $|\psi,t>$ is called the "raw" statevector, and Eq. (3.1) is called the raw equation.

Eq. (3.1) can be rewritten in the basis $<a_n|$, and its solution found:

$$d<a_n|\psi,t>/dt = [a_nw(t)-\lambda a_n^2]<a_n|\psi,t> \tag{3.4}$$

$$<a_n|\psi,t> = \alpha_n(0)exp[a_nB(t)-\lambda ta_n^2]$$

$$= \alpha_n(0)\{expB^2/4\lambda t\} exp-[B(t)-2\lambda ta_n]^2/(4\lambda t) \tag{3.5}$$

The "physical" statevector $|\psi,t>_p$ is simply the normalized raw statevector:

$$<a_n \mid \psi,t>_p = \frac{\alpha_n(0)\exp-[B(t)-2\lambda ta_n]^2/(4\lambda t)}{\{\Sigma_m |\alpha_m(0)|^2 \exp-[B(t)-2\lambda ta_m]^2/(2\lambda t)\}^{1/2}} \tag{3.6}$$

The second thing to observe is that the solution (3.6) for $\mid \psi,t>$ depends upon the brownian motion $B(t)$ that drives it. However, we do not know what is the actual $B(t)$ we encounter, so a rule must be given which assigns a probability to each $B(t)$ and its associated $\mid \psi,t>$. Usually in the case of stochastic differential equations like (3.1) or (3.4) one assigns the probability $\rho(B)dB$, where $\rho(B)$ is given by Eq. (3.3). However, here we shall declare the following *probability rule*, the physical postulate that the probability is

$$<\psi,t \mid \psi,t>\rho(B)dB = (2\pi\lambda t)^{-1/2}dB\Sigma_n |\alpha_n(0)|^2 \exp-[B-2\lambda ta_n]^2/(2\lambda t) \tag{3.7}$$

(clearly the integral of this probability is 1).

What $B(t)$'s are most probable? From Eq. (3.7) we see that these are such that

$$|B(t)-2\lambda ta_n| < c(\lambda t)^{1/2}, \quad n=1,2 \tag{3.8}$$

where we may choose the number of standard deviations c so that the remaining probability is as small as we like. Now consider a time satisfying $\lambda t(a_2-a_1)^2>>c^2$, which is large enough so that the two ranges of $B(t)$ given by Eq. (3.8) do not overlap. When this no-overlap condition is reached, the physical statevector is essentially reduced! For example, if $B(t)=2\lambda ta_1$, then, according to Eq. (3.6),

$$<a_1 \mid \psi,t>_p = \alpha_1(0)/\{|\alpha_1(0)|^2 + |\alpha_2(0)|^2\exp-2\lambda t[a_2-a_1]^2\}^{1/2} \approx 1 \tag{3.9a}$$

$$<a_2 \mid \psi,t>_p = \alpha_2(0)/\{|\alpha_1(0)|^2\exp+2\lambda t[a_2-a_1]^2 + |\alpha_2(0)|^2\}^{1/2} \approx 0 \tag{3.9b}$$

Moreover, by integrating Eq. (3.7) over those values of B (given by Eq. (3.8) for $n=1$) which lead to the result (3.9), we see that the probability associated with all these solutions is essentially the quantum theory prediction $|\alpha_1(0)|^2$.

To summarize, the evolution equation (3.1) and the probability rule (3.7) are the mathematical basis of the theory. They work by giving a high probability to large norm raw statevectors, with values of $B(t)$ that create a disparity in the relative magnitudes of the amplitudes $<a_n \mid \psi,t>$.

To conclude this discussion, I want to briefly mention the density matrix, the evolution equation for the physical statevector, and the generalization of this simple example. The density matrix is

$$D_{nm}(t) = \int<a_n \mid \psi,t>_{pp}< \psi,t \mid a_m><\psi,t \mid \psi,t>\rho(B)dB = <<a_n \mid \psi,t><\psi,t \mid a_m>>$$

$$= \alpha_n(0)\alpha_m(0)^*\exp-(\lambda t/2)[a_n-a_m]^2 \tag{3.10}$$

the last expression being obtained by performing the integral using Eqs. (3.3), (3.5). Thus the density matrix satisfies the differential equation

$$dD_{nm}(t)/dt = -(\lambda/2)[a_n-a_m]^2D_{nm}(t) \tag{3.11}$$

From Either Eq. (3.10) or (3.11) the decay of the off-diagonal element associated with reduction can readily be seen. The evolution equation of the density operator D, that follows from Eq. (3.11), has the Lindblad form

$$dD(t)/dt = \lambda\{AD(t)A - (1/2)(A^2D(t)+D(t)A^2)\} \tag{3.12}$$

It is possible to write down the nonlinear equation of evolution[24,10] for the physical statevector $\mid \psi,t>_p$

$$d| \psi,t>_p/dt = [(A - {}_p<\psi,t |A| \psi,t>_p)w(t) - \lambda(A - {}_p<\psi,t |A| \psi,t>_p)^2]| \psi,t>_p \tag{3.13}$$

where the probability weighting associated with $| \psi,t>_p$ is the usual one, Eq. (3.3).

The simple example discussed above is generalized by introducing a family of commuting operators A_j and associated independent white noise functions w_j, as well as a hamiltonian H:

$$d| \psi,t>/dt = \{-iH + \Sigma_j[A_j w_j(t)-\lambda A_j^2]\}| \psi,t> \tag{3.14}$$

3.2 Single Particle

The CSL equation of evolution of the statevector describing a single particle moving in one dimension is

$$d| \psi,t>/dt = \{-iH + \int dz f(X-z)w(z,t)-\lambda\}| \psi,t> \tag{3.15}$$

Here H is the usual hamiltonian. Following GRW, $f(X-z)$ is taken to be a gaussian function of width $\alpha^{-1/2}\approx 10^{-5}$ cm., depending upon the position operator X

$$f(X-z) = (\alpha/\pi)^{1/4}\exp-(\alpha/2)(X-z)^2 \tag{3.16}$$

and the characteristic reduction rate per particle is taken to be $\lambda\approx 10^{-16}$ sec^{-1}. $w(z,t)$ is a white noise field

$$<w(z,t)> = 0, \quad <w(z,t)w(z',t')> = \lambda\delta(z-z')\delta(t-t') \tag{3.17}$$

It will prove useful to write $w(z,t)=\partial B(z,t)/\partial t$, where the stochastic field $B(z,t)$ is brownian in t but white in z, with probability density given in the functional integral form

$$\rho(B(z,t)) = C\exp-\int dz B(z,t)^2/2\lambda t \tag{3.18}$$

Eq. (3.15) has the form of Eq. (3.14), except the continuous index z replaces the discrete index j ($A_j \to f(X-z)$, $\Sigma_j A_j^2 \to \int dz f(X-z)^2=1$).

Consider a particle in the initial state (2.1), where $| a_n>$ now is taken to describe a well-localized (to a distance much less than $\alpha^{-1/2}$) wavepacket centered at position a_n, and $|a_2-a_1| >> \alpha^{-1/2}$: I shall call this the "canonical two-packet state". Eq. (3.15) works as follows. Suppose the white noise fluctuations average out to nearly zero or to negative values at the sites a_n of both packets. Then the norm of each packet exponentially decreases with time constant λ^{-1} so, although the statevector remains in a superposition of packets with comparable amplitudes, the statevector eventually becomes improbable. When the fluctuations are much more positive than negative at only one site, that packet grows while the other diminishes, and this situation occurs with overwhelming probability.

This reduction behavior can be seen from the solution of Eq. (3.15). Neglecting the hamiltonian motion of the packets, the solution can be written analogously to Eq. (3.5) as

$$<x | \psi,t> = <x | \psi,0>\exp[\int dz f(x-z)B(z,t)-\lambda t]$$

$$= <x | \psi,0>\{\exp\int dz B(z,t)^2/4\lambda t\} \exp-\int dz[B(z,t)-2\lambda t f(x-z)]^2/(4\lambda t) \tag{3.19}$$

Also similarly to Eqs. (3.6), (3.7), the physical statevector's components and the probability associated with this solution can be written respectively as

$$<a_n \mid \psi,t>_p \approx \frac{\alpha_n(0)\ exp\text{-}\int dz[B(z,t)\text{-}2\lambda tf(a_n\text{-}z)]^2/(4\lambda t)}{\{\Sigma_m|\alpha_m(0)|^2\ exp\text{-}\int dz[B(z,t)\text{-}2\lambda tf(a_m\text{-}z)]^2/(2\lambda t)\ \}^{1/2}} \tag{3.20}$$

$$<\psi,t \mid \psi,t>\rho(B)DB \approx CDB\Sigma_n|\alpha_n(0)|^2\ exp\text{-}\int dz[B(z,t)\text{-}2\lambda tf(a_n\text{-}z)]^2/(2\lambda t) \tag{3.21}$$

with the usual functional integral notation that $DB \equiv \Pi_z dB(z)$. In these equations, the approximation has been made that x can be replaced by a_n in f(x-z) (because of the narrowness of the wavepacket). The no-overlap condition can now be applied. The most probable B(z,t)'s occur when either or both exponents in Eq. (3.21) are small, a condition I will write analogously to Eq. (3.8) as[29]

$$\int dz[B(z,t)\text{-}2\lambda tf(a_n\text{-}z)]^2 < c^2(\lambda t), \quad n=1,2 \tag{3.22}$$

However, when t is sufficiently large $(\lambda t >> c^2)$, there is no overlap: a B(z,t) that makes one exponent small makes the other large. Thus a probable B(z,t) makes only one of the exponentials in Eq. (3.20) large, of magnitude>$exp\text{-}c^2$, while the other decays with time constant λ^{-1}. This results in reduction precisely as in Eqs. (3.9) of the simple example.

Of course, the reduction may also be seen in the behavior of the density matrix. The density operator evolution equation for a single particle turns out to be identical to that of GRW[8]

$$dD(t)/dt = \text{-}i[H,D] + \lambda\{\int dzf(X\text{-}z)D(t)f(X\text{-}z) - D(t)\} \tag{3.23}$$

with solution (neglecting the hamiltonian H)

$$<x \mid D(t) \mid x'> = \{exp\text{-}\lambda t[1\text{-}\Phi(x\text{-}x')]\}<x \mid D(0) \mid x'> \tag{3.24}$$

where $\Phi(x\text{-}x')=\int dzf(x\text{-}z)f(x'\text{-}z)=exp\text{-}\alpha(x\text{-}x')^2/4$. When $x \approx x' \approx a_n$ are in the same packet, then $\Phi(x\text{-}x') \approx 1$ and there is little change in $<x \mid D(t) \mid x'>$, but when x and x' are in different packets, then $\Phi(x\text{-}x') \approx 0$ and $<x \mid D(t) \mid x'>$ decays as $exp\text{-}\lambda t$.

To close this discussion I will display the evolution equation of the state-vector, and of the density operator, in the nonrelativistic CSL theory for many identical particles[8,10]:

$$d\mid \psi,t>/dt = \{\text{-}iH + \int dxdzn(x)f(x\text{-}z)w(z,t) \\ \text{-}\lambda\int dxdx'n(x)n(x')\Phi(x\text{-}x')\}\mid \psi,t> \tag{3.25}$$

$$dD(t)/dt = \text{-}i[H,D] + \lambda\int dxdx'dz\Phi(x\text{-}x')\cdot \\ \{n(x)Dn(x')\text{-}(1/2)[n(x)n(x')D+Dn(x)n(x')]\} \tag{3.26}$$

where $n(x)= \phi^*(x)\phi(x)$ is the particle number density operator and $\phi^*(x)$, $\phi(x)$ are respectively the creation and annihilation operators of a particle at position x. In the next section it will be shown that the CSL nonhermitian and time-dependent "hamiltonian" in Eq. (3.25) is an appropriate generator of time translations for a Galilean invariant theory.

4. GALILEAN INVARIANCE

What is meant by time translation invariance? In classical physics it means that if you perform a measurement now and get a certain result, then if you perform the identical measurement at a later time you must get identically the same result, within experimental error. But that is frequently not what happens in nature, which is not classical. We can get quite different results when we repeat measurements. What is meant by time translation invariance is that when we perform an experiment consisting of many identical measurements (either taken simultaneously or sequentially), and then repeat that multi-measurement experiment, the *statistics* of the results must be identical. This can be called *stochastic time translational invariance* .

Thus it is quite reasonable to construct a theory which is *stochastic relativistic invariant,* meaning that the theory predicts identical statistical results for such multi-measurement experiments in all inertial coordinate systems. One might expect the evolution equation of that theory to depend upon time in a stochastic manner. Indeed, from this point of view, it is remarkable that ordinary quantum theory, which is stochastic time translational invariant, *doesn't* have any explicit time dependence in its hamiltonian! Let us now consider what is sufficient for a *stochastic Galilean invariant* theory.

4.1 Galilean Group Requirement

Pick an arbitrary fiducial reference frame, and go to any other reference frame by first performing a boost to the velocity v, followed by a time translation through the interval t and by a space translation through a (for simplicity only two spacetime dimensions will be considered throughout this paper). The statevector in such a reference frame will be called | ψ,a,t,v>.

Call the infinitesimal generators of the boost, time translation and space translation K-iL, H-iV, P-iR respectively, where K, H, P are the usual generators. These are the mathematical tools necessary to transform the statevector in a frame characterized by parameters (a,t,v) to the statevector in another frame whose parameters are infinitesimally close. In ordinary quantum theory these generators are independent of the frame they act upon, i.e. they do not depend upon a, t and v. In a stochastic relativistic theory, where events occur due to a fluctuating spacetime field, the generators could depend upon a, t and v, and we will suppose that L, V and R may have this dependence.

In order to have a stochastic Galilean invariant theory, both a Galilean group requirement and stochastic requirements must be fulfilled. The Galilean group requirement is that when one transforms the statevector from one reference frame to another, regardless of the route taken (e.g. first boost and then time translate, or first time translate and then boost) one must end up with the same statevector. Let us impose this first. Of course, when this has been accomplished, all that we will have is the rule to tell us what the universe looks like from different coordinate systems: there will as yet be no guarantee that the universe looks the same, in the sense of stochastic relativity.

The action of the generators is

$$\{1 + da[iP+R(a,t,v)]\}| \psi,a,t,v> = | \psi,a+da,t,v> \tag{4.1a}$$

$$\{1 - dt[iH+V(a,t,v)]\}| \psi,a,t,v> = | \psi,a,t+dt,v> \tag{4.1b}$$

$$\{1 + dv[iK+L(a,t,v) -t(iP+R(a,t,v)) +ima]\}| \psi,a,t,v> = | \psi,a,t,v+dv> \tag{4.1c}$$

Note that in Eq. (4.1c) the generator of velocity translations has been used. For the purposes of this article I want to make a distinction between a "reference frame", characterized by fixed values of a, t and v, and a "coordinate system" characterized by fixed values of a and v, but with variable t: a reference frame is an instantaneous photograph of a coordinate system. Now, the pure boost generator K takes | ψ,a,t,v> into | ψ,a-tdv,t,v+dv>, and the parameter a-tdv characterizes a sequence of different coordinate systems as t increases. It is necessary to use a velocity translation if one wishes to transform from one coordinate system to another.

By equating the results of two successive different transformations (4.1) in either order one obtains

$$[iP+R,iH+V] = \partial_a V + \partial_t R \tag{4.2a}$$

$$[iP+R,iK+L] = \partial_a L + \partial_v R - t\partial_a R + im \tag{4.2b}$$

$$[iK+L,iH+V] = \partial_t L + t\partial_a V - \partial_v V - (iP+R) \tag{4.2c}$$

If we define $P' \equiv P + i\partial_a$, $H' \equiv H - i\partial_t$, $K' \equiv K + it\partial_a - i\partial_v$, Eqs. (4.2) can be written in a form similar to the usual commutation relations

$$[iP'+R,iH'+V] = 0, \quad [iP'+R,iK'+L] = im, \quad [iK'+L,iH'+V] = -(iP+R) \tag{4.3a,b,c}$$

By utilizing the usual commutation relations of the generators P,H,K we obtain

$$i[P',V]-i[H',R]=[V,R], \quad i[P',L]-i[K',R]=[L,R], \quad i[K',V]-i[H',L]+R=[V,L] \tag{4.4a,b,c}$$

For the CSL theory, V is hermitian. If I choose R and L to be hermitian too (although a more general class of theories could be explored), then the left sides of Eqs. (4.4) are hermitian while the right sides are antihermitian, so both sides must vanish. I will choose R=L=0 and the right sides will vanish. The remaining restrictions on V

$$i[P',V] = 0, \quad i[K',V] = 0 \tag{4.5a,b}$$

have nontrivial solutions. With an eye toward the CSL theory, assume that V has the form

$$V(a,t,v) = \int dx\phi^*(x)\phi(x)G(x,a,t,v) + \int dxdx'\phi^*(x)\phi(x)\phi^*(x')\phi(x')G'(x,x',a,t,v) \tag{4.6}$$

where G and G' are so far arbitrary functions of their arguments. From Eq.(4.5a), using $i[P,\phi(x)]=-\partial_x\phi(x)$, we obtain the conditions $\partial_x G-\partial_a G=0$, $\partial_x G'+\partial_{x'}G'-\partial_a G'=0$, so $G=G(x+a,t,v)$ and $G'=G'(x-x',x+a,t,v)$. From Eq.(4.5b), using $i[K,\phi(x)]=0$ we obtain the conditions $t\partial_a G-\partial_v G=0$, $t\partial_a G'-\partial_v G'=0$, so $G=G(x+a+vt,t)$, $G'=G'(x-x',x+a+vt,t)$.

This Galilean group requirement can be fulfilled by taking

$$G=-\int dz f(x+a+vt-z)w(z,t)=-\int dz f(x-z)w(z+a+vt,t), \quad G'=\lambda\Phi(x-x') \tag{4.7a,b}$$

With these choices, the space translation and boost generators are the usual ones, while the time translation equation is

$$d|\psi,a,t,v>/dt = \{-iH + \int dxdz\phi^*(x)\phi(x)f(x-z)w(z+a+vt,t) \\ -\lambda\int dxdx'\phi^*(x)\phi(x)\phi^*(x')\phi(x')\Phi(x-x')\}|\psi,a,t,v> \tag{4.8}$$

Suppose we are given the particular hamiltonian of our universe, depending upon the white noise function $w(z_f,t_f)$, in the fiducial coordinate system (parameters a=v=0 and coordinates z_f,t_f). Eq. (4.8) tells us that the hamiltonian in another coordinate system (parameters a,v and coordinates $z=z_f-a-vt_f$, $t=t_f$) depends upon the sample function $w_{a,v}(z,t) \equiv w(z_f,t_f)$. Eq. (4.8) differs from the quoted non-relativistic CSL equation (3.5) only in the dependence of the white noise function on its spatial argument. We will now see that this is an inessential difference.

4.2 Stochastic Requirements

A stochastic relativistic invariant theory requires the ensemble of hamiltonians to be identical in all inertial frames, and requires the probability rule to be applicable under a change of inertial frame.

But, as seen from a particular coordinate system, The History of Our Universe (THOU) is unique, with a hamiltonian dependent upon a unique white noise function w(z,t). Why then do we need to talk about an ensemble of hamiltonians with an ensemble of white noise functions $\{w(z,t)\}$ (by the ensemble $\{w(z,t)\}$ is meant both the functions and their associated probabilities) and an associated probability rule?

Consider the single statevector, as defined in some coordinate system at some time in THOU. The ensemble of histories that evolves out of this statevector is driven by the ensemble of white noise functions, and weighted by the probability rule. We hypothesize that THOU chosen from among these histories is not an improbable one.

Suppose that during THOU a succession of identical measurements is performed in a coordinate system. This gives an ensemble of results whose statistical distribution is the same as would be obtained if one considered a single measurement performed in that coordinate system in the ensemble of universes driven by $\{w(z,t)\}$. (The elements behind this observation are that white noise is a stationary process and that the rule associating the probability $<\psi,a,t,v \mid \psi,a,t,v>P(w)$ ($P(w)$ is the probability of $w(z,t)$) with the response of the system to the measurement is a stationary rule.) The successive measurements during THOU in effect sample the ensemble of hamiltonians and the ensemble of system responses, so the ensemble tells us what happens in each of its members, of which THOU is typical.

According to Eq. (4.8), the ensemble of hamiltonians will be identical in all inertial frames provided $\{w(z+a+vt,t)\}=\{w(z,t)\}$. But, using Eq. (3.17),

$$<w(z+a+vt,t)> = 0, \quad <w(z+a+vt,t)w(z'+a+vt',t')> = \lambda\delta(z-z')\delta(t-t') \qquad (4.9)$$

and these two moments of the gaussian process completely determine the ensemble. Since (4.9) shows that the moments are identical in all inertial frames, so also the ensemble of white noise functions and their associated hamiltonians must be identical in all inertial frames. Then the probability rule works in all coordinate systems as its functioning depends upon the ensemble of hamiltonians being of CSL form. This ensures stochastic relativistic invariance from the active point of view (i.e. a succession of measurements performed in any coordinate system will yield the same statistical results) for the ensemble and thus for THOU.

The passive point of view for THOU and for the ensemble must be examined separately. The passive point of view for THOU requires that a measurement evolution in one coordinate system be viewed from another coordinate system as a possible evolution. This is guaranteed by the form invariance of the equation of evolution (4.8) under coordinate transformations. The passive point of view also requires the result of the measurement in the two coordinate systems to be identical (up to a Lorentz transform of meter readings). This will be so because, sufficiently long after a measurement has been completed, the large component of the statevector corresponding to the actual outcome and the small tail corresponding to the nonoccurring outcomes do not have their relative disproportionate magnitudes reversed by the coordinate transformation operation.

The passive point of view for the ensemble describing a single measurement in one coordinate system requires that it be viewed from another coordinate system as a feasible ensemble, with the same probabilities associated to the outcomes of the measurement. Sufficient conditions for this, in addition to the identity of the ensemble of hamiltonians in both coordinate systems, are that the probability associated to each measurement evolution be invariant under change of coordinate system:

$$\partial_a<\psi,a,t,v \mid \psi,a,t,v>P(w)=0, \quad \partial_v<\psi,a,t,v \mid \psi,a,t,v>P(w)=0, \qquad (4.10a,b)$$

(This is sufficient but not necessary: it is only necessary that *all* of the measurement evolutions leading to the same result have the same total probability in each coordinate system.) However <ψ,a,t,v | ψ,a,t,v> is invariant under translations and boosts because they are generated by unitary transformations. Also,

$$P(w(z,t)) = CDwexp-\iint dzdtw(z,t)/2\lambda, \quad Dw = \Pi_z\Pi_t dw(z,t) \qquad (4.11)$$

is unchanged when w undergoes a translation of its spatial argument from z to z+a+vt. Therefore <ψ,a,t,v | ψ,a,t,v>P(w) is unchanged.

This concludes the demonstration that the CSL theory is Galilean invariant. As is to be expected, the density matrix evolution equation (3.26) that the theory produces is (manifestly) Galilean invariant.

4.3 On Tails, Localization, and Retrodiction

This is as good a place as any to call particular attention to the fact that in the SL and CSL theories, the statevectors are never completely reduced. There is always a little bit of "what might have been" present along with "what is". In complete reversal of the position I previously took[30], I am going to argue that, rather than an embarassment to be eventually overcome in a better theory, tails are an absolute necessity.

First, however, note that tails already exist in ordinary quantum theory: for example, the tails of atomic wavefunctions spread out over all space. Although the SL and CSL theories are realist theories, they are only committed to representing reality by the statevector, they are not committed to a realist interpretation of the components of the statevector in a particular basis. Nevertheless, because the position representation is more important in these theories than any other representation, it is tempting to explore the spacetime picture given by taking the one-particle position density

$$\rho(x,t) = <x|Tr_{x'\neq x}\{| \psi,t>_{PP}<\psi,t |\}|x> \qquad (4.12)$$

(I am ignoring here the issues involved with identical particles) and regarding it as "really" representing the spatial extension of the particle. I will here succumb to this temptation.

The picture we get is Schrodinger's original picture, before Bohr talked him out of it. An individual particle can be "mostly localized" as in an atomic bound state, in the sense that $\int_V \rho(x)dx$ is almost 1 when V is "small" ($<10^{-5}$ cm., say). However, an individual particle may be "very spread out" in space, as in a scattering state. The "only mystery" of quantum theory referred to by Feynman, the two-slit interference pattern of a single particle, is simply explained by the particle really being in both slits at the same time.

The new feature of the SL and CSL theories, compared to quantum theory, is that they make a distinction between microscopic and macroscopic. Unlike an individual uncorrelated particle, the center of mass of a macroscopic object is always "mostly localized", it is never "very spread out" in space. This latter state is prevented by the reduction dynamics which rapidly turns a would-be superposition of distinctly spatially separated macroscopic states of comparable amplitude (allowable by quantum theory) into one state plus a tail, before such a superposition can appreciably develop. As I will later argue, this has the corollary in the context of a possible relativistic CSL theory, that the spacetime localization region of a macroscopic object is an absolute concept (i.e. agreed upon by all inertial observers), but the spacetime localization region of a microscopic object is a relative concept.

Now I come to the necessity for tails. First, they are necessary for retrodiction. If one knows the white noise function w(z,t) and the present statevector, tail and all, one can use the hamiltonian to correctly run the time evolution (3.25) backwards. When running the statevector of a completed measurement backwards, the insignificantly sized tail will grow and the portion of the statevector describing the correct outcome will diminish in amplitude until both are comparably sized. If one had cut off the tail, the result of this calculation could be spectacularly wrong. Because of the time direction built into the CSL equation (the term proportional to λ in (3.25) makes the evolution equation not time-reversal invariant), a tail that has a negligible effect on future evolution will have an enormous effect on evolution into the past.

However, their utility for retrodiction does not make tails absolutely necessary because retrodiction is not an absolutely necessary capability that a theory must provide. But, if we want to have a Lorentz invariant theory, the boost operation requires a transformation, to a new hyperplane, that is partly a sweep backwards in time. Just as in the above paragraph, a boost to a new frame can dramatically enhance the size of the tail. One gets the wrong statevector in the new reference frame without having the tail in the original frame. I believe the tail is absolutely necessary because I cannot see how to make a Lorentz invariant theory of statevector reduction without it.

5. LORENTZ INVARIANT CSL THEORY?

It is natural to try to construct a Lorentz invariant theory by following the procedures presented in the previous section that resulted in a Galilean invariant theory. I will show that it is possible to fulfill the Poincarè group requirements and construct the generators in terms of local quantum fields. The new feature is that not only the hamiltonian but also the boost generator depend upon the white noise function. Furthermore, active Lorentz invariance is satisfactory as the ensemble of hamiltonians is the same in all inertial frames, and passive Lorentz invariance for THOU is likewise satisfactory. What fails is the statistical requirement leading to passive Lorentz invariance for the ensemble. I shall argue that this is to be expected. However, if one relaxes the demand of passive Lorentz invariance on hyperplanes in favor of spacelike surfaces which do not extend arbitrarily far into the past, one can recover passive Lorentz invariance in the framework of the Tomonaga-Schwinger formalism. So, it appears that a reasonably satisfactory framework for a relativistic quantum field CSL theory exists.

The problem then becomes to choose a hamiltonian that will give a satisfactory dynamical theory. I will consider the simplest possible nonhermitian hamiltonian, and suggest a mechanism that reduces the statevector. With each fermion there is associated a scalar "meson" field. When the fermion is in the canonical two-packet state, the statevector is reduced to one of the two meson configurations, bringing along with it the attached fermion. However, this model has a grave difficulty: there is an infinite meson production rate per unit volume! So there is not yet a satisfactory relativistic CSL theory. However, I will close with a discussion of a few examples of how I think such a theory will behave, and the pictures it might give of the evolution of particles in spacetime during the reduction process.

5.1 Poincarè Group Requirement

Exactly parallel to the discussion in section 3, the action of the infinitesimal generators is

$$\partial| \psi,a,t,v>/\partial a = [iP+R(a,t,v)]| \psi,a,t,v> \tag{5.1a}$$

$$\partial| \psi,a,t,v>/\partial t = -[iH+V(a,t,v)]| \psi,a,t,v> \tag{5.1b}$$

$$\partial| \psi,a,t,v>/\partial v = -[iK+L(a,t,v) -t(iP+R(a,t,v)) \\ +a(iH+V(a,t,v))]| \psi,a,t,v> \tag{5.1c}$$

(The velocity of the boosted frame with respect to the fiducial frame is tanhv).
By equating the second derivatives of $|\psi,a,t,v\rangle$ in either order one obtains the conditions

$$[iP+R,iH+V] = \partial_a V + \partial_t R \tag{5.2a}$$

$$[iP+R,iK+L] = \partial_a L + \partial_v R - t\partial_a R - a\partial_t R + iH+V \tag{5.2b}$$

$$[iK+L,iH+V] = \partial_t L + t\partial_a V + a\partial_t V - \partial_v V - (iP+R) \tag{5.2c}$$

If we define $P'\equiv P+i\partial_a$, $H'\equiv H-i\partial_t$, $K'\equiv K+it\partial_a+ia\partial_t-i\partial_v$, Eqs. (5.2) can be written in a form similar to the usual commutation relations

$$[iP'+R,iH'+V] = 0, \quad [iP'+R,iK'+L] = iH+V, \quad [iK'+L,iH'+V] = - (iP+R) \tag{5.3a,b,c}$$

By utilizing the usual commutation relations of the generators P,H,K we obtain

$$i[P',V]-i[H',R]=[V,R], \quad i[P',L]-i[K',R]=[L,R]+V, \quad i[K',V]-i[H',L]=[V,L]-R \tag{5.4a,b,c}$$

The new feature is that we cannot choose R=L=0 because, from Eq. (5.4b), this means that V will vanish. So we will take R=0 and look for a solution where both V and L are nonvanishing. We conclude from the hermitian nature of the left side of (5.4c) and the nonhermitian nature of the right side that [V,L]=0. This can be achieved if V and L are functions of commuting operators. For definiteness, I will take the operators to be a scalar field $\phi(x)$ at different points x,

$$\phi(x) \equiv (4\pi)^{-(1/2)}\int dk\{a(k)\exp ikx + a^*(k)\exp -ikx\}/k^0 \tag{5.5a}$$

$$[a(k),a^*(k')]=k^0\delta(k-k') \tag{5.5b}$$

although scalar combinations of other fields could be used. Assume that V and L have the local[31] forms

$$V(a,t,v)=\Sigma_n\int dxg_n[\phi(x)]G_n(x,a,t,v), \quad L(a,t,v)=\Sigma_n\int dxg_n[\phi(x)]H_n(x,a,t,v) \tag{5.6a,b}$$

where g_n, G_n, and H_n are so far arbitrary functions of their arguments. The constraints (5.4)

$$i[P',V] = 0 , \quad i[P',L] = V, \quad i[K',V] = i[H',L] \tag{5.7a,b,c}$$

together with the commutation relations of the usual generators with the field

$$i[P,\phi(x)] = -\partial\phi(x)/\partial x, \quad i[H,\phi(x)] = \dot\phi(x), \quad i[K,\phi(x)] = \dot\phi(x)x \tag{5.8a,b,c}$$

imply the following restrictions on Eqs. (5.6). Applying (5.7a) to the form (5.6a) for V, and integrating by parts to throw the derivatives of g_n onto G_n, we obtain the conditions $\partial_x G_n-\partial_a G_n = 0$, so $G_n = G_n(x+a,t,v)$. From Eq. (5.7b) comes the conditions $\partial_x H_n-\partial_a H_n = G_n$, which can be satisfied with $H_n = xG_n$. Finally, from (5.7c) we obtain the constraint $t\partial_a G_n+(x+a)\partial_t G_n-\partial_v G_n = 0$, which is fulfilled if G_n is an arbitrary function of the two arguments

$$x_f \equiv (x+a)\cosh v + t\sinh v, \quad t_f \equiv t\cosh v + (x+a)\sinh v \tag{5.9a,b}$$

These restrictions allow the fulfillment of the Poincarè group requirements where the time translation generator has a CSL form, by replacing G_1 by the white noise function $-w(x_f,t_f)$ and G_2 by λ:

$$\partial|\psi,a,t,v\rangle/\partial a = iP|\psi,a,t,v\rangle \tag{5.10a}$$

$$\partial|\psi,a,t,v\rangle/\partial t = \{-iH + \int dxg[\phi(x)]w(x_f,t_f) - \lambda\int dxg^2[\phi(x)]\}|\psi,a,t,v\rangle \tag{5.10b}$$

$$\partial| \psi,a,t,v>/\partial v = \{-iK + \int xdxg[\phi(x)]w(x_f,t_f) - \lambda\int xdxg^2[\phi(x)] + tiP$$
$$+ a(-iH + \int dxg[\phi(x)]w(x_f,t_f) - \lambda \int dxg^2[\phi(x)])\}| \psi,a,t,v> \qquad (5.10c)$$

5.2 Stochastic requirements

Active Lorentz invariance of this theory immediately follows using the argument given in section 4.2: in particular, according to Eq. (3.17), the moments of the white noise ensemble are invariant under the Lorentz transformation (5.9)

$$< w(x_f,t_f) > = 0, \quad < w(x_f,t_f) \, w(x'_f,t'_f) > = \lambda\delta(x-x')\delta(t-t') \qquad (5.11)$$

Passive Lorentz invariance of THOU likewise follows according to the argument in section (4.2).

However, passive Lorentz invariance for the ensemble cannot be obtained because the total ensemble probability is not conserved under boosts. To see this, we use Eq. (5.10c) to calculate the change of total probability associated with the transformation of a single statevector in the fiducial frame into an ensemble under an infinitesimal boost:

$$\partial<<\psi,0,0,v \,|\, \psi,0,0,v>>/\partial v|_{v=0} = -4\lambda<\psi,0,0,0 \,|\int_{-\infty}^{0}xdxg^2[\phi(x)]|\, \psi,0,0,0> \qquad (5.12)$$

It is clear from Eq. (5.12) that the problem comes from the region x<0, the part of the reference frame that goes backwards in time during the boost. Why is this?

First, let me return to the subject of retrodiction. Suppose one has an ensemble at time t, and one wishes to run it backwards to a time <0. What this means is that one looks for an ensemble at the time <0 that will evolve into the ensemble at time t. It can be found if the ensemble at the time <0 actually evolved out of an ensemble prepared at an earlier time yet. But suppose the ensemble at time t evolved out of a single statevector at time 0. There is no way that an ensemble at a time earlier than 0 can evolve into the single statevector at time 0, so it is not reasonable to ask the theory to construct such an ensemble. (A precisely analogous situation occurs in ordinary diffusion, where the probability distribution for an ensemble of brownian motions cannot be run backwards in time indefinitely without eventually encountering an obstacle, a distribution that no earlier distribution could have evolved into.)

But an unreasonable request just like this is what we are making if we ask the theory to produce an ensemble of statevectors in a boosted frame that corresponds to an ensemble in the fiducial coordinate system that arose from a single statevector given at fiducial time 0. The preparation of a state in the fiducial coordinate system at, or earlier than, some instant of fiducial time is an operation that interferes with the autonomous evolution of the statevector, and this operation takes forever from the point of view of any boosted coordinate system. Thus it is quite reasonable that the theory balks, as expressed in Eq. (5.12), at producing a probability measure for the ensemble in a boosted frame.

5.3 Tomonaga-Schwinger Equation

We do not have to give up passive Lorentz invariance for the ensemble if we are willing to utilize a relativistic formalism that permits transformations to spacelike hypersurfaces that are arbitrary except for the following imposed restriction. No hypersurface is to cross backwards in time beyond the "preparation hypersurface", earlier than which the system was prepared and the statevector is unknown. The evolution equation is the Tomonaga-Schwinger equation[32,33] adapted to the CSL formalism

$$d| \psi,s>/ds=\iint dtdx\delta(\sigma(x,t)-s)\{g[\phi_{int}(x,t)]w(x,t)-\lambda g^2[\phi_{int}(x,t)]\}| \psi,s> \qquad (5.13)$$

Here $\sigma(x,t)=s$ describes a family of appropriate hypersurfaces, and $\phi_{int}(x,t)$ is the scalar field in the interaction picture. The use of this formalism is explored elsewhere[34], where it is shown that $d<<\psi,s \mid \psi,s>>/ds=0$, so probability is conserved under a transformation of such spacelike hypersurfaces.

6. REDUCING ON VIRTUAL MESONS

It looks so far that there is no formal barrier to constructing a relativistic CSL theory, so the next step is to choose a hamiltonian. In this section I will investigate (approximations to) the model with evolution equation

$$d\mid \psi,t>/dt = \{-iH_F -iH_M + i\eta(4\pi)^{1/2}\int dz \Psi*(z)\Psi(z)\phi(z)$$
$$+ \int dz\phi(z)w(z,t) - \lambda\int dz\phi^2(z)\}\mid \psi,t> \qquad (6.1)$$

In Eq. (6.1), H_F and H_M are the hamiltonians for fermion ($\Psi(z)$) and meson fields, η is the coupling constant for these two fields, and the simplest choice $g[\phi(z)]=\phi(z)$ has been made.

6.1 Unmoving Model

We take the nonrelativistic limit in which the fermion has infinite mass, and $\Psi*(x)$ is just the creation operator for the fermion at point x (i.e. $\Psi*(x)\mid 0>=\mid x>$). All I will consider is a one-fermion state, for which the evolution equation is

$$d<x \mid \psi,t>/dt = \{-i\int dka*(k)a(k) + i\eta(4\pi)^{1/2} \phi(x) +$$
$$\int dz\phi(z)w(z,t) - \lambda\int dz\phi^2(z)\}<x \mid \psi,t> \qquad (6.2)$$

As is well known[33], the meson field in this model "dresses" the fermion, and the renormalized mesons uncouple from the fermion. Upon defining the renormalized meson annihilation operator $b(k)\equiv a(k)-(\eta/k^0)\exp-ikx$, Eq. (6.2) becomes

$$d<x \mid \psi,t>/dt = \{-i\int dkb*(k)b(k) + \int dz\phi(z)w(z,t) - \lambda\int dz\phi^2(z)\}<x \mid \psi,t> \qquad (6.3)$$

(where the constant energy of the dressed fermion has been omitted). Finally, I shall omit the hermitian part of the hamiltonian in Eq. (6.3) entirely, obtaining the easily treated evolution equation

$$d\mid \psi,t>/dt = \{\int dz\phi(z)w(z,t) - \lambda\int dz\phi^2(z)\}\mid \psi,t> \qquad (6.4)$$

This means that mesons which are created will not move away from the site of their creation.

The ground state of the hermitian part of the hamiltonian in (6.3) is the state of the virtual mesons surrounding the fermion located at x. This state $\mid m,x>$ satisfies $b(k)\mid m,x>=0$, and can be written as a coherent state

$$\mid m,x> = e^{\eta\int k_0^{-2}dka*(k)\exp-ikx} \mid 0>e^{-\eta^2\int k_0^{-3}dk/2} \qquad (6.5)$$

where $\mid 0>$ is the virtual meson vacuum state, $a(k)\mid 0>=0$.

Consider the fermion in an initial canonical two-packet state where the packets are centered at a_1 and a_2 and no free renormalized mesons are present. This state can be written approximately as

$$\mid \psi,0> = \alpha_1(0)\mid \alpha_1> + \alpha_2(0)\mid \alpha_2>, \quad \mid \alpha_n> \equiv \mid m,a_n>\mid a_n> \qquad (6.6)$$

where $<x \mid a_n>$ is a narrow wavepacket of the fermion centered at a_n, $\mid \alpha_n>$ is the state consisting of this fermion packet and its associated meson field, and the approximation consists in replacing $\mid m,x>$ by $\mid m,a_n>$ for x in the neighborhood of the packet centered on a_n.

The solution of (6.4) subject to the initial condition (6.6) is

$$| \psi,t> = \exp[\int dz\phi(z)B(z,t) - \lambda t\int dz\phi^2(z)]| \psi,0>$$

$$= \{\exp\int dzB(z,t)^2/4\lambda t\}\Sigma_n\alpha_n(0)| a_n>$$
$$\cdot\exp-\int dz[B(z,t)-2\lambda t\phi(z)]^2/(4\lambda t)| m,a_n> \qquad (6.7)$$

6.2 Crude Model of Meson Field

Before continuing with this problem, I want to solve an even easier problem, with a very crude model of the meson field, to indicate how the reduction mechanism could be expected to work. The expectation value of the meson field strength in the state | m,x> is readily evaluated with the help of Eqs. (5.5), (6.5):

$$<m,x |\phi(z)| m,x> = (\eta\pi^{1/2}/m)\exp-m|x-z| \equiv f(x-z) \qquad (6.8)$$

Now suppose that, instead of the expression (6.5) for | m,x>, we model the state | m,x> as an eigenstate of the meson field at each point of space, with eigenvalue f(x-z):

$$\phi(z)| m,x> = f(x-z)| m,x> \qquad (6.9)$$

Of course this neglect of the dispersion of $\phi(z)$ in the state | m,x> is a gross simplification. However, the solution (6.7), with | m,x> defined as in Eq. (6.9),

$$<\alpha_n| \psi,t>= \{\exp\int dzB(z,t)^2/4\lambda t\}\alpha_n(0)$$
$$\cdot\exp-\int dz[B(z,t)-2\lambda tf(a_n-z)]^2/(4\lambda t) \qquad (6.10)$$

is a solution we have encountered before. It is precisely the solution given in Eqs. (3.19) to the single particle nonrelativistic problem (except there f(x-z) was the gaussian (3.16), while here it is the exponential (6.8)). As was shown in section (3.2), this solution describes statevector reduction. What is the physical mechanism here that brings reduction about?

Since the two fermion packets are centered at different locations, the meson field at each point z of space is in a superposition of two differing amplitudes, $\approx f(a_1-z)$ or $\approx f(a_2-z)$. Just as in the simple example of section 3.1, the reduction mechanism chooses one of these two amplitudes. The overwhelmingly probable statevectors in the ensemble are those for which the field amplitudes chosen at all z are either $f(a_1-z)$ or $f(a_2-z)$. The reduction of this meson field "brings along" the fermion packet to which it is correlated.

The density matrix can be found from the solution (6.10) using the prescription $D = < | \psi,t><\psi,t | >$ given in Eq. (3.10):

$$<\alpha_n| D(t)|\alpha_m> = \alpha_n(0)\alpha_m^*(0)\exp-(\lambda t/2)\int dz[f(a_n-z)-f(a_m-z)]^2 \qquad (6.11)$$

from which the reduction behavior can also be surmised.

6.3 Solution of the Unmoving Model

Returning now to the solution (6.7) with the meson field given by Eq. (6.5), it is useful to introduce the eigenbasis | $\varphi(z)$> of the meson field $\phi(z)$ at each point z:

$$\phi(z)| \varphi(z)> = \varphi(z)| \varphi(z)> \quad , \qquad \int D\varphi(z)| \varphi(z)><\varphi(z) | = 1 \qquad (6.12)$$

The meson state expressed in this basis, $<\varphi(z) | m,x>$, is a Gaussian in $\varphi(z)$ and so it is completely characterized by the mean f(x-z) given by (6.8) and covariance

$$<m,x |[\phi(z)-f(x-z)][\phi(z')-f(x-z')]| m,x> =(4\pi)^{-1}\int dkk_0^{-1} \cos k(z-z') \qquad (6.13)$$

whose inverse $\sim \int dk_0 \cos k(z-z')$ appears in the gaussian's exponent:

$$<\varphi(z) \mid m,x> = C\exp\text{-}(4\pi)^{-1}\iint dzdz'[\phi(z)\text{-}f(x\text{-}z)][\phi(z')\text{-}f(x\text{-}z')]\int dk_0 \cos k(z\text{-}z') \quad (6.14)$$

It is now possible to calculate the probability $<\psi,t \mid \psi,t>\rho(B)DB$ associated with a particular solution $\mid \psi,t>$ from (6.7) and (6.14):

$$<\psi,t \mid \psi,t>\rho(B)DB = \rho(B)DB \int D\varphi(z)\Sigma_n<\psi,t \mid \varphi(z)> \mid a_n><a_n \mid <\varphi(z) \mid \psi,t>$$
$$= CDB\Sigma_n \mid \alpha_n(0) \mid^2$$
$$\cdot\exp\text{-}\iint dzdz'[B(z,t)\text{-}2\lambda tf(a_n\text{-}z)]g(z\text{-}z',t)[B(z',t)\text{-}2\lambda tf(a_n\text{-}z')]/(2\lambda t) \quad (6.15a)$$

$$g(z\text{-}z',t) \equiv (2\pi)^{-1}\int dk_0(k_0+2\lambda t)^{-1}\cos k(z\text{-}z') \quad (6.15b)$$

For λt less than the meson mass, we can approximate $(k_0+2\lambda t)$ in (6.15b) by k_0, and $g(z\text{-}z',t) \approx \delta(a_n\text{-}z)$. The expression (6.15a) then becomes identical to the expression (3.21), and the reduction description is identical to that in the non-relativistic model of section (3.2) and in the crude model of section 6.2.

For large t, $g(a_n\text{-}z,t) \sim (\lambda t)^{-1}$, making the variance of the Gaussian in Eq. (6.15a) no longer proportional to λt, but instead proportional to $(\lambda t)^2$. This has the consequence that the non-overlap of two $B(z,t)$'s leading to two different outcomes ceases to improve with time as it did in the previous two examples. Instead, each outcome contains a tail of constant (not continually reducing) size. This is because there is a small probability that any meson field configuration surrounding one packet will also be possessed by a distant packet, and the reduction mechanism has no way to distinguish them.

These considerations indicate that the meson field reduction mechanism may work for the relativistically invariant model.

6.4 Meson Production

The meson energy production rate can be directly calculated for the relativistic model (6.1), with no approximations. The density matrix evolution equation is

$$dD(t)/dt = \text{-}i[H,D] + \lambda\int dz\{\phi(z)D\phi(z) - D\phi^2(z)/2 - \phi^2(z)D/2\} \quad (6.16)$$

We use this to calculate $<H>=TrHD(t)$:

$$d<H>/dt = \text{-}(\lambda/2)\int dz[\phi(z),[\phi(z),H]]D \quad (6.17)$$

Since $[\phi(z),H]=id\phi(z,t)/dt|_{t=0}$ and $[\phi(z),d\phi(z,t)/dt|_{t=0}]=i\delta(0)$, we obtain for the rate of energy production per unit volume

$$d(<H>/\int dz)/dt = (\lambda/2)\delta(0) \quad (6.18)$$

What can be done about this infinite result? It is independent of the state of the system, which might be construed as indicating that it may be removed by some renormalization procedure. It would not be infinite if the interaction is nonlocal, i.e. if Eq. (6.1) was replaced by

$$d\mid \psi,t>/dt = \{\text{-}iH_F \text{-}iH_M + i\eta(4\pi)^{1/2}\int dz\Psi*(z)\Psi(z)\phi(z)$$
$$+ \int dz\int dz'\phi(z)f(z\text{-}z')w(z',t) - \lambda\int dz\int dz'\phi(z)\Phi(z\text{-}z')\phi(z')\}\mid \psi,t> \quad (6.19)$$

but then the theory would not be relativistically invariant: can we make a nonlocal relativistic theory? It would not be infinite if the white noise function $w(z,t)$ was more realistic in that it did not have arbitrarily short wavelengths and high frequencies. These suggest avenues of exploration in the search for a satisfactory relativistic theory of statevector reduction, which is still elusive.

7. Speculations

I want to comment on two aspects of the incompleted program for a relativistic theory of statevector reduction outlined above. The first comment argues that, if the meson field reduction mechanism is to be pursued, the "mesons" ought to be "gravitons." The second comment examines some aspects of the spacetime picture of individual particle behavior that I expect the completed theory would provide.

7.1 Reduction on the Gravitational Field

What properties ought to be possessed by the meson field responsible for reduction?

If we follow the suggestion of GRW, the characteristic reduction length is 10^{-5} cm. In the meson theory, this is the Compton wavelength of the meson, which translates to a meson mass of about 1 eV.

In order for the reduction rate to increase as the number of particles contained in an apparatus increases, the meson field strength must increase as particles are brought together. If there were opposite signs of the fermion charge, as in electromagnetism, it would be possible to make an object possessing negligible meson field strength out of a macroscopic collection of such fermions. A negligible reduction rate would result for a superposition of states of this object separated by a macroscopic distance. To prevent this, the meson should be associated with only one sign of charge.

The coupling η between the meson and the fermion fields modeled in Eq. (6.1) should be either extremely weak or already known or both. This is because the meson field is not only responsible for reduction. The meson coupling to the fermions results in interaction between the fermions. For example, it should play a role in fermion-fermion scattering. Either the coupling is so weak that no anomaly has yet been detected in predicted scattering amplitudes based upon the known interactions, or the interaction is one of the known interactions.

The coupling of the meson field to other particles should be universal: it ought not be just a coupling to fermions, but a coupling to every other form of matter. For, suppose someone figures out how to make an apparatus out of bosons. It is reasonable to expect that the statevector would reduce for a measurement made by such an apparatus.

So, we need a candidate for a meson field with a mass about 1 eV., coupled to particles with only one sign of charge, with a universal coupling that is either weak or already known or both. What else could this be but the gravitational field?

7.2 Spacetime Pictures

I want to consider a few simple measurement situations, as viewed (i.e., calculated, not necessarily observed) from different Lorentz frames.

In the first example, a particle is emitted from a site at x=0, at time t=0. It has equal likelihood of traveling to left or right, so it is described by wavepackets $\psi_L(x,t)$ and $\psi_R(x,t)$. An apparatus capable of detecting the particle sits to the left at x=-a. Suppose that the white noise function is such that the particle is actually detected by the apparatus. What is the spacetime picture from the point of view of the x-coordinate system, and how would this be viewed from a boosted x'-coordinate system?

In the x-system, the two packets trace world lines to left and right from the initial event (0,0), until the left-packet hits the apparatus and is detected, say at time t_1. The reduction mechanism immediately commences, and within the characteristic reduction time τ of the apparatus, the wavefunction consists largely of a piece describing the detected particle:

$$\psi(t_1) = A_L{}^U\{(1/\sqrt{2})\psi_L(x,t_1) + (1/\sqrt{2})\psi_R(x,t_1)\} \rightarrow \tag{7.1a}$$

$$\psi(t_1+\tau) = C(t_1+\tau) A_L{}^D\psi_L(x,t_1+\tau) + c(t_1+\tau) A_L{}^U\psi_R(x,t_1+\tau) \tag{7.1b}$$

where $A_L{}^U$, $A_L{}^D$ refer to the apparatus states of undetection and detection respectively, and C is a big number and c is a small number. The corresponding spacetime picture (using $\rho(x,t)$, the single particle probability density defined in Eq. (4.12)) is that the right-packet fades away substantially during the time interval t_1 to $t_1+\tau$.

Two further points are worth mentioning here.

The first is that the reduction process once started never stops, so the amplitude c of the right-packet continues to exponentially decline for time $>t_1+\tau$. Thus, although the dramatic difference in the amplitudes C and c occurs in the time interval t_1 to $t_1+\tau$, there is every τ seconds an equally dramatic proportionate decline in the amplitude c.

The second is that, along with the fading away of the right-packet, $\rho(x,t)$ for any particle in the apparatus's undetected state likewise fades away. The difference between an apparatus particle and the measured particle is that, although the former are also in a superposed state just before reduction, their wavefunctions are localized in a small (of order 10^{-5} cm.) connected region of space, whereas before reduction the measured particle occupies two widely separated regions of space.

Now consider this experiment as viewed from the x'-coordinate system. The history is qualitatively exactly the same: the collision of the left-packet with the apparatus triggers the reduction process, which dramatically reduces the amplitude of the right-packet during a short time interval, say t_1' to $t_1'+\tau'$. However, since the hyperplane t_1 is not the same as the hyperplane t_1', the observers from the two systems will differ as to the part of the right-packet worldline where the rapid reduction in amplitude commenced.

In both coordinate systems, the particle occupies the same region of spacetime before, during and after the measurement, in the sense that $\rho(x,t)$ has "bumps" along the whole length of the right and left worldlines. However, when the time occurs in one's coordinate system that the amplitude for the particle's being in one region is much larger than the amplitude for its being in another region, one may legitimately make the distinction of section 4.3, that the particle is thereafter "mostly localized" in the region of overwhelmingly greater amplitude.

So, to all the other things that used to be thought of as absolute, but which relativity has taught us are reference frame dependent, a relativistic CSL theory adds the region of spacetime in which a particle is "mostly localized."

As a second example, consider a different experimental setup with the same particle, that of an interference experiment. Suppose the packets sent to left and right encounter reflectors so that they are returned to the origin x=0 where sits a detector. The point I wish to make here is that this measurement looks the same from all coordinate systems. In each system the particle is mostly localized along both worldlines. So, there does not *have* to be an ambiguity due to the relativity of localization, even for an uncorrelated particle.

For the last example, consider again the setup of the first example, except that an additional apparatus is placed to the right at x=b, where b is slightly larger than a.

From the point of view of the x-system, the left-packet hits the left-apparatus first and so triggers the reduction. Let us suppose as before that the particle is actually detected by the left-apparatus at time t_1. The right-packet's amplitude is rapidly diminished during the interval t_1 to $t_1+\tau$. Although it is not particularly important from the point of view of the x-system, the right-packet's amplitude continues to diminish with characteristic time τ until time t_2, when the right-packet hits the right-apparatus and puts it into a superposition. Thereafter, because there is now an even larger discrepancy in the two macroscopic states, the right-packet continues to diminish in amplitude even faster, with characteristic time $\tau/2$. (The probability of a reversal of the result is the same as that predicted by ordinary quantum theory, $|c|^2$, and is negligibly small.) The sequence of wavefunctions describing this history is

$$\psi(t_1) = A_L{}^U A_R{}^U\{(1/\sqrt2)\psi_L(x,t_1) + (1/\sqrt2)\psi_R(x,t_1)\} \rightarrow \qquad (7.2a)$$

$$\psi(t_1+\tau_1) = A_R{}^U\{C(t_1+\tau_1)A_L{}^D\psi_L + c(t_1+\tau_1)A_L{}^U\psi_R\} \rightarrow \qquad (7.2b)$$

$$\psi(t_2+\tau_2) = C(t_2+\tau_2)A_R{}^U A_L{}^D\psi_L + c(t_2+\tau_2)A_R{}^D A_L{}^U\psi_R \qquad (7.2c)$$

By proper choices of the numerical values of the parameters in this example (a, b, t_1, t_2 and the relative velocity of the two coordinate systems), the view from the x' system will be one in which the right-packet is seen to first hit the right-apparatus at time t_1', and thereby trigger the reduction. From this point of view the amplitude of the right-packet diminishes rapidly during the time interval t_1' to $t_1'+\tau'$. It continues to diminish at the same rate until the left-packet (which is now known to be the "mostly" location of the particle) hits the left-detector, and then the reduction process proceeds even more rapidly. The sequence of wavefunctions here is

$$\psi(t_1') = A_L{}^U A_R{}^U\{(1/\sqrt2)\psi_L(x,t_1) + (1/\sqrt2)\psi_R(x,t_1)\} \rightarrow \qquad (7.3a)$$

$$\psi(t_1'+\tau_1') = A_L{}^U\{C'(t_1'+\tau_1')A_R{}^U\psi_L + c'(t_1'+\tau_1')A_R{}^D\psi_R\} \rightarrow \qquad (7.3b)$$

$$\psi(t_2'+\tau_2')=C'(t_2'+\tau_2')A_R{}^U A_L{}^D\psi_L+c'(t_2'+\tau_2')A_R{}^D A_L{}^U\psi_R \qquad (7.3c)$$

The outcome of the measurement (7.2c) or (7.3c) is independent of coordinate system[36]. Also, both observers agree that, upon each collision of a wavepacket with an apparatus, there is an enhanced reduction rate. However, the observers disagree as to the cause of the "important" reduction, where the onset of the dramatic difference between the amplitudes is first triggered. One says the cause is the collision of the left-packet with the left-apparatus (see Eq. (7.2b)). The other says it is the collision of the right-packet with the right-apparatus (see Eq. (7.3b)). So, we have a relativity of causation. This occurs because the theory is nonlocal. However, the equation of motion is local. The nonlocality resides in the probability rule, which says that the white noise tends to fluctuate in such a way as to enhance the norm of the statevector. The norm of the statevector is a nonlocal quantity.

Tails, the associated concept of "mostly localized" and the idea that a particle (but not a particle in an apparatus) can in some sense be truly in two or more places at the same time, the relativity of localization and the relativity of causation are some of the realistic notions I believe one may think of as entailed by a relativistic theory of statevector reduction.

ACKNOWLEDGMENTS

I would like to express my deep gratitude to GianCarlo Ghirardi and Renata Grassi for their substantial and ongoing contributions to this problem and to my understanding. I would also like to thank John Bell, Lajos Diosi, Nicolas Gisin, John Hannay, Roger Penrose and Abner Shimony for stimulating conversations.

REFERENCES AND REMARKS

1. E. Schrodinger, Die Naturwissenschaften **23**, 807, 823, 844 (1935).

2. D. Bohm and J. Bub, Reviews of Modern Physics **38,** 453 (1966).

3. P. Pearle, Physical Review D **13**, 857 (1976).

4. P. Pearle, International Journal of Theoretical Physics **48**, 489 (1979).

5. F. Karolyhazy, Il Nuovo Cimento **52**, 390 (1966) and in this volume;
F. Karolyhazy, A. Frenkel and B. Lukacz, in *Physics as Natural Philosophy*, edited by A. Shimony and H Feshbach (M.I.T. Press, Cambridge Mass.,1982) and in *Quantum Concepts in Space and Time*, edited by R. Penrose and C. J. Isham (Clarendon, Oxford, 1986).

6. R. Penrose in *Quantum Concepts in Space and Time*, edited by R. Penrose and C. J. Isham (Clarendon, Oxford, 1986).

7. L. Diosi, Physics Letters A **120**, 377 (1987); Physical Review A **40**, 1165 (1989).

8. P. Pearle, Physical Review A **39**, 2277 (1989).

9. P. Pearle and J. Soucek, Foundations of Physics Letters **2,** 287 (1989).

10. G. C. Ghirardi, P. Pearle and A. Rimini, Trieste preprint IC/89/44.

11. G. C. Ghirardi and A. Rimini, in this volume.

12. G. C. Ghirardi, A. Rimini and T. Weber, Physical Review D **34**, 470 (1986) and in Foundations of Physics **18**, 1, (1988).

13. J. S. Bell in *Schrodinger-Centenary celebration of a polymath*, edited by C. W. Kilmister (Cambridge University Press, Cambridge 1987).

14. J. S. Bell, Physics **1**, 195 (1965).

15. J. S. Bell in *The Ghost In The Atom*, edited by P.C.W. Davies and J.R. Brown (Cambridge University Press, Cambridge 1986).

16. L. Diosi, Journal of Physics A **21**, 2885 (1988).

17. N. Gisin, Physical Review Letters **52**, 1657 (1984).

18. P. Pearle in *Quantum Concepts in Space and Time*, edited by R. Penrose and C. J. Isham (Clarendon, Oxford, 1986).

19. P. Pearle, Physical Review D **29**, 235 (1984).

20. A. Zeilinger, R. Gaehler, C. G. Shull and W. Treimer in *Symposium on Neutron Scattering,* edited by J. Faber Jr. (Amer. Inst. of Phys., 1984); A. Zeilinger in *Quantum Concepts in Space and Time*, edited by R. Penrose and C. J. Isham (Clarendon, Oxford, 1986).

21. P. Pearle in *New Techniques in Quantum Measurement theory* , edited by D. M. Greenberger (N.Y. Acad. of Sci., N.Y., 1986). This was first brought to my attention by Y. Aharonov.

22. N. Gisin, Physical Review Letters **53**, 1776 (1984).

23. P. Pearle, Physical Review Letters **53**, 1775 (1984).

24. N. Gisin, Helvetica Physica Acta, **62**, 363 (1989).

25. G. Lindblad, Communications in Mathematical Physics **48**, 119 (1976).

26. P. Pearle, Physical Review D **33**, 2240 (1986).

27. P. Pearle, Foundations of Physics **12**, 249 (1982).

28. Eq. (3.1) is a Stratonovich stochastic differential equation, which means for our purposes that the white noise and brownian motion can be manipulated as if they were ordinary functions.

29. Eq. (3.22) is only meant to be suggestive, as B(z,t) is not an integrable function. The condition is better expressed as an inequality for the functional integral $\int_R DB \exp{-\int dz[B(z,t)-2\lambda t f(a_n-z)]^2/(2\lambda t)} > \int_R DB \exp{-c^2}$, where the subscript R denotes that the functional integral is restricted to a suitable range of B(z,t).

30. P. Pearle, Journal of Statistical Physics **41**, 719 (1985), and in references 23, 26, 21, and even as recently as reference 8. My conversion is due to discussions at Erice with Bell, Ghirardi, Gisin and Shimony, which made me realize that a realist can cheerfully survive with tails in his world picture, and a recent discussion with Penrose which helped me realize that tails are not only tolerable, but they are a necessity for boost invariance.

31. The nonlocal form $V(a,t,v) = \Sigma_n \int dx' dx g_n[\phi(x),\phi(x')] G_n(x,x',a,t,v)$, $L(a,t,v) = \Sigma_n \int dx' dx h_n[\phi(x),\phi(x')] H_n(x,x',a,t,v)$ is found not to satisfy the constraint (5.7c) unless G_n and H_n are proportional to $\delta(x-x')$, i.e. unless the form is local.

32. S. Tomonaga, Progress in Thoretical Physics **1**, 27 (1946). Eq. (5.13) contains no term describing the usual interaction of the quantum field theory because it is presumed that $\phi_{int}(x,t)$ is the solution of that theory. If one wishes, the usual term can be added and $\phi_{int}(x,t)$ taken to describe the free field.

33. S. Schweber, *An Introduction To Relativistic Quantum Field Theory* (Row Peterson, Illinois, 1961).

34. G. C. Ghirardi, R. Grassi and P. Pearle, in preparation.

35. I would like to thank Roger Penrose for urging these considerations.

36. Y. Aharanov and D. Albert, Physical Review D **21**, 3316 (1980) and Physical Review D **24**, 359 (1981) emphasize this point. However, these authors considered ordinary relativistic quantum theory, with an instantaneous reduction, and concluded that the notion of a statevector itself had to be abandoned, for it is incompatible with the additional requirements of relativity and probability conservation. Because the norm of the statevector is not constrained to be 1 in the CSL theory, there is no conflict, and no need to get rid of the notion of statevector.

THE BREAKDOWN OF THE SUPERPOSITION PRINCIPLE

F. Károlyházy

R. Eötvös University
Budapest

1. INTRODUCTION

First of all, let me express my sincere gratitude to Prof. Arthur I. Miller for his kind invitation to this course. I am happy to have the opportunity to present an approach to the problem of the reduction of the wave function which I have long since cherished but also kept feeling coy about.

There are other men to whom I am grateful. I started to work on what will be the subject of my lecture during my stay in Chapel Hill with Bryce S. DeWitt. I had completed my analysis in 1972 but I published the main results only in Hungarian (Károlyházy, 1974). The fact that in spite of that the content of this work leaked through to a wider circle of interested people I owe completely to the efforts of A. Frenkel, who is a Hungarian physicist and to Prof. A. Shimony, who, incidentally, has a Hungarian name. They kept encouraging me and gave help in every respect. It is my pleasant duty to say thanks to them.

In that early work I arrived at and exploited the idea that in the behaviour of the wave function of any system (simple or composite) the deterministic (Schrödinger equation) and the stochastic (reduction) aspects are both inherent and reveal themselves hand in hand, essentially simultaneously, with no cause to distinguish between "ordinary" and "measurement" processes and that the time evolution of the wave function can be best visualized as going through expansion—reduction cycles. During the period of expansion the wave function of the system, written somewhat symbolically $\psi(x, t)$, obeys the Schrödinger equation, whereas at the end of each expansion period it will undergo a stochastic reduction $\psi \rightarrow \psi_i$ with $\psi_i = N_i \psi(x, t) f_i(x)$ where N_i takes care of the norm and $f_i(x)$ is a "projection function" of the type

$$f_i(x) = C \cdot \exp\left(-\frac{(x - x_i)^2}{R^2}\right),$$

chosen stochastically by Nature itself, with the proper probability, out of a set $f_n(x)$, $n = 1, 2, \ldots$ of similar functions. As a consequence, superpositions of the type | cat dead ⟩ + | cat alive ⟩ cannot develop and the wave function always remains in a close connection with the actual, physically real state of the single system it describes. Thus my results have much in common with the GRW theory (Ghirardi, Rimini and Weber, 1986).

However, by no means do I claim to have worked out the beautiful formalism of the GRW theory to obtain the exact details of the effect of the perpetual reductions on the wave function in special cases. My interest lay elsewhere. I didn't want to use arbitrary parameters to prescribe how often and how drastically the positions of the various parts of the system have to be localized, instead, I sought to establish the generally valid nature and "intensity" of the reduction that can be expected to occur spontaneously in nature without injecting any new parameter.

This seems a very presumptuous program and I am ready to confess that what I have constructed is merely a heuristic model, falling short definitely of being a complete theory. Still I hope that it is worth looking into. Beside offering a generally applicable picture about how the stochastic aspect of the evolution of the wave function depends on the structure of

Sixty-Two Years of Uncertainty
Edited by A. I. Miller
Plenum Press, New York, 1990

the system under investigation (giving insight e.g., among other things, into the principal difference of the motion of an electron and a macroscopic body) the calculations based on this heuristic model suggest a new type of experiments with numerically (!) fairly definite predictions concerning tiny anomalies in the motion of macroscopic bodies. Moreover, the numerical values are such that it seems possible to confirm or to reject the existence of such anomalies with present day tools.

One should get eager about such experimental possibilities. Still, in order to get numerical estimates, I had to rely upon a model comprising admittedly speculative ideas. It was for fear that they appear less than half-baked that I was so timid to "come out into the open" with my theory.

2. THE PROBLEM OF THE SUPERPOSITION PRINCIPLE

1. After several enjoyable introductory talks at this course it seems perfectly superfluous to give a background to the problem. But one has to start his speech somehow so please let me give my version in telegram style.

The average "working view" can be stated as follows.

For an electron,

$$\varphi_1 + \varphi_2 \qquad \text{yes}$$

— because, if ψ refers to the surrounding and Ψ refers to the total system surrounding + electron, we can have $\Psi = \psi \cdot (\varphi_1 + \varphi_2)$. Interference between φ_1 and φ_2 can be pursued and it turns out that the state $\varphi_1 + \varphi_2$ is very definitely different from either the state φ_1 or the state φ_2, or, by the same token, from the state $\varphi_1 - \varphi_2$.

For a macroscopic body,

$$\varphi_1 + \varphi_2 \qquad \text{no}$$

— because then
$$\Psi = \psi_1 \varphi_1 + \psi_2 \varphi_2 \tag{1}$$

The simple superposition $\varphi_1 + \varphi_2$ cannot develop, due to the inevitable interactions with the surrounding, only the entangled superposition (1) can emerge from the Schrödinger equation. But then we can have no interference between φ_1 and φ_2, so we can regard "the state of the body" as a mixture of the states φ_1 and φ_2, that is, we can safely assume that either φ_1 or φ_2 —with the proper probability—corresponds to reality. The content of the superposition (1) wouldn't change a bit if we wrote a minus sign between its two members: the superposition principle breaks down FAPP.

This "indolent" attitude finds support in the everyday experience that a pointer of an apparatus or any other macroscopic body has a practically definite position. (There are views that deny that to be a fact.)

2. However, I think that there hides something important, namely a subconscious urge of psychological rather than logical nature behind our readiness to acquiesce to the above reasoning. To wit, when we abandon the sum in (1) in favor of one of its members, we have the illusion that we turn from a situation where a body in itself *doesn't possess a state at all* to a situation where it *has* a state (either φ_1 or φ_2). And that makes us feel better for we abhor the idea that an entangled superposition can objectively correspond to an actual, physically real state much more than the idea that a simple superposition like e.g. $\varphi_1 + \varphi_2$ for an electron in a two slit interferometer can do so. One might wonder why this is so when an electron that "doesn't know its own position" is just as far from a classical picture as can be.

I risk a thought as a possible answer, a thought which, incidentally, I sometimes find useful while teaching QM to my students.

There is an analogy between the state of the electron right behind the two slits and the mental state of some young Harry who is being attracted both by Mary and Judy but could not yet make up his mind.

It is very enlightening to discover that the "uncertain (sometimes painful) state" of Harry is as objective as any of the two possible "eigenstates" with definite commitments to either Mary or Judy. Analogies like that (even if not stated explicitly) help us to get friendly to the objective uncertainty in the electron state $\varphi_1 + \varphi_2$ (and strengthens, by the way, my personal conviction that the wave function should always stand in a one to one

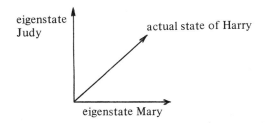

Figure 1. Mental state of Harry

correspondence with the single system it describes). (It makes fun to speculate about a "measurement" on Harry's state. A huge inheritance, subject to the condition of a prompt marriage, probably would do.)

As to entangled superpositions, it would be much more difficult to think up a similarly convincing every day analogy.

However, the illusion that "killing the monster structure (1)" yields a remnant of the form of a simple product (either $\psi_1 \varphi_1$ or $\psi_2 \varphi_2$) and thus enables the body (e.g. pointer) to possess a state and by that to come close to reality is absolutely misleading.

First: entangled superpositions are – especially nowadays, after the advent (for an early review, see J. F. Clauser and A. Shimony, 1978) of two particle correlation experiments – and have been quite common in microscopic systems, where the particles don't possess separate states of their own yet the superposition principle has proven valid. There is, therefore, no a priori reason to object to the form (1).

Second: Superpositions of this type *must in fact exist* (= describe actuality) whenever two macroscopic systems, say A and B come in contact, if QM has to have any validity at all. A state $\psi_A \cdot \psi_B$ wouldn't survive for any time whatsoever, it would immediately go over, because of single microscopic interactions, into a mess

$$\psi_{A1} \psi_{B1} + \psi_{A2} \psi_{B2} + \psi_{A3} \psi_{B3} + \ldots$$

Only, the components ψ_{A1}, ψ_{A2}, etc. wouldn't then represent "very" different states.

It remains true that we never observe such monsters involving macroscopic bodies with a "truly large" difference between its component states to stand in a one to one correspondence to reality. The conclusion seems inevitable: Nature prevents it in some specific way *not to be seen* immediately from the *form* of the wave function. The wave function, obeying solely the Schrödinger equation, detours only gradually from reality. I went to all this agony to expose the fallacy of our a priori aversion to superpositions of the form (1) to let it dawn upon ourselves that the absence of certain situations that should occur if the Schrödinger equation had unlimited validity has to be due to some definite trait in nature, not just to the intricate look of a wave function. The stochastic outcome of the typical measurements in microphysics, combined with the fact that the position of a macroscopic body always remains "classical", suggest that nature curtails the validity of the Schrödinger equation by a more or less continuous series of stochastic reductions.

But how? Along what lines? What is it that matters? Distance? Complexity? Mass? None of them in itself, to be sure.

We have Michelson interferometers with arms many meters long. Successful experiments show that the distant parts of a spatially extended wave function for an elementary particle do remain coherent (if undisturbed): large distance in itself is no sign for the breakdown of the superposition principle.

Superconductivity and superfluidity can be interpreted in terms of superpositions of very complicated components.

Finally, the "classical" motion of a body with however large mass does not only suggest that there are limitations to the superposition principle, it also proves that the superposition principle *remains valid* at least to the extent to enable the wave packet for the center of mass of the body to build up. (In the early days of QM the thought eventually came up that the motion of a body with a large enough mass M is "perfectly classical", even to the extent that the uncertainties Δx and Δv in the position and velocity fail to satisfy the relation

$$\Delta x \cdot \Delta v \geqslant \frac{\hbar}{2M}$$

but it was prompty shown that such idea would be incompatible with the two slit interference experiments.)

In short, even though the stochastic influence on the time evolution of the wave function shows drastically in the *final* outcome of some processes (Schrödinger's cat etc.), it seems to be too weak to make an easy guess at how it works.

3. To find a clue I looked around for another unsolved problem in contemporary physics and tried to find a connection. My candidate was quantum gravity. During so many years everybody came off second best when tackling this problem. In the following speculation I shall — possibly quite erroneously — adopt the view that this failure indicates that the gravitational field should not be quantized according to the usual pattern of quantum field theories, in other words, that general relativity is essentially valid in its original classical form invented by Einstein.

The possible fertility of this assumption lies in the fact that classical GR cannot be exactly valid. It could claim unlimited validity only if one ignored QM completely. In classical GR spacetime is curved, yet it has an absolutely sharp structure because the curvature depends on the mass distribution and classically there is nothing uncertain about the position or motion of the various bodies. But we don't want to ignore QM. Then we must concede at least a slight uncertainty, some sort of a smear in the structure of spacetime, to "absorb" the consequences of the inevitable quantum mechanical uncertainty in the state of the various bodies.

On the other hand, a spacetime with a slightly blurred structure — whatever that means — is apt to be generally responsible for the curtailment of the validity of the Schrödinger equation.

Of course, to say that spacetime structure is classical for all practical purposes is, in a sense, not an assumption, it is a fact. We know that it is the large masses that contribute perceptibly to the curvature of spacetime and we also know from experience that it is the large masses that "move classically", i.e. that refuse to develop appreciable uncertainties in position and velocity. Our assumption is that it couldn't be otherwise, that nature, in some (not yet fully understood) way tries to reconcile the classical Einstein equations with QM as much as possible.

How can a hazy spacetime contribute to the reduction problem?

To see this imagine an isolated body, once an electron, once a rifle bullet, moving in space and time. (A bullet, completely isolated? Ridiculous! True, but it is also true that there is nothing in the present formalism of QM that forbids us to visualize an ideal isolation for any — however complex — system. If spacetime haziness is a decisive factor, that should show up in this simple comparison.)

There is no difference in the form of the wave equation for the electron and the center of mass coordinate of the bullet. The wave function spreads out in both cases, the bullet "wants to behave" the same way as the electron. In a spacetime with a slight smear in its structure one can imagine (and will be seen explicitly in our model) at least that much that the relative phase between the distant parts of the wave function gradually (with increasing distance) becomes indefinite. To use a short term to characterize this situation we shall say that the coherence between the distant parts of the wave function gets destroyed. (What we have in mind is an "objective uncertainty" of the relative phase, not just some rapid, erratic variation of it, which would, non at the less, give a definite value every moment.)

This decay of the coherence between the parts of the extended wave function can then be interpreted as a signal for a stochastic reduction of the extended wave function to one of its smaller (coherent) parts, executed by nature itself, i.e. without reference to any observer.

Now the density of the periods of the phase of the wave function along the time axis increases with increasing mass (the time dependence of the wave function is something like $\exp(-iMc^2 t/\hbar)$), therefore it is plausible that the greater the mass the more sensitive the wave function to the imprecision in the spacetime structure. The spatial domain within which the wave function can be regarded coherent gets smaller and smaller. We may expect that for large enough masses a "classical world line" will be imitated by the propagation of the wave function for the center of mass. It will try again and again to spread out to large spatial domains but by successive reductions it will be forced to remain small. The bullet should certainly fall into this cathegory, for an electron it may turn out that the impairment of the coherence is altogether negligible. If we are able to estimate somehow the "extent of haziness" in spacetime structure, we can investigate the "transition region" between microscopic (unlimited validity of the Schrödinger equation) and macroscopic (classical world lines) motion.

However, we can investigate the uncertainty in the relative phase of the wave amplitude at two different points due to the blurred spacetime quite generally, also when the wave function of the system extends in the configuration space of its constituent particles rather than in ordinary space, i.e. also in the case of arbitrarily composite systems. (In fact, a bullet is a composite system but it turns out that its center of mass can be treated separately.) The idea that a staccato of reductions in configuration space should parallel the Schrödinger equation in the time evolution of the wave function to prevent it from spreading out in configuration space so much as to develop strongly incoherent parts, is generally applicable. This is important because there are situations, e.g. the blackening of a photographic plate at a tiny spot either here or there, due to a single photon, where no obviously macroscopic difference in the position of some "pointer" is involved, and one may wonder, whether or not the superposition survives until someone looks at the plate. (I got reassuring results in my model.)

The idea (without elaboration) that a smear of some sort in the spacetime structure can eventually wash away the phase relations between distant parts of a wave function occurred to Feynman (1962). But if we really want to capitalize on that idea we need a specific model for our spacetime.

Decay of the coherence (complete uncertainty of the relative phase) between distant parts of the wave function is not exactly the same thing, to be sure, as a stochastic reduction. It would be miraculous to have an understanding of the relation of wave propagation and spacetime as deep as to see something about the perpetual dying away of parts of the wave function in a stochastic manner. The spacetime model I have constructed can't in itself afford that. The decay of the coherence does only indicate that the situation is ripe for a reduction. We have to put it in the theory as a postulate: stochastic reductions have to occur to keep the wave function fairly coherent. However, we at least know when (how often and with what vigor) the reductions have to take place. The tendency of the wave function to spread out more and more in configuration space is practically always present, therefore we can safely say that the stochastic aspect of the time evolution is as inherent in nature as the deterministic aspect, expressed by the Schrödinger equation. Because spacetime imprecision kills only the coherence and not (stochastically) the superfluous members of the superposition, I used to be inclined to say that God sends Saint Peter to do the job. But I admit that "spontaneous localization" is a much better expression.

Accepting the decay of coherence as a guide we see that it is the large masses that are not permitted to develop appreciable uncertainties in the position. As a consequence, they don't seem to be able to add more uncertainty to the spacetime structure than was already in it. This is very satisfactory from the point of view of the hypothesis that nature, *by a slight mutual restriction of validity*, tries to reconcile QM and classical GR as much as possible.

Our task is then

a) to find a model for a spacetime with the proper amount of haziness in it,

b) to investigate the wave propagation on the smeared spacetime.

3. HAZY SPACETIME

1. When imagining a full theory of quantum gravity, one visualizes the metric tensor (the gravitational field) as a quantum mechanical dynamical variable, just like electromagnetic field in quantum electrodynamics. There would be probability amplitudes (complex numbers) for every conceivable configuration of spacetime metric + source with complicate but principally definite phase differences. That would mean an extension of the realm of the Schrödinger equation, not a restriction of it. Nothing in the formalism would, presumably, prevent large bodies from developing large uncertainties in position, accompanied by accordingly large uncertainties in the spacetime structure. If we were provided such a complete theory, then there would be no immediate justification for seeking the origin of stochasticity in uncertain metric (but at least we could calculate the extent of that uncertainty in any given situation).

But suppose nature wants to keep the metric unique, opposing somehow a free response of spacetime uncertainty to the quantum mechanical "play" in the distribution of matter. We shall make use of this idea by simply neglecting completely the quantum dynamical response of spacetime to the actual processes occurring in it. Instead, we shall put, once for all, sort of a "rigid" amount of uncertainty into it.

(This procedure is, in all probability, an over simplification. It is quite conceivable that some "extra fringes" with quantum mechanical probability amplitudes can be superimposed

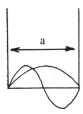

Figure 2. Pointer confined to interval a

on our spacetime with stable haziness, due to material processes. Gravitons may turn out as real as photons, but they wouldn't be expected to do much harm to our cause.)

Now, what amount of uncertainty? The only way to avoid unnecessary arbitrariness is to look for some kind of a minimum.

Let us ask the following simple question. Suppose we wish to talk about a world-line segment of length s = cT, say, along the x_0 axis of our coordinate system, in some flat space-time. With what accuracy can we implement ("realize") such a world-line segment by some-thing "graspable"?

We are, of course, not interested at all in the *practical* possibilities. Though we are talking about "realization", what we have in mind is rather a "conceptual probe" into the principal limits. Therefore we deliberately disregard any restriction on our "devices" other than those posed by the very basic features of QM and classical GR. If we can get an estimate of the imprecision Δs that we can be convinced is an irrefutable minimum, then we can also say that any spacetime that has an even sharper structure than expressed by Δs (e.g. is com-pletely classical) is an idealization that has been pushed too far. Naturally, the thought can-not be excluded that the imprecision in the spacetime structure is *greater* than the minimum obtained by thought experiment stressed to the limit. In that case, however, the conse-quences to be expected would be even more pronounced.

The most straightforward way to "build up" the desired world-line segment is to use a "clock", consisting of two bodies ("pointer" and "dial") that oscillate against each other and produce there by distinguishable configurations. Let the pointer have the mass M and be con-fined to an interval a. In order to oscillate to and fro its quantum mechanical state must be a superposition of the, say, "ground state" and "first excited state" (thinking in one dimension at this detail). The energy $E = p^2/2M$ of the pointer is uncertain ($p = h/2a$ in the ground state and twice as large in the other state), we have $h\nu = \Delta E \approx h^2/a^2 M$, ν being the frequency of the tick-tacking of the pointer.

The smallest measurable time interval is $\tau = 1/\nu \approx a^2 M/h$ and so

$$\Delta s = c\tau = \frac{a^2 Mc}{h}$$

This is also the maximal dimension ℓ of the pointer (the time for action in making the pointer oscillate is τ, it would be unreasonable to regard "distant" parts that can in no way be reacted within this time to belong to the pointer. From the above equation $\Delta E \approx hc/\Delta s$, therefore the mass uncertainty of the pointer is $\Delta M \approx h/c\Delta s$.

The mass M acts on spacetime structure. If the length of the world-line segment is s = cT when projected via light signals on a fictitious distant world-line, its length s' in the immediate vicinity of (and within) the clock is

$$s' \approx \left(1 - \frac{r_s}{2r}\right) cT,$$

where $r \approx \ell \approx \Delta s$ and $r_s = 2GM/c^2$ is the Schwarzschild radius of the clock. The uncertainty of M produces an uncertainty $\Delta s'$ in s':

$$\Delta s' \approx \frac{G}{c^2} \Delta M \frac{1}{\ell} cT \approx \frac{Gh}{c^3 \Delta s^2} s.$$

The optimum is reached when $\Delta s' \approx \Delta s$. Then

$$\Delta s^3 \approx \frac{Gh}{c^3} s = \Lambda^2 s, \qquad (2)$$

where we use the symbol Λ for the Planck length.

The relation (2) expresses the ultimate precision with which a world-line segment can be physically implemented. Note how small this limitation in the precision of s is. For $T = s/c = 1$ second, $\Delta T \approx 10^{-29}$ second! But it seems to be absolute, with no alternative: Δs cannot be reduced by increasing the uncertainty of some complementary quantity. Note also that (2) is a nonlinear relation between Δs and s and that the coefficient contains only the universal constants characterizing the two theories involved. Of course, (2) loses its meaning when the value of s approaches that of the Planck length, but we shall be concerned with much larger values of s, in fact, it turns out that when dealing with the problem of reduction we need not invade the "province of high energy particle physics" (which we couldn't do anyway), i.e. the region $cT < 10^{-13}$ cm or $T < 10^{-23}$ s.

One might state that the concentration of matter in our "clocks" is of unreasonable degree, even for the purpose of a conceptual analysis. It is indeed possible to take the view that using "material parts" is not the only permissible way to "build up" our world-line segment, that, instead, while discussing the *concept of spacetime,* we may accept the *"conceptual reality"* of a wave function. We may wish to realize the world-line segment of length s simply by the corresponding segment of the center of mass wave function of a large body. This certainly means a much more liberal attitude toward the question of what constitutes a "realization", because the width of the body can be much larger than the width of the wave function for its center of mass. In fact, we shall want to increase both the dimension and the mass of the body in order to diminish the rate of the spreading out of its wave function and to reduce the increase of its gravitational influence on the structure of spacetime. Interestingly enough, when we combine the various factors to get an optimum, we again obtain the relation (2).

Now we contend that the "rigid amount" of haziness in spacetime structure we have been talking of should be such as to reflect the imprecision expressed by (2).

2. We set up the following simple *mathematical model* for a physical, matter-free space-time domain of (essentially) Minkowskian metric.

We introduce a set or family $\{(g_{ik})_\beta\}$ of metric tensors, each $(g_{ik})_\beta$ representing a matter-free spacetime, slightly deviating from the Minkowskian spacetime:

$$(g_{ik})_\beta = \bar{\delta}_{ik} + (\gamma_{ik})_\beta, \ \Box(\gamma_{ik})_\beta = 0, \ \bar{\delta}_{00} = -\bar{\delta}_{11} = -\bar{\delta}_{22} = -\bar{\delta}_{33} = 1, \ \bar{\delta}_{ik} = 0 \text{ if } i \neq k.$$

The index β labels the various members of the family. No physical meaning will be attached to any single member of the family, only to the whole set. Only $(\gamma_{00})_\beta \equiv \gamma_\beta$ will be needed.

We choose arbitrarily a large box of volume V in three-space and develop γ_β into the Fourier series

$$\gamma_\beta(\mathbf{x}, t) = \frac{1}{\sqrt{V}} \sum_\mathbf{k} (c_\mathbf{k} e^{i(\mathbf{kx} - \omega t)} + c_\mathbf{k}^* e^{-i(\mathbf{kx} - \omega t)}) \tag{3}$$

with

$$\mathbf{k} = \frac{2\pi}{V^{1/3}} \mathbf{n}, \ (n_x, n_y, n_z \text{ integers}) \text{ and } \omega = ck,$$

the latter equation corresponding to the assumption that our spacetime is matter-free.

We identify the index β with the set of specific values for every complex coefficient $c_\mathbf{k}$. To make the definition of the family of metrics complete, we shall assign a weight function to every $c_\mathbf{k}$. We shall assume that when moving through the family, the $c_\mathbf{k}$'s will vary around the average value zero and will take on their values independently from each other. In short, we make the simple assumption that

$$\overline{c_\mathbf{k}} = 0 \ (\text{also } \overline{\prod_\mathbf{k} c_\mathbf{k}} = 0), \ \overline{c_\mathbf{k} c_{\mathbf{k}'}^*} = 0 \text{ for } \mathbf{k} \neq \mathbf{k}', \text{ and } \overline{|c_\mathbf{k}|^2} = F(k), \tag{4}$$

F(k) yet to be determined, and the bar denoting average over the whole family.

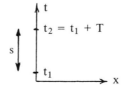

Figure 3. World-line segment

221

The "physical" world-line segment between the time coordinates t_1 and $t_2 = t_1 + T$ along the x_0 axis is represented by the set $\{s_\beta\}$ with

$$s_\beta = \int \left[(g_{ik})_\beta \frac{dx^i}{dt} \frac{dx^k}{dt}\right]^{\frac{1}{2}} dt = cT + \frac{c}{2} \int_{t_1}^{t_2} \gamma_\beta \, dt. \tag{5}$$

Obviously, $\overline{s_\beta} = cT$. We *define* the uncertainty in the length $s \equiv \overline{s_\beta}$ by $\Delta s^2 \equiv \overline{(s - s_\beta)^2}$. Using (4) we have

$$\Delta s^2 = \frac{1}{8\pi^3} \int dk \frac{F(k)}{\omega^2} (1 - \cos \omega T), \tag{6}$$

and it is easily shown that (2) is reproduced if we make the choice

$$F(k) = \Lambda^{\frac{4}{3}} k^{-\frac{5}{3}}. \tag{7}$$

Our derivation of the relation between s and Δs remains valid if the world-line segment correspond to a slow motion ($v \ll c$) in our coordinate system. This is not a Lorentz-invariant restriction (neither is the condition that the deviation of the components of the metric tensor from the Minkowskian values should be small). There is a nonrelativistic aspect smuggled in into our model. But it doesn't seem preposterous to assume that the haziness of spacetime is somehow connected with the matter distribution of the universe. We shall assume that we are in a coordinate system in which all macroscopic bodies move slowly.

Thus, a *single physical spacetime domain,* with a rigid amount of smear in its structure, is represented in our mathematical model by a bundle of spacetimes, all of them having a sharp structure but differing slightly from each other. We repeat that no physical significance should be attached to the single "leaves" of the bundle. They have no "dynamical status" at all. The only justification for their introduction is that they invite a very easy treatment of the wave propagation. (Perhaps a last remark here: one may be tempted to regard the c_k's as stochastic variables, representing some stochastic disturbance on the spacetime structure of as yet unknown origin. Though such view is harmless "practically"—averages have to be calculated anyway—it would be a misinterpretation in our case. What we have in mind is not a complicate, but an *uncertain* spacetime.) For the weight function $P(c_k)$ for the coefficients c_k we shall make the simplest possible choice:

$$P(c_k) = \frac{l_k^2}{\pi} \exp(-l_k^2 |c_k|^2) \quad \text{with} \quad l_k = \Lambda^{-\frac{2}{3}} k^{\frac{5}{6}}, \tag{8}$$

with a cutoff $c_k = 0$ if $k \gg 10^{13}$ cm^{-1}, the details of which are unimportant.

4. WAVE PROPAGATION

The most natural way to treat wave propagation on our hazy spacetime is to introduce a family $\{\psi_\beta\}$ of wave functions, ψ_β corresponding to the metric $(g_{ik})_\beta$. We want to deal with many particle Schrödinger equations and it is not immediately clear how to generalize them for non-Minkowskian metrics, but we can strike out the following roundabout way. For a single (say, scalar) elementary particle with mass m we have the general relativistic wave equation

$$\frac{1}{\sqrt{-g}} (\sqrt{-g}\, g^{ik} \varphi_{,i})_{,k} - \left(\frac{mc}{\hbar}\right)^2 \varphi = 0.$$

This gives in the case of slow motion and nearly Minkowskian metric, after factorizing out $\exp(-imc^2 t/\hbar)$, the nonrelativistic approximation

$$i\hbar\dot{\psi}_\beta(\mathbf{x}, t) = \left[-\frac{\hbar^2}{2m} \Delta + V_\beta\right]\psi_\beta(\mathbf{x}, t), \tag{9}$$

where the small "perturbation" V_β is given by

$$V_\beta(\mathbf{x}, t) = mc^2[g_{00}(\mathbf{x}, t))_\beta - 1] = mc^2\gamma_\beta(\mathbf{x}, t). \tag{10}$$

Now the generalization into an N particle nonrelativistic Schrödinger equation is straightforward. We shall have for every β the equation

$$i\hbar\dot{\Psi}_\beta(x_i, t) = [H + U_\beta(x_i, t)]\Psi_\beta, \tag{11}$$

with, for example,

$$H = \sum_{i=1}^{N} -\frac{\hbar^2}{2mi}\Delta_i + \sum_{i \neq k} V_{ik} \tag{12}$$

and

$$U_\beta(x_i, t) = c^2 \sum_{i=1}^{N} m_i\gamma_\beta(x_i, t). \tag{13}$$

The particles may be the nucleons and electrons of a body, or the molecules of a gas, depending on the nature of the system and on the approximation used. The V_{ik} are ordinary interaction energies between pairs of particles (and are supposed to be, naturally, much smaller than the energies belonging to the rest mass of the constituent particles). The effect of the slight deviation from the Minkowskian metric is the appearance of the small perturbation U_β in the wave equation. From eqn (3) we see that U_β oscillates rapidly in time.

Now comes the decisive step. Although it would be very difficult to find the solution $\Psi(x_i, t)$ of an "ordinary" N particle Schrödinger equation (i.e. without the perturbation U_β) explicitly, we can tell how such a solution will be modified by the perturbation U_β. As a solution for eqn (11) we shall simply write

$$\Psi_\beta(x_i, t) = \Psi(x_i, t)\exp[i\phi_\beta(x_i, t)] \tag{14}$$

with

$$\phi_\beta(x_i, t) = -\frac{1}{\hbar}\int^t U_\beta(x_i, t')\,dt'. \tag{15}$$

(The perturbation (13) is switched on adiabatically.)

Though the smallness and the rapidly oscillating character of U_β is not in itself sufficient to make (14) a necessarily good approximation, it can be shown that we are entitled to use the solution (14) for our purposes in all cases of practical interest. (In writing down (14) we have dropped a small phase term which doesn't depend appreciably on the coordinates x_i, and, furthermore, we have made use of the mild assumption that $\Psi(x_i, t)$ doesn't change unreasonably wildly with x_i.)

Using (3) and (13) in (15) we can write down $\phi_\beta(x_i, t)$ explicitly:

$$\phi_\beta(x_i, t) = -\frac{1}{\hbar}\sum_{i=1}^{N} m_i c^2 \sum_k \frac{1}{\sqrt{V}}\left(\frac{c_k}{-i\omega}e^{i(kx_i - \omega t)} + \frac{c_k^*}{i\omega}e^{-i(kx_i - \omega t)}\right) \tag{16}$$

(Excuse me that $i = \sqrt{-1}$ also appears in the equation, but I hope it cannot be confused with the summation index.)

At this stage we can even redefine the c_k's at every moment t so as to absorb the factor $i\exp(-i\omega t)$ and use the time independent phase function

$$\phi_\beta'(x_i) = -\frac{1}{\hbar}\sum_{i=1}^{N} m_i c \sum_k \frac{1}{\sqrt{V}}k^{-1}(c_k'\exp(ikx_i) + c_k'^*\exp(-ikx_i) \tag{17}$$

The family $\{\phi_\beta'(x_i)\}$ is identical with the family $\{\phi_\beta(x_i, t)\}$, therefore, from now on, we shall drop the apostrophes in (17) and use it in place of (16).

Thus, "solving the Schrödinger equation on our hazy spacetime" simply means that the "ordinary" solution $\Psi(x_i, t)$ goes over into the bundle $\{\Psi_\beta(x_i, t)\}$ of the functions $\Psi_\beta(x_i, t) = \Psi(x_i, t)\exp[i\phi_\beta(x_i)]$, differing from each other only in a phase factor. If the additional phase

$\phi_\beta(x_i)$ wouldn't depend on x_i (only on β), this change wouldn't mean a thing, the bundle would correspond to a single ray in Hilbert space. But because ϕ_β depends on x_i, we can say that the bundle is the mathematical model of a „*single physical wave function*" with a smaller or larger uncertainty of the *relative phase* between two points (x_i) and (x_i') of the configuration space. (We don't attach any physical meaning to a single member of the mathematical set $\{\Psi_\beta\}$.)

The average of the excess phase $\phi_\beta(x_i)$ over β (i.e. over the c_k's) is obviously zero in every point of the configuration space. Therefore, the relevant quantity is

$$\overline{\Delta \phi_\beta^2} = \overline{[\phi_\beta(x_i) - \phi_\beta(x_i')]^2} . \tag{18}$$

This quantity suggests the notion of *coherence distance* in configuration space. The postulate of the successive reductions, whatever specific form we choose for it, should "follow the lead" of this expression: it has to have the tendency to trim the wave function $\Psi(x_i, t)$ in such a way as to prevent it from developing parts with an unreasonably large coherence distance between them.

However, before turning to the question of how to formulate the postulate for the reduction reasonably, let us orient ourselves about the general nature of the difference between two situations giving rise to a large uncertainty in the relative phase.

Let us introduce the quantities

$$q_k = \sum_{i=1}^N m_i e^{ikx_i}$$

for every k. These quantities are essentially the Fourier coefficients of the mass distribution of the system in the configuration given by x_i. From (17) we can express the phase ϕ_β by these quantities:

$$\phi_\beta(q_k) = - \frac{c}{\hbar} \sum_k \frac{1}{\sqrt{V}} \, k^{-1} (c_k q_k + c_k^* q_k^*). \tag{19}$$

If the two configurations to be compared correspond to mass distributions described by the Fourier coefficients q_k and $q_k + a_k$, then, using (4) and (7), we see that

$$\overline{[\phi_\beta(q_k) - \phi_\beta(q_k + a_k)]^2} = \frac{1}{V} \sum_k \frac{|a_k|^2}{\lambda_k^2} \tag{20}$$

with

$$\lambda_k = \frac{\hbar}{c} \Lambda^{-\frac{2}{3}} k^{\frac{11}{6}} .$$

In other words: *it is just the difference in the mass distribution, corresponding to the configuration* (x_i) *and* (x_i') *that matters* in the question of the uncertainty in the relative phase between the two points (x_i) and (x_i') of the configuration space. $\Delta \phi_\beta^2$ is a simple quadratic expression of the difference in the Fourier coefficients of the density, containing no free parameter. (This feature of the uncertainty of the relative phase is very convenient when discussing the behaviour of gases or systems which consist partly of gases.)

Thus we see the general nature of the "decay of the coherence" between the "distant parts" of the wave function, to use our earlier expression. However, the special case of the (nearly) free motion of a single body is of more interest from the point of view that it offers experimental possibilities. It is also the easiest case to see the stochastic aspect enter the wave propagation. We look at first at this issue.

5. FREE MOTION

1. We investigate the free motion of a macroscopic body, for example a solid ball of mass M, radius R, volume Ω and density ρ.

At first we shall assume that the ball is completely isolated, in spite of the fact that this is a practical impossibility. The results will provide an a posteriori justification for this assumption. It will turn out that the smear in the spacetime structure has, if M is large enough, a bigger effect on the motion of the ball, than, say, some gas around it under customary circumstances.

When dealing with the free motion of a massive body as a whole, one usually doesn't hesitate to separate the center of mass coordinate from the other degrees of freedom and associate the whole mass with the former. This cannot be done without circumspection in our case because the small "gravitational perturbation" resulting in the extra phase (17) is picked up individually by the, say, N atoms of the body. Strictly speaking, one should investigate the phase in the whole configuration space. However, if the wave function $\Psi(x_i, t)$ describes a state of the ball that is compatible with the assumption that the ball is in thermal equilibrium, then for all configurations (x_i) that deviate appreciably from a regular arrangement of the x_i (corresponding to the crystal structure) around the center of mass coordinate x, the value of Ψ (and, of course, all Ψ_β) is zero because of the binding forces. That means that we need $\phi_\beta(x_i)$ only for that regular arrangements of x_i. Then the sum in (17) can be converted into an integral over a volume Ω around x, with the result

$$\phi_\beta(x) = -\frac{Mc}{h}\frac{1}{\sqrt{V}}\sum_k d(k)k^{-1}[c_k \exp(ikx) + c_k^* \exp(-ikx)], \tag{21}$$

where

$$d(k) = 3\left(-\frac{\cos kR}{(kR)^2} + \frac{\sin kR}{(kR)^3}\right). \tag{22}$$

Now we are entitled to separate the center of mass coordinate x from the internal degrees of freedom, because ϕ_β depends – apart from β, i.e. the set of c_k's – only on x, and discuss the "physical" wave function

$$\{\psi_\beta(x, t)\} = \{\psi(x, t)\exp(i\phi_\beta(x))\} \tag{23}$$

of the center of mass. $\psi(x, t)$ will spread out in space and the set of the ϕ_β's will provide the uncertainty in the relative phase between the points x and x'. From (21) and (22) we see that, on the one hand, the mass has a prominent role, as expected, on the other hand, the effect of the spacetime uncertainty has a tendency (mirrored by the factor $d(k)$) to average out over extended bodies.

If we want to trace the decay of the coherence of the physical wave function we have to calculate the spread of the relative phase between the points x and $x' = x + a$. We use (4) and (7), then we set

$$\frac{1}{V}\sum_k = \frac{1}{(2\pi)^3}\int d^3k$$

and integrate over the directions of k to get

$$\overline{\Delta\phi_\beta^2} = \overline{[\phi_\beta(x') - \phi_\beta(x)]^2} \approx \Lambda^{\frac{4}{3}}\left(\frac{Mc}{h}\right)^2\int_0^\infty dk\, k^{-\frac{5}{3}}d^2(k)\left(1 - \frac{\sin ka}{ka}\right). \tag{24}$$

(It has been written out to see the irrelevance of the cutoff in c_k.)

The uncertainty in the relative phase between x' and x increases with increasing a. At a critical value a_c the relative phase uncertainty will reach the value π. Until the spatial extension of the wave function remains smaller or equal to a_c, the bundle $\{\psi_\beta\}$ may be thought of as representing, at least approximately, a single ray in Hilbert space. We may call a_c coherence length and a domain in space of the linear dimension a_c a coherence domain. The simplest and most straightforward way to introduce the postulate for reductions in this type of wave motion is the following prescription of expansion—reduction cycles. Let us start with a moment when the wave function is confined to only one coherence domain. Let $\{\psi_\beta\}$ obey the Schrödinger equation until it develops incoherent parts with comparable weight, i.e. until it spreads out to cover two (thinking in one dimension) neighboring coherence domains. Then let us perform a stochastic reduction (with the corresponding weight) to one of the coherent parts so that $\{\psi_\beta\}$ will again be confined to a single coherence domain.

Apart from the "breathing" between the values a_c and $2a_c$, the width of the wave function always remains close to a_c. If a_c is very small, then, as time passes, the wave function

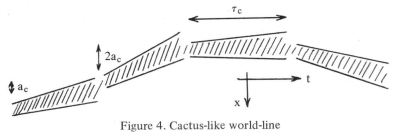

Figure 4. Cactus-like world-line

draws a "world-line" on spacetime. The time τ_c necessary for the wave function to spread from the width a_c to $2a_c$ is easily seen to be of the approximate value

$$\tau_c \approx \frac{Ma_c^2}{\hbar}.$$ (25)

This is the duration of a cycle. The uncertainty in the momentum remains practically the same through out, but the mean value of the momentum makes a little stochastic jump at the end of each cycle of the order

$$|\Delta p_c| \approx \frac{\hbar}{a_c}.$$ (26)

(The fore part of a spreading wave function "contains" a little more momentum than the hind part.) Because of the small kicks (26), the world tube for the wave function will show a cactus-like shape in spacetime.

If we put π^2 on the left hand side of equation (24), we can solve it for a_c. The formula for a_c depends on whether a_c turns out to be larger or smaller than the radius of the ball. We find

$$a_c \approx \frac{1}{\Lambda^2} \left(\frac{\hbar}{Mc}\right)^3 = \frac{\hbar^2}{G} \frac{1}{M^3} \quad \text{if } a_c \gg R,$$ (27a)

$$a_c \approx \left(\frac{R}{\Lambda}\right)^{\frac{2}{3}} \frac{\hbar}{Mc} = \left(\frac{\hbar^2}{G}\right)^{\frac{1}{3}} \frac{R^{\frac{2}{3}}}{M} \quad \text{if } a_c \ll R.$$ (27b)

It is not without some solemnity that we take a closer look at these relations. We see that if the body is some tiny grain so that the wave function of its center of mass is permitted to spread out to much larger dimensions than the dimension of the body itself—a typical habit of atomic particles—then the extension R of the body doesn't enter the expression for its coherence length at all, only its mass and the coherence length increases very rapidly with decreasing mass. Incidentally, we may ask what (27a) would give for an elementary particle. For an electron, using also (25), we obtain

$$a_c \approx 10^{35} \text{ cm}, \quad \tau_c \approx 10^{70} \text{s},$$ (28)

for a proton, the corresponding values would be $\approx 10^{25}$ cm and $\approx 10^{53}$s. That suggests that for systems consisting of only one or a few microparticles spacetime haziness remains unnoticed, the superposition principle is not endangered and there is no hope to observe its breakdown in split beams interference experiments, or in two-particle correlation experiments involving large spatial separation between the particles. Needless to say, such experiments become the more important. The beautiful results showing that Bell's inequality is violated in such experiments (Aspect et al, 1982) corroborates our view that "distance in itself is not enough".

For an "ordinary" ball of $R \approx 1$ cm and normal density it turns out that $a_c \ll R$ and (27b) yields

$$a_c \approx 10^{-16} \text{ cm}, \quad \text{with } \tau_c \approx 10^{-4}\text{s},$$ (29)

the value of τ_c coming again from (25).

Hazy spacetime doesn't allow the center of mass of the ball to develop an uncertainty larger than $\approx 10^{-16}$ cm. For all practical purposes, the center of mass seems to be a classical point as expected. However, let me point out right away that *the interference of the stochastic aspect with the causal evolution* induced by the Schrödinger equation is *very mild even in this case*. Remember that the wave length of the wave function of the center of mass of our ball would be only $\approx 10^{-20}$ cm even "at rest", i.e. when moving only with the evanescent velocity of its thermal agitation! (29) means that the wave function can more or less freely spread out over a domain accommodating 10^4 wave lengths, which is probably more than was ever drawn in any text book on QM. (Also the number of "unperturbed" time oscillations of the wave function during the time interval τ_c, as estimated with the help of the term $\exp(-iEt/\hbar)$ is reassuringly large.) Stochasticity is very magnanimous to the Schrödinger equation indeed. Still, spacetime imprecision seems to be mounted properly to prevent situations where a really drastical *collapse* of the wave function would be needed. The term collapse becomes an exaggeration, the series of little stochastic jumps should be regarded as a model to follow the suggestion made by the decay of the coherence of the physical wave function. The jumps can even be replaced by a continuous stochastic process as we have learned from Gisin (1984), Caves and Milburn (1987), Diósi (1989), Ghirardi, Pearle and Rimini (1989), Ghirardi, Grassi and Rimini (1989).

It is tempting to see in (27) a natural division of the various grains and balls into three categories. We may say that for $a_c \gg R$ microbehaviour dominates, for $a_c \ll R$ macrobehaviour prevails. $a_c \approx R$ corresponds to the "transition region" between the microscopic and macroscopic behaviour. (These relation correspond to

$$\frac{\hbar^2}{G} \gg M^3 R, \quad \frac{\hbar^2}{G} \ll M^3 R \quad \text{and} \quad \frac{\hbar^2}{G} \approx M^3 R \tag{30}$$

as was pointed out to me by B. Lukács (see Károlyházy and al., 1982)). We find (assuming customary densities) for the transition region the values

$$a_c \approx R \approx 10^{-5} \text{ cm}, \quad \tau_c \approx 10^3 \text{s}, \quad M \approx 10^{-14} \text{g}. \tag{31}$$

This value of radius and mass corresponds to a colloidal grain, containing about 10^9 molecules.

While of theoretical interest, the actual significance of this division is lessened by the circumstance that the effect of the surrounding — especially for masses in the transition region — can't very well be neglected.

6. EXPERIMENTAL POSSIBILITIES

That brings us to the question of experimental possibilities. That was my main motivation all along but now I shall confine myself to a few remarks. (For a more detailed discussion of one of the several types of experiments that suggest themselves, see the former reference.)

The little stochastic changes given by (26) in the mean value of the momentum occurring once in every time interval τ_c causes the kinetic energy to creep up slowly, by an amount of $\Delta E_c \approx \hbar^2/Ma_c^2$ per cycle. Such violation of the conservation laws is not appalling in itself in a hazy spacetime and also its extent is harmless from the point of view of everyday observation. Our ball of $R \approx 1$ cm would, for example, collect only the tiny amount $\approx 10^{-3}$ erg of energy from spacetime even if it remained completely isolated during the whole life of the universe. Practically, there is always a damping force $F = -\kappa \dot{x}$ (e.g. the ball moves in some gas). The motion of the ball will be "pumped up" to an average velocity v_a at which the gain ΔE_c in time τ_c equals the dissipation $F\dot{x}\tau_c$. Then

$$\frac{\hbar^2}{Ma_c^2} \approx \kappa v_a^2 \tau_c, \quad \text{i.e.} \quad v_a^2 \approx \frac{\hbar^2}{Ma_c^2 \kappa \tau_c}. \tag{32}$$

The velocity v_a has to be compared with the thermal velocity v_t, satisfying

$$Mv_t^2 \approx kT. \tag{33}$$

It turns out that in ordinary air of room temperature, for a ball with density $\rho \approx 1 \text{g}/\text{cm}^3$ and radius $R_{crit} \approx 0,1$ cm,

$$v_t \approx v_a. \tag{34}$$

For smaller R, v_t (= normal Brownian motion) dominates over v_a. For $R > R_{crit}$, $v_y > v_t$, i.e. the *anomalous Brownian motion* is larger. With increasing mass the anomalous Brownian motion predicted by our theory will surpass the normal one by many orders of magnitude. (This supplies the a posteriori justification for first neglecting the effect of the surrounding gas on the heavy ball.) In a given surrounding v_t decreases rapidly, v_a increases slowly with the mass M. However, it would be of no advantage to increase the mass of a body too much in order to detect the tiny anomaly in its motion. It takes time for the anomalies to build up, the longer time the larger mass. One has to wait long between observations and one has to keep the body during all these times practically under the *sole* influence of, say, the surrounding gas. Therefore, bodies with a radius of ≈ 1 cm seem to be the best candidates. The ball might be suspended somehow (if by a thin thread, then microfriction is forbidden within the thread) or put aboard a spacecraft. In a pendulum-type experiment anomalous elongations of the order of 10^{-5} cm can readily be expected (using balls of a few cm-s of diameter). Microseizmic disturbances have to be excluded for several hours but this doesn't seem too difficult. There are other sources of unwanted perturbations but all in all the experimental check of our theory looks feasible.

I have coined the term "anomalous Brownian motion" for this phenomenon in my early paper, because I wanted to emphasize that complete isolation from the surrounding is neither possible nor necessary.

Anomalous Brownian motion manages to be a good candidate to expose the breakdown of the superposition principle because for a sufficiently isolated solid body

a) the tiny energy-increasing effect of the perpetual stochastic "pinching" of the wave function concentrates on a single coordinate that "can be seen" (we couldn't do much with a similarly small oscillation of the center of mass of a certain amount of gas in a container),

b) the tiny gains in energy can accumulate, for the damping force acting on a body in a gas is proportional to the velocity of the body and can therefore be very small indeed.

One may come across the idea that some similar effects might be expected in various types of extended systems other than just nearly free solid bodies. But in all situations I could think up (ensemble of superconducting electrons or superfluid He atoms, phase transitions, etc.) either the wave function doesn't have the tendency to develop parts corresponding to appreciably different mass distributions at all (like in the case of superconducting electrons) or the eventual effects of the successive reductions are superseded by far by a mess of interactions between the various parts of the system.

7. DECAY OF SUBMACROSCOPIC SUPERPOSITIONS

The procedure described in Chapter 5 to "take the hint" of the decaying coherence and set the rule for the stochastic reductions is somewhat crude. We can make it a little more formal, without, of course, changing the basic idea.

I go into this shortly because I want to say a few words about what we can term "the breakdown of the superposition principle on a submacroscopic level."

We saw that uncertainty in the relative phase is connected with the difference in the density distribution, generally, and that, specifically, a very small ambiguity in the position of a big body already represents a large enough difference in the density distribution for a reduction to take effect. But what about superpositions involving only very slight differences in the mass distribution?

Originally, I investigated the hypothetical situation in which we have a gas in a container and a very small droplet of water (consisting eventually of a few hundred molecules only) that doesn't know its own position. That is, I assumed that the initial wave function of the total system gas + droplet is a distinct superposition |droplet here⟩ and |droplet there⟩, the difference between "here" and "there" being ≈ 1 cm. This distance is much smaller than a_c for the droplet, because of the very small mass of the latter, therefore, if the droplet were alone, the superposition would survive indefinitely. (Its components would overlap, though.) I was able to show that the thermal noise in the surrounding gas is sufficient to kill the superposition (of the total system!) within ≈ 1s. This is so because the gas keeps developing "intolerably" large differences in its density distribution and that triggers reductions in the mass distribution of the *total system,* affecting slowly (and stochastically) the relative weight of the states |droplet here⟩ and |droplet there⟩ in the surviving wave function. After a long enough time we have either the one or the other definite situation and the probability for either outcome could be shown to equal the initial weight of the corresponding state in the superposition. (See in this context the beautiful presentation of the "gambler's ruin" by Pearle, 1984).

Here I shall take a glimpse at the following "didactical" situation.

Let us have a tiny grain (and nothing else) in the state

$$\varphi(x_g) = c_1\varphi_1(x_g) + c_2\varphi_2(x_g) \tag{35}$$

where φ_1 and φ_2 are narrow states, separated, say, by a few millimetres, this distance being much smaller than the coherence length for the grain. (I am thinking in one dimensional motion and, moreover, try to drop the habit of referring to the "correct" wave function $\{\varphi_\beta\}$ as much as possible.) Now suppose we put a massive ball (say, R ≈ 1 cm) in the close vicinity of the grain. (There should be no tangible interaction between them.) Will the ball "teach the grain its own position"?

As an introduction to this problem, let us first return to Chapter 5 and resume talking about a single isolated body, placed near the origin of the coordinate system. We shall suppress two out of the three coordinates of the center of mass. Because $a_c \ll R$, we can regard $\phi_\beta(x)$ as being linear in x in the small domain $\psi(x, t)$ occupies. We can drop $\phi_\beta(0)$ and replace the coefficient of x, containing the set of the c_k's by a single real variable c_β so that in place of (21) and (23) we can simply write

$$\phi_\beta(x) = c_\beta x \quad \text{and} \quad \psi_\beta(x, t) = \psi(x, t) \exp(ic_\beta x). \tag{36}$$

Inspection of (21) shows that c_β has a Gaussian weight function with $\overline{c_\beta} = 0$ and (readjusting slightly the definition of the coherence length a_c by a numerical factor)

$$\overline{c_\beta^2} = \frac{1}{a_c^2}. \tag{37}$$

To assess the "quality of the coherence" of the "physical wave function" $\{\psi_\beta(x, t)\}$, we introduce the time dependent matrix

$$\rho(x, x', t) = \psi(x, t)\psi^*(x', t) \overline{\exp[ic_\beta(x - x')]} = \psi(x, t)\psi^*(x', t) \exp\left[-\frac{(x - x')^2}{2a_c^2}\right] \tag{38}$$

(which I could say is the average of $\langle x | \psi_\beta \rangle \langle \psi_\beta | x' \rangle$, but I don't want to go into questions of definition), and form the trace of its square

$$D(t) \equiv \text{Tr } \rho^2 = \int dx\, dx'\, |\psi(x, t)|^2\, |\psi(x', t)|^2 \exp\left[-\frac{(x - x')^2}{a_c^2}\right]. \tag{39}$$

It's no use trying to call ρ anything else than density matrix, even though that name is rather misleading in our case. For permanently isolated systems the equation Tr $\rho^2 < 1$, in everyday thinking, is connected with something *we* don't know, in our case, Tr $\rho^2 < 1$ reflects something *nature* doesn't know (namely, the exact relative phases between various points). Because of the hazy spacetime the "quality of coherence" of a wave function is never perfect, not even when the extension of the wave function is being tolerated by nature.

When $\psi(x, t)$ is spreading out, the coherence of $\{\psi_\beta\}$ deteriorates, D(t) decreases.

Let ψ evolve obeying the ordinary Schrödinger equation for a short, but otherwise arbitrary time interval τ and let Δ_{det} denote the change of D(t) during this time. We write simply $\psi(x)$ for $\psi(x, t + \tau)$.

We may try to counteract this deterioration by prescribing a stochastic reduction at the moment $t + \tau$, relying on a series of "pinching functions"

$$f_i(x) = C \exp\left[-\alpha\frac{(x - x_i)^2}{a_c^2}\right], \tag{40}$$

i.e., relying on functions that contain some small parameter α but are, apart from that, *sensitive to configurations* (x, x') *exactly in the same proportion as the latter are damaging to the coherence*. The parameter α must be small if τ is small, otherwise the "deterministic aspect" in the wave propagation would get wiped out. (The requirements concerning the set x_i of centers and the value of C are rather trivial, the point being that the equation

$$\sideset{}{'}\sum_{i=-\infty}^{+\infty} f_i^2(x) = 1 \quad \text{for all} \quad x$$

should be satisfied.) The question is just what value of α is reasonable.

Now the *"degree of interference"* with the deterministic time evolution at the moment of the occurrence of the reduction

$$\psi(x) \to \psi_i(x) = N_i \psi(x) f_i(x) \tag{41}$$

to a $\psi_i(x)$ with a particular value of i is best estimated with the help of the "auxiliary quantity"

$$\rho_{\text{aux}} = \sideset{}{'}\sum_i |\psi(x)f_i(x)\rangle \langle \psi(x)f_i(x)| \tag{42}$$

by forming the expression

$$\Delta_{\text{sto}} = \text{Tr}\,\rho_{\text{aux}}^2 - 1. \tag{43}$$

This is a tricky thing and it should be properly understood. Through the actual transition $\psi \to$ some definite ψ_i the coherence of $\{\psi_\beta\}$ improves, $D(t + \tau + 0)$ is closer to unity than $D(t + \tau)$. But the *unpredictability* of the outcome of the reduction — and that is meant by interference with the deterministic evolution — is expressed by (43).

Δ_{sto} depends on α. The only sound partner α for any choice of τ is that that makes $\Delta_{\text{det}} \approx \Delta_{\text{sto}}$. That is the most reasonable way to *accept the hint of the decaying coherence to take steps to prevent it.* $|\Delta_{\text{sto}}| < |\Delta_{\text{det}}|$ would mean "negligence" in saving the coherence, $|\Delta_{\text{sto}}| > |\Delta_{\text{det}}|$ would mean that we do more harm than good to the "ideal" time evolution.

Pursuing this line further in the case of a single body, incidentally, we would reproduce the results of Chapters 5 and 6. Instead, we return to our grain and ball.

We go through exactly the same steps as with the single ball, only, in place of $\psi(x, t)$ we shall have

$$\Psi(x_b, x_g, t), \quad \text{with} \quad \Psi(x_b, x_g, 0) = \psi(x_b)[c_1\varphi_1(x_g) + c_2\varphi_2(x_g)]$$

The crucial point is that $\phi_\beta(x_b, x_g)$ turns out, as is readily seen from (21), to depend, to a good approximation, *only on the center of mass coordinate X of the total system*. The factor responsible for the deterioration of the coherence (c.f. equation (39)) will be, as a consequence, of the form $\approx \exp[-(X - X')^2/a_c^2]$ and the same expression will appear in the localization functions f_i. In other words, the reductions will go via the center of mass of the total system. The reductions will be triggered by the quick spreading out of the massive ball, but they will affect also the grain. For a grain with a mass of $\approx 10^{-17}$g the superposition of the said kind survives only for ≈ 1s (instead of $\approx 10^6$s) in this didactical example.

8. CONCLUDING REMARKS

It is a widely accepted view that if the breakdown of the superposition principle is bound somehow to macroscopic systems, then the search for its exact whereabouts is hopeless because of the irreproducibility and complexity of the wave functions involved. In the foregoing I wanted to argue that the opposite may be true.

The intriguing paradoxon of the overwhelming successes of QM on the one hand and the thing on the other hand that "it makes absolutely no sense" (Penrose, 1986) chases us to look for liberators under the face of reduction postulates. However, without something to connect them with, they play a role somewhat similar to that of the Newtonian absolute space.

Hazy spacetime did look attractive to me as a candidate for a connection. There is no doubt that if in the future our present day spacetime picture turns out to be some sort of a phenomenological approximation, to day we are far from a real insight into that.

What, then, can be the excuse for a model as crude as the one just presented? Well, the bonus, perhaps, of its naivety is that while seeking estimates, it deviates as little as possible from well-established relations.

REFERENCES

Aspect, A., Dalibard, J. and Roger, G., 1982, *Phys. Rev. Lett.,* 49:1804.

Caves, C. M. and Milburn, G. J., 1987, *Phys. Rev. A,* 36:5543.

Clauser, J. F. and Shimony, A., 1978, *Rep. Prog. Phys.,* 41:1881.

Diósi, L., 1989, *Phys. Rev. A,* 40:1165.

Ghirardi, G. C., Grassi, R. and Rimini, A., 1989, preprint IC/89/105.

Ghirardi, G. C., Pearle, Ph. and Rimini, A., 1989, preprint IC/89/44.

Ghirardi, G. C., Rimini, A. and Weber, T., 1986, *Phys. Rev. D,* 34:470.

Gisin, N., 1984, *Phys. Rev. Lett.,* 52:1657.

Károlyházy, F., 1974, *Magyar Fizikai Folyóirat,* 12:24.

Károlyházy, F., Frenkel, A. and Lukács, B., 1982, On the possibility of observing the eventual breakdown of the superposition principle, *in:* "Physics as natural philosophy, essays in honor of László Tisza on his seventy-fifth birthday", A. Shimony and H. Feshbach, ed., MIT press, Cambridge.

Pearle, Ph., 1984, *Phys. Rev. D,* 29:235.

Penrose, R., 1986, Gravity and state vector reduction, *in:* "Quantum Concepts in Space and Time", R. Penrose and C. J. Isham, ed., Oxford University Press, New York.

MATHEMATICAL AND PHILOSOPHICAL QUESTIONS IN THE THEORY OF OPEN AND MACROSCOPIC QUANTUM SYSTEMS

Hans Primas

Laboratory of Physical Chemistry
ETH-Zentrum
CH-8092 Zürich
Switzerland

1. INTRODUCTION

It is a curious fact that our popular text books and most discussions of conceptual and philosophical problems of quantum mechanics are still based on the very first attempts to formalize this theory which have been worked out more than 50 years ago. These formulations of nonrelativistic quantum theory do, however, not represent the state of the art.

On the basis of the historical codifications of quantum mechanics one may be tempted to presume that it makes no sense to speak of individual quantum objects and that a realistic interpretation of quantum mechanics amounts to the introduction of hidden variables. Fortunately, such a conclusion is premature, no changes in the original ideas of the pioneers of quantum theory are necessary. All we have to do is to work out the mathematical formalization properly, to give up our preconceptions about the localizability of matter, of the separability of the objects of scientific inquiry, and the associated ontology of classical physics.

We know today that the historical Hilbert-space formalism (as codified by von Neumann [28]) is of rather limited validity and not appropriate for the discussion of foundational problems of quantum theory. It refers to *strictly closed systems* with only finitely many degrees of freedom. *Such systems do not exist in nature.* In the generic case, the approximation of a real physical system by a strictly closed finite system gives *qualitatively* inadequate results. That is, we are not allowed to ignore the effects of the environment of a physical system. The mathematical tools which avoid the limitations of the historical Hilbert-space formalism of quantum mechanics and which allow an efficient discussion of open quantum systems − both in their individual and statistical descriptions − have been available for many years and will be discussed here from the viewpoint of algebraic quantum theory. At present, I do not see good reasons to consider a program of "completing" quantum mechanics by some so-called «hidden variables» that should more comprehensively specify the state of the system, or any other ad hoc modification of quantum mechanics. In any case, I never will consider so-called «hidden variables» or ad hoc modifications of the first principles of quantum mechanics. *Quantum mechanics is a much richer theory than initially expected,* and at present there is no indication that quantum theory could be inadequate for the description and the explanation of the behavior of matter. But quantum mechanics is also an *intrinsically holistic theory* which goes far beyond the initial ideas of its creators. Furthermore, the modern conception of quantum theory is much broader than our narrow-minded and old-fashioned textbooks may suggest.

Sixty-Two Years of Uncertainty
Edited by A. I. Miller
Plenum Press, New York, 1990

I will therefore adopt the *working hypothesis* that in a proper codification quantum mechanics has universal and unlimited validity at the level of atoms, molecules and non-cosmological macroscopic bodies. Subatomic, genuinely Lorentz-relativistic phenomena, and problems related to men's free will and consciousness will, however, be excluded. While I take the first principles of quantum mechanics for granted, I will refuse obstinately to accept uncritically any of the additional working rules of the physicist's trade. Likewise I do not feel to be bound by any regulations philosophers may require. To refuse to submit myself to any authority does not mean that I will try to be smarter than the experimenters or wiser than the philosophers but only that I would like to avoid hidden rules and preconceptions. We have to distinguish carefully between physics as a trade, and physics as a fundamental theory. While the first principles of quantum mechanics can be codified, it is hard to characterize the art of physical research. The working physicist certainly uses first principles but in addition also well-established working rules borrowed largely from everyday experience. As emphasized by Michael Polanyi [34, chapt.4], there are rules which are not known as such to the person following them, hence we are able to do things without necessarily knowing what we are doing thereby. If such rules are not codified as tentative working hypotheses, the working scientist feels he is doing just the natural thing. Even the most convincing visions of scientific truth may contain elements of basic error. We cannot avoid that but intellectual honesty requires that we at least *try* to include in full statement those premises logically prior to the first principles of the theory we consider as basic.

Habitually physics deals with simple systems or with simple aspects of complex systems. In their experimental work physicists try hard to *create* almost isolated systems. However, strictly isolated systems do not exist and certainly a fundamental theory has to be more than a theory of isolated systems. If we claim that quantum mechanics plays a fundamental role for our understanding of nature, we cannot restrict its basic concepts to esoteric systems which have little relevance for the other natural sciences.

In the development of physics, the status of a law may change because "deeper" levels of physical reality are disclosed. What was once considered to be a fundamental law may later lose this status and become more or less a phenomenological law with a restricted range of validity. For example, thermodynamics has been put forward as one of the fundamental theories of physics (e.g. for some time by Planck and Einstein, more recently also by Stueckelberg, von Weizsäcker and Prigogine). However, we regard thermodynamics as a description with a rather limited range of validity – in fact, it is impossible even to define its basic concepts temperature and entropy for arbitrary states of matter. All known first principles refer to situations with high intrinsic symmetry. We may adopt the point of view that the invariance principles and the associated symmetries are the really fundamental laws of nature. But it is necessary to break fundamental symmetries, as clearly recognized by Pierre Curie [7]: *"C'est la dissymétrie qui crée le phénomène"*.

Measurements belong to the everyday domain with its many broken symmetries. For that reason *the foundation of a fundamental theory should not be built on measurements*. In order to understand measuring processes we have to understand first the spontaneous symmetry breaking of the time-inversion symmetry, the reasons for preferring the retarded solutions, the existence of classical phenomena, the emergence of irreversibility, and the causality principle of the engineers («response comes after the stimulus»). The operational approach presupposes that all these problems are already solved. I prefer to think that the crowning of the development of quantum theory will be an understanding of the measuring process in terms of first principles which represent the undivided wholeness of the material reality. The most interesting phenomenon of fundamental physics is the *spontaneous breaking of symmetries* which, however, cannot be described in a rational manner in terms of the usual Hilbert-space formalism of quantum mechanics.

Algebraic quantum mechanics gives us the tools to come to grips with the measurement problem. Furthermore one can define rigorously an *object* as an open quantum system which is characterized by a complete set of intrinsic potential properties and which is distinguished from arbitrary open quantum systems by its *individuality*. This concept paves the way for an *individual and ontic interpretation* of quantum theory. Such a move is crucial for a general theory of matter. In the molecular domain there is an essentially continuous transition from atoms to small molecules, to macromolecules, to grains, crystals and macroscopic bodies, so that the distinction between microscopic and macroscopic becomes obsolete. It would be altogether unreasonable to require that different languages and ontologies should be used for describing small and large molecules since in general there is no qualitative difference between these systems. Everyone believes in some kind of realism. In the macroscopic domain even the most hard-boiled positivists take on a metaphysical ontological commitment, they certainly believe in the reality of measuring instruments and in the reality of observations. No scientist adopts the view that single macroscopic crystals are not real until observed. Similarly, all molecular biologists claim that single DNA-molecules *really exist* and that these molecules *have* at every instant a well-defined tertiary structure, quite independently from the question whether we know it or not. A scientist likes to find out what reality is and how he possibly can grasp it. Or in the words of Alfred Tarski [55]: *"the search of truth is the essence of scientific activities"*.

A working scientist never will accept the idea that theories are just tools for expressing empirical regularities, and that interpretations that go beyond the level of observation are idle baggage. Fundamental science is concerned to *explain* what happens in nature, so a scientist can hardly avoid to adopt Albert Einstein's [11] ontological thesis: *"Es gibt so etwas wie den 'realen Zustand' eines physikalischen Systems, was unabhängig von jeder Beobachtung oder Messung objektiv existiert und mit den Ausdrucksmitteln der Physik im Prinzip beschrieben werden kann"* [1]. If we posit that algebraic quantum mechanics is valid for the atomic, the molecular *and* the macroscopic domain, then we should choose an interpretation which essentially agrees with the usual realistic view of our everyday life. Such an ontological commitment is permissible *provided* we restrict our discussion to objects which are not entangled with their environments (to be discussed in section 5), and if we adopt an appropriate ontic interpretation of quantum mechanics (compare section 6).

If we adopt the working hypothesis that quantum mechanics is universally valid, we have to cut a Gordian knot. All universal first principles we know refer to *strictly closed* systems. But by definition, closed systems have no operational meaning. Nevertheless, an external observer seems to be an indispensable element for a complete description of physical reality. Clearly this situation poses conceptual problems which cannot be solved in a trivial manner (compare sections 7 and 8).

2. *THE LIMITATIONS OF VON NEUMANN'S CODIFICATION*

Von Neumann's [28] formalization of quantum mechanics is of rather limited validity, so that it would be premature to draw philosophical conclusions from this codification. For example, the so-called measurement problem and the "reduction of the wave packet" are not philosophical enigmas but technical questions which simply go beyond von Neumann's codification.

Von Neumann's codification of quantum mechanics is based on the following *uniqueness theorem* [52, 27, 25, 26]:

> *Every representation of Weyl's canonical commutation relations over a locally compact phase space is up to unitary equivalence a direct sum of Schrödinger representations.*

Von Neumann's restriction to locally compact phase spaces excludes the description of irreversible processes, of dissipative processes, of classical systems, of spontaneous

---

[1] English translation : There is such a thing as the "real state" of a physical system that exists objectively, independently of any observation or measurement, and which, in principle, can be described by the concepts of physics.

breakdown of symmetries, and of Hamiltonian systems having infinitely many degrees of freedom (like the electromagnetic field).

From our experience we know that there are many physical systems which allow (at least in an excellent approximation) a description in terms of *nonlinear classical physics*. If we assume that quantum mechanics is a universally valid theory we have to face the following questions: How is it possible that in a quantum world there can be domains admitting classical descriptions? How can we describe the interactions between purely quantal systems and classical systems? How can we describe the dynamical behavior of classical measurement devices quantum-mechanically?

Closely related with these questions is the impossibility of a correct description of the measuring tools within von Neumann's codification. The very possibility of confronting a scientific theory with experience presupposes a domain where we have a language based on the classical two-valued predicate logic, and where in addition there exist irreversible processes which lead to facts which can be stored in memories. In von Neumann's codification there are no such systems. Detection and registration processes are necessarily irreversible, terminating in a record which can be understood in terms of two-valued Boolean logic. Such genuinely irreversible processes *can* be described rigorously in quantum mechanics but every non-phenomenological Hamiltonian description of a truly irreversible process requires *noncompact local observables* which are banned in von Neumann's codification.

That is, already on the basis of formal considerations, *von Neumann's Hilbert-space codification of quantum mechanics is too narrow and not qualified as a universally valid theory*. Fortunately, many of these difficulties fade away in a more recent formalization of quantum mechanics: *algebraic quantum mechanics*. The algebraic codification is a straightforward generalization of traditional quantum mechanics, classical point mechanics and statistical mechanics, it is valid for systems with finitely or infinitely many degrees of freedom. In the case of purely quantal finite systems it is physically equivalent to von Neumann's codification of quantum theory. Yet, algebraic quantum mechanics allows in a natural way the description of superselection rules and the emergence of classical observables.

3. THE VARIOUS STATE CONCEPTS IN PHYSICAL THEORIES

There are many conceptually different state concepts which all play a crucial role in physical theories and which must not be confused. We distinguish between *ontic states* (which refer to intrinsic properties individual objects have), *epistemic states* (which express the state of our knowledge concerning an individual object), *statistical states* (which refer to the results of statistical experiments), and *system-theoretic states* (which describe the past of a system). The fact that the mathematicians have borrowed the term *state* from algebraic statistical quantum mechanics as a synonym for the *purely mathematical concept* of a «normalized positive linear functional» adds fuel to the fire. This unfortunate terminology has become so common in the theory of *-algebras that we are forced to accept it.

The various state concepts of physics and mathematical system theory all go back to classical point mechanics. In the Newtonian formulation, the equations of motion for a system of N point particles are expressed by a system of $3N$ second-order differential equations in terms of the spatial coordinates of the particles. In the system-theoretical version of Lagrange these equations are rewritten as a system of $6N$ first-order differential equations for the $3N$ coordinates and the $3N$ velocities. In the Hamiltonian reformulation the $6N$ variables are taken as the $3N$ coordinates

$$q(t) = \{ q_1(t),...,q_{3N}(t) \} \ ,$$

and the $3N$ canonically conjugated momenta

$$p(t) = \{ p_1(t),...,p_{3N}(t) \} \ .$$

If we know all the forces acting on the particles, and if we know the initial conditions $q(0)$ and $p(0)$, then by solving the equations of motion we can *predict* the values of $q(t)$

and $p(t)$ for all future $(t > 0)$, and *retrodict* these values for the past $(t < 0)$. In classical point mechanics, the $6N$-tuple $\omega(t) = \{p(t), q(t)\}$ is called the *individual state* of the N-particle system at time t.

The phase space Ω of classical mechancis is the set of all feasible individual states and is uncountable. The family of all subsets of the phase space is an atomic Boolean algebra which is much too large as to qualify for a Boolean algebra of experimentally distinguishable events. There is a one-to-one correspondence between the atoms and the individual states, but these do not correspond to experimentally decidable questions. The only reasonable choice for the finest Boolean algebra $\mathcal{B}$ of experimentally decidable events is isomorphic to the σ-algera Σ / Δ, where Σ is the σ-algebra of Borel sets of Ω and Δ is the σ-ideal of Borel sets of Lebesgue measure zero (compare [29]). This choice is closely related to Kolmogorov's [20] foundation of mathematical probability theory. A countably additive probability measure μ on the measurable space $(\Omega, \mathcal{B})$ is called a *statistical state* of the mechanical system with the phase space Ω.

The individual state and the statistical state of classical mechanics are in the first place mathematical concepts, they may be associated to various physical interpretations. There are ontic interpretations which relate the state to actual being, there are epistemic interpretations which refer to our knowledge, there are system-theoretical interpretations which characterize the state by the history of the system, and there are operational interpretations which identify a state by a preparation procedure. The following variants are frequently used in classical physics:

(i)   *Ontic interpretation of individual states*
      The individual state $\omega(t)$ describes in an exhaustive manner all the properties the N-particle system *has* at time t.

(ii) *System-theoretical interpretation of individual states*
      The individual state $\omega(t)$ specifies the equivalence class of all histories (for $t < 0$) of the system which give rise to the same predictions for all conceivable future experiments on the system.

(iv) *Epistemic interpretation of statistical states*
      The probability measure μ refers to our knowledge. We do not know the individual state ω but we know that it is more likely to be in some Borel-subsets of Ω than others. That is, $\mu(B)$ is interpreted as the *probability* that the individual state ω *is* in the Borel set $B \in \mathcal{B}$.

(v)   *System-theoretical interpretation of statistical states*
      The statistical state μ specifies the equivalence class of all histories of the system which give rise to the same *statistical predictions* for all conceivable future experiments on the system.

(vi) *Operational interpretation of statistical states*
      The statistical state μ refers to a statistical ensemble representing an equivalence class of *preparation procedures leading to the same statistical predictions*.

Note that from a formal point of view, using only the formalism of classical mechanics, there is no argument against further interpretations which replace the concept of "prediction" by "retrodiction".

In classical mechanics there are clear-cut interrelations between these state concepts. However this situation is not generic; in general these notions are very different both from a conceptual and from a mathematical point of view and one should be very careful not to confuse them. For example, in classical point mechanics, every individual system-theoretical state admits an ontic interpretation while in general it makes no sense to assign an ontological meaning to a system-theoretical state.

It is generally accepted that quantum mechanical predictions are *intrinsically* of statistical character. This circumstance, however, neither implies that an individual interpretation of quantum mechanics is impossible, nor that a purely statistical interpretation is unproblematic. In contradistinction to classical statistical mechanics, a statistical quantum state does *not* specify an ensemble in the sense of a mixture of individual systems in pure states. Moreover, the statistical interpretation *presupposes* the existence of classical domains, the existence of irreversible events, and a preferred direction of time such that the concept of prediction makes sense. If we do not want to include these presuppositions into our list of postulated first principles, we cannot start with a statistical interpretation but we have to *derive* the statistical description from the individual description.

If we do not specify clearly the kind of state we are discussing we cannot communicate. For example, the following statements: "it is very unlikely that a given complicated system will have a definite wave function" and "a pure case is all too easily converted into a mixture by any small erratic disturbance" by Willis E. Lamb Jr. [24], tacitly presume an operationalistic view, namely that to say «a system is in a certain state» can only mean «that an experimentalist has prepared the system to be in that state». Lamb's objectivistic way of speaking is very misleading since his statement refers to our *knowledge,* and not to an objective statement about nature. Of course, one can never get the information necessary for the specification of a pure state of a macroscopic object. However, this question has nothing to do with the question whether or not a macroscopic object can have an ontic state, a problem which by definition is independent of our information about this state. Even if the dynamics depends sensitively on the initial conditions, the ontic state always is decribed by a pure state while for an epistemic description one may be forced to use mixed states. If an individual description is available, one always can go to a statistical description by forming an ensemble of uncorrelated replicas of individual systems.

4. *ALGEBRAIC QUANTUM MECHANICS*

INTRINSIC AND CONTEXTUAL DESCRIPTIONS

Algebraic quantum mechanics is nothing else but a precise and complete codification of the heuristic ideas of quantum mechanics of the pioneer days. Moreover it is a straightforward generalization both of classical point mechanics and traditional quantum mechanics. In the domains proper of these theories, nothing is changed. Nevertheless, both theories are treated in the same framework. In contradistinction to von Neumann's codification, the algebraic formulation avoids the limitations of the Stone-von Neumann uniqueness theorem and rejects the unfounded irreducibility postulate.

Algebraic quantum mechanics provides a mathematically rigorous basis for the study of finite and infinite systems, which may be either purely quantal, purely classical or mixed quantal/classical. In this formulation, the bounded observables correspond to the selfadjoint elements of an abstract *-algebra of observables. A *-algebra is just a collection of mathematical objects $A, B, \ldots$ that can be combined linearly, multiplied in a bilinear and associative way, and mapped by the conjugate linear *-operation $A \to A^*$ which satisfies $A^{**} = A$ and $(AB)^* = B^*A^*$. These operations correspond to the familiar operations performed with linear operators acting on a Hilbert space. If a *-algebra $\mathcal{A}$ is endowed with a Banach-space norm $\| \ \|$ with the properties $\|AB\| \leq \|A\| \cdot \|B\|$ and $\|A^*A\| = \|A\|^2$, then $\mathcal{A}$ is called a C*-algebra. If a C*-algebra is the dual of a Banach space, then it is called a W*-algebra[1].

Depending on the circumstances, the algebra of observables can be chosen as a C*-algebra $\mathcal{A}$ of *intrinsic observables,* or as a W*-algebra $\mathcal{M}$ of *contextual observables.* It turns out that the individual states of an isolated system are in a one-to-one correspondence with the extremal normalized positive elements of the C*-algebra $\mathcal{A}$ of intrinsic observables. The statistical states of an epistemic description correspond to the positive normalized elements of the predual of the W*-algebra $\mathcal{M}$ of contextual observables.

---

[1] The C*- and W*-algebraic formalism is a bit technical but there are excellent texts. For all mathematical questions we refer to Sakai [47], Dixmier [9], Pedersen [31], Bratteli and Robinson [4,5], Kadison and Ringrose [18,19].

Algebraic quantum mechanics has a *descriptive part* and a *constructive part*. For a given universe of discourse, the C*-algebra $\mathcal{A}$ and the associated time-evolution group is *intrinsically* given, while the W*-algebra $\mathcal{M}$ depends on the observer's context. Provided the observer is able to describe precisely what he considers as relevant and what as irrelevant, then he can specify a particular topology on the dual of the algebra $\mathcal{A}$ of intrinsic observables which allows by the so-called GNS-construction to constitute the appropriate W*-algebra of contextual observables. However, from a practical point of view such a procedure can be exceedingly difficult. Nevertheless, one can continue in a purely descriptive phenomenological way and describe the context-dependent W*-algebra in terms of algebraic structures. For example, classical systems are characterized by commutative algebras, thermodynamic systems are characterized by W*-algebras of type III, and so on.

From a purely formal point of view we characterize quantum systems according to the *structure of their algebra of observables*. The most important algebraic characterization of an algebra refers to its center. The *center* $Z(\mathcal{M})$ of an algebra $\mathcal{M}$ is defined as the set of all operators in $\mathcal{M}$ which commute with every operator in $\mathcal{M}$,

$$Z(\mathcal{M}) := \{ Z \mid Z \in \mathcal{M}, ZM = MZ \text{ for every } M \in \mathcal{M} \} \ .$$

The center of a W*-algebra is a commutative W*-algebra. The corresponding sub-system with $Z(\mathcal{M})$ as algebra of observables is called the *classical part* of the whole system with $\mathcal{M}$ as algebra of observables. The selfadjoint elements of the center $Z(\mathcal{M})$ are called *classical observables*. If the algebra of observables is commutative, then it equals its center, $\mathcal{M} = Z(\mathcal{M})$, so that all observables are classical, and accordingly the corresponding system is called *classical*. If the center is trivial (i.e. if the center consists at most of the scalar multiples of the identity), the corresponding system has no genuine classical observables, hence no classical part, and is therefore called *purely quantal*.

Our use of the word «classical» *has nothing to do with Planck's constant*. In fact, important features of classical quantum systems depend crucially on the value of Planck's constant. Moreover, *quantum mechanics does not reduce to classical mechanics as a special case*, say in the limit of macroscopic systems. From a theoretical viewpoint, the distinction between microphysics and macrophysics has become obsolete for a long time. There is a practically continuous transition between microphysics and macrophysics; there are microsystems with classical properties, and there are macrosystems without classical properties.

Paradigmatic examples for the algebraic classification of physical theories are the well known extreme cases of traditional quantum mechanics and of classical Hamiltonian point mechanics.

FIRST EXAMPLE: TRADITIONAL QUANTUM MECHANICS

The irreducible Hilbert-space representation of traditional quantum mechanics is a *statistical description* and starts with a separable Hilbert space $\mathcal{H}$, the associated algebra $\mathcal{B}(\mathcal{H})$ of all bounded linear operators acting on $\mathcal{H}$ plays the role of the algebra $\mathcal{M}$ of bounded *contextual* observables. The algebra $\mathcal{B}(\mathcal{H})$ is a factorial W*-algebra of type I; it has a trivial center, so that traditional quantum mechanics describes *purely quantal* systems. Traditional quantum mechanics uses the concept of statistical states which refer to the outcome of statistical experiments; they enjoy the property of σ-additivity and are described by *normal* normalized positive linear functionals on $\mathcal{B}(\mathcal{H})$, that is by normalized positive elements of the predual[1] $(\mathcal{B}(\mathcal{H}))_*$ of the algebra $\mathcal{B}(\mathcal{H})$ of contextual observables. The predual $(\mathcal{B}(\mathcal{H}))_*$ is isomorphic to the Banach space $\mathcal{B}_1(\mathcal{H})$ of all nuclear operators, so that the statistical states of traditional

---

[1] The *predual* $\mathcal{A}_*$ of a W*-algebra $\mathcal{A}$ is characterized by the fact that the dual $(\mathcal{A}_*)^*$ of the predual $\mathcal{A}_*$ of $\mathcal{A}$ equals $\mathcal{A}$.

quantum mechanics can be represented by density operators and pure statistical states by idempotent density operators.

The less known *individual description* of traditional quantum mechanics is based on the C*-algebra $\mathcal{A}$ of bounded *intrinsic* observables which is the algebra $\mathcal{B}_\infty(\mathcal{H})$ of all compact operators[1] in $\mathcal{B}(\mathcal{H})$. It is a separable[2] nuclear C*-algebra which is exceptional in the sense that all its irreducible *-representations are equivalent. Furthermore, the converse is also true: if $\mathcal{A}$ is a separable C*-algebra which has only one equivalence class of irreducible *-representations, then $\mathcal{A}$ is C*-isomorphic to $\mathcal{B}_\infty(\mathcal{H})$ with an appropriate Hilbert space $\mathcal{H}$ [44]. This fact corresponds to the uniqueness theorem for the canonical commutation relations by Stone [52] and von Neumann [27]. The conjugate space $\mathcal{B}_\infty(\mathcal{H})^*$ of the C*-algebra $\mathcal{B}_\infty(\mathcal{H})$ of intrinsic observables is isomorphic to the Banach space $\mathcal{B}_1(\mathcal{H})$ of all nuclear operators.

From the perspective of algebraic quantum mechanics the formal status of traditional quantum mechanics is exceptional in so far as the dual $\mathcal{B}_\infty(\mathcal{H})^*$ of the algebra $\mathcal{B}_\infty(\mathcal{H})$ of intrinsic observables is isomorphic to the predual $\mathcal{B}(\mathcal{H})_*$ of the algebra $\mathcal{B}(\mathcal{H})$ of contextual operators. In abstract terms: in traditional quantum mechanics the bidual $\mathcal{A}^{**}$ of the C*-algebra $\mathcal{A}$ of intrinsic observables equals the W*-algebra $\mathcal{M}$ of contextual operators, $\mathcal{A}^{**} = \mathcal{M}$. In general, however, we have $\mathcal{A}^{**} \supseteq \mathcal{M}$. Except in the case of traditional quantum mechanics the bidual $\mathcal{A}^{**}$ of the algebra $\mathcal{A}$ of intrinsic observables is always *much* larger than the algebra $\mathcal{M}$ of contextual observables. In the algebraic language, the Stone-Neumann uniqueness theorem is expressed by the exceptional relation $\mathcal{A}^{**} = \mathcal{M}$. That is, in traditional quantum mechanics the intrinsic observables determine uniquely the contextual observables.

To summarize: The algebra $\mathcal{A}$ of intrinsic observables of traditional quantum mechanics in its individual description is given by a C*-algebra of compact operators, a separable and noncommutative algebra. The individual states are represented by the normalized extremal positive elements of the dual $\mathcal{A}^*$. The statistical description of the same system is based on the noncommutative W*-algebra $\mathcal{M}$ of contextual observables, a factor of type I with separable predual $\mathcal{M}_*$, $\mathcal{M} = (\mathcal{M}_*)^*$. The statistical states are represented by the normalized positive elements of the predual $\mathcal{M}_*$. Traditional quantum mechanics is characterized by the exceptional fact that the bidual of the algebra of intrinsic observables is the algebra of contextual observables, $\mathcal{A}^{**} = \mathcal{M}$. As a consequence, there is an unusual one-to-one correspondence between individual states and extremal statistical states (the so-called pure states).

SECOND EXAMPLE: CLASSICAL MECHANICS

Classical point mechanics with finitely many degrees of freedom can be characterized in the Hamiltonian formalism by a locally compact phase space Ω, a smooth symplectic manifold like $\mathbb{R}^{2n}$. In the usual *individual description* of classical point mechanics the points ω of the phase space Ω represent the individual states of the system. The algebra $\mathcal{A}$ of bounded *intrinsic* observables of the corresponding algebraic description is the separable, nuclear and *commutative* C*-algebra $C_\infty(\Omega)$, the algebra of all complex-valued continuous functions on Ω which vanish at infinity, endowed with the supremum norm. On the other hand, by the Gelfand representation theorem every commutative C*-algebra $\mathcal{A}$ can be canonically represented as an algebra $C_\infty(\Omega)$ on a suitable locally compact space Ω; if $\mathcal{A}$ is separable then the space Ω is second countable and metrizable as a separable complete metric space.

The *statistical description* of classical mechanics uses a different formalism. A statistical ensemble can be specified by a normalized positive distribution function on the phase space Ω,

---

[1] Recall that a compact operator is a norm limit of finite rank operators. The linear set of all compact operators on a Hilbert space $\mathcal{H}$ where the bound of an operator is considered as its norm, furnishes a Banach space $\mathcal{B}_\infty(\mathcal{H})$.

[2] The C*-algebra of all compact operators acting in a Hilbert space $\mathcal{H}$ is separable if and only if $\mathcal{H}$ is separable.

$$\int_\Omega f(\omega)\, d\omega \ = \ 1 \ ,$$

which defines a probability measure μ

$$\mu(B) \ = \ \int_B f(\omega)\, d\omega \ \ , \ \ B \in \Sigma \ ,$$

where Σ is the σ-algebra of Borel sets of the phase space Ω. Classical statistical mechanics is based on the Kolmogorov probability space (Ω,Σ,μ), and the Gibbsian ensemble interpretation, whereby the members of the ensemble are *individual* objects in *individual* states. Note that this statistical ensemble interpretation *presupposes* the individual interpretation of classical mechanics. The shortcut to introduce statistical ensembles by specifying a distribution function is admissible in classical mechanics, since in *classical* theories the set of all statistical states is a *simplex* so that every mixed state (i.e. probability measure) has a *unique* decomposition into pure states (i.e. Dirac measures on Ω). That is, a classical distribution function in fact specifies a unique ensemble[1]. In classical statistical mechanics, an observable is defined as a Borel-measurable real-valued function on the phase space Ω. The space of equivalence classes of complex-valued μ-essentially bounded Borel-measurable functions on Ω is a Banach space under the norm $\|f\| = \mu\text{-ess sup } |f|$, and is denoted by $L^\infty(\Omega,\Sigma,\mu)$. It is a commutative W*-algebra and represents the algebra $\mathcal{M}$ of contextual observables. Since its elements are equivalence classes of functions (and not point functions), the W*-algebra $L^\infty(\Omega,\Sigma,\mu)$ does not contain enough information to reconstruct the phase space Ω. In the terminology of algebraic quantum mechanics, the W*-algebra $L^\infty(\Omega,\Sigma,\mu)$ is the algebra $\mathcal{M}$ of *contextual* observables, while the experimentally accessible statistical states are represented by the normalized positive elements of the Banach space $L^1(\Omega,\Sigma,\mu)$ which is the predual of $L^\infty(\Omega,\Sigma,\mu)$. It is remarkable that there are no pure states in $L^1(\Omega,\Sigma,\mu)$, all pure states are in the dual $(L^\infty)^*$ and not in the predual $L^1 = (L^\infty)_* \subset (L^\infty)^*$. The dual $(L^\infty)^*$ is a monster which contains myriads of unphysical states. This W*-formulation of classical statistical mechanics has a canonical Hilbert-space representation which is generally known under the name *Koopman formalism* [22].

To summarize: In its individual description classical point mechanics is characterized by a separable and commutative C*-algebra $\mathcal{A}$ of intrinsic observables. The individual states are represented by the normalized extremal positive elements of the dual $\mathcal{A}^*$. The statistical description of the same system is based on a nonseparable commutative W*-algebra $\mathcal{M}$ of contextual observables. The statistical states are represented by the normalized positive elements of the separable predual $\mathcal{M}_*$ of the algebra of contextual observables, $\mathcal{M} = (\mathcal{M}_*)^*$. In classical statistical mechanics there are no pure statistical states. This implies that *the individual states of classical mechanics are experimentally inaccessible* [36, chapt. 3.5 and 4.6, 37, chapt. 5.4]. In contradistinction to traditional quantum mechanics, the bidual of the algebra of intrinsic observables is much larger than the algebra of contextual observables, $\mathcal{A}^{**} \supset \mathcal{M}$.

GENERAL QUANTUM SYSTEMS

The basic entity in the algebraic codification of quantum theory is an abstract C*-algebra $\mathcal{A}$ which has to be chosen "as small as possible". This means in more technical terms, the algebra $\mathcal{A}$ is chosen as a separable nuclear[2] C*-algebra. The

---

[1] In nonclassical theories every nonpure state has infinitely many decompositions into pure states so that in this case a so-called mixed state does not characterize the mixture.

[2] A C*-algebra is called *nuclear* if there is a unique way of forming its tensor product with any other C*-algebra. This property is important for the unique composition of quantum systems. A C*-algebra $\mathcal{A}$ is nuclear if and only if the enveloping W*-algebra $\mathcal{A}^{**}$ is injective. A W*-algebra $\mathcal{M}$ acting on a Hilbert space $\mathcal{H}$ is said to be injective if there is a retraction of norm one from $\mathcal{B}(\mathcal{H})$ onto $\mathcal{M}$.

selfadjoint elements of this algebra represent the so-called intrinsic observables which are in one-to-one correspondence to the intrinsic potential properties of the system. Those intrinsic potential properties which are actualized at time t characterize the intrinsic individual state of the system at time t. It can be shown that these individual states are in one-to-one correspondence with the so-called pure states on the algebra $\mathcal{A}$ of intrinsic observables [42]. It is very important to note that the concept of a "pure state" is a strictly mathematical one, it just means an extremal normalized positive linear functional on the C*-algebra $\mathcal{A}$. It follows that a property is actualized at time t if and only if the corresponding observable has a dispersionfree value with respect to the individual state. That is, if $\rho_t \in \mathcal{A}^*$ represents the individual state at time t, and if $A \in \mathcal{A}$ is the observable corresponding to the property under discussion, then A is dispersionfree with respect to ρ_t if

$$\rho_t(A^2) = \{\rho_t(A)\}^2.$$

In this case we adopt Dirac's [8] regulative principle[1] and say that the observable *has* the value $\rho_t(A)$. If a property is not actual, we do not attribute any value to the corresponding observable.

The perfect isolation of a quantum system is reflected in the bidirectional deterministic character of its time evolution. In the simplest cases[2], the dynamics in algebraic quantum mechanics is given directly by a one-parameter group $\{\alpha_t : t \in \mathbb{R}\}$ of automorphisms α_t of the C*-algebra $\mathcal{A}$ of intrinsic observables. The time evolution $\rho_0 \to \rho_t$ of the individual state is then given by

$$\rho_t(A) := \rho_0\{\alpha_t(A)\} \quad \text{for all} \quad A \in \mathcal{A} \quad \text{and all} \quad t \in \mathbb{R}.$$

This reversible dynamics is the mechanism which actualizes potentialities. From our viewpoint, the idea that "the only processes... in which potentialities are actualized are measurements" [50], is not acceptable. While in the usual interpretation of classical physics all potential properties are taken to be always actualized, in quantum physics such a premise would be contradictory. But we are allowed to posit (as we do in classical physics) that always the potential properties of a *maximal subset* of the set of all potential properties are actualized. If by the time evolution some potential properties become actual some other ones must necessarily disappear into potentiality, a principle which Piron [33] attributes to Aristotle.

With the only exception of traditional quantum mechanics, a general C*-algebraic quantum system has always *infinitely many physically inequivalent* W*-representations. This is not a pathology but a most important feature which has to be expected on physical grounds. However, the "correct" choice of the representation is a non-trivial problem. A particular representation can be fixed mathematically by specifying a privileged state on the intrinsic C*-algebra $\mathcal{A}$. Such a privileged state can be used as reference state in the so-called *GNS-construction* (according to Gelfand, Naimark and Segal), giving a unique faithful W*-representation.

Every state ρ (in the mathematical sense of a normalized positive linear functional) on a C*-algebra $\mathcal{A}$ gives rise to a cyclic representation $(\pi_\rho, \mathcal{H}_\rho, \Psi_\rho)$ with a complex Hilbert space $\mathcal{H}_\rho$ and a normalized vector $\Psi_\rho \in \mathcal{H}_\rho$, and the properties

(i) $\rho(A) = \langle \Psi_\rho | \pi_\rho(A) | \Psi_\rho \rangle$ for every $A \in \mathcal{A}$,

(ii) the closed subspace of $\mathcal{H}_\rho$ spanned by $\pi_\rho(\mathcal{A})\Psi_\rho$ equals $\mathcal{H}_\rho$.

---

[1] Compare the *first* edition of Dirac's "The Principles of Quantum Mechanics" of 1930, § 11. In the later editions Dirac adopts the nowadays more popular operational interpretation.

[2] In general the dynamics is not well defined for *all* observables in $\mathcal{A}$. In this case, the time evolution is defined by a one-parameter group acting only on a dense set of observables.

The triple $(\pi_\rho, \mathcal{H}_\rho, \Psi_\rho)$ is called the *GNS-representation* of $\mathcal{A}$ induced by ρ. Every cyclic representation arises this way. The GNS-representation is unique in the sense that every cyclic representation $(\pi, \mathcal{H}, \Psi)$ such that

$$\rho(A) = \langle \Psi | \pi(A) | \Psi \rangle \quad \text{for all } A \in \mathcal{A} ,$$

is unitarily equivalent to the GNS-representation $(\pi_\rho, \mathcal{H}_\rho, \Psi_\rho)$. The weak closure of $\pi_\rho(\mathcal{A})$ in the algebra $\mathcal{B}(\mathcal{H}_\rho)$ of all boundend operators acting in the Hilbert space $\mathcal{H}_\rho$ is a von Neumann algebra $\mathcal{M}_\rho$ which conceptually corresponds to the W*-algebra of contextual observables whereby the context is characterized by the reference state ρ on the algebra $\mathcal{A}$ of intrinsic observables.

A standard choice in many exactly soluble models (say for superfluidity, superconductivity, ferromagnet, phase transitions, laser models, spontaneous symmetry breakings and so on[1]) are either the ground state or the thermal equilibrium states (the so-called β-KMS states). However, these are exceptional cases – in general physically interesting systems are not in stationary states, and in the absence of such overly idealized situations we do not have a workable recipe to find the appropriate W*-representation. This may be disappointing but is inevitable. It just means that no comprehensive universal *operational* description for the whole material reality can be found. The reason is that *nothing can be said about nature unless some abstractions are made.* There is no science without abstractions but abstractions are context-dependent, they do not falsify our description of the material reality but *they create the patterns of reality.* All concepts of empirical science refer to observations obtained by some pattern recognition methods which distinguish between *relevant and irrelevant features.* What has to be considered as relevant, and what as irrelevant, is not written down in the universally valid C*-algebraic description of material reality but has to be introduced into quantum theory by making the very same abstractions as the pattern recognition methods of empirical science do. In algebraic quantum mechanics these abstractions can be introduced by specifying an appropriate new topology on the state space of the intrinsic C*-algebraic formalism, or, equivalently, by specifying a privileged state which can be used as reference vector for the relevant GNS-representation.

5. QUANTUM OBJECTS

QUANTUM MECHANICS IS A HOLISTIC THEORY

A fundamental preconception of all science is the idea that one has not to consider the whole universe at once but that one can advance by compartmentalization. Classical science defends in addition the idea that the analysis of nature in terms of interacting but independently existing objects is in accordance with the empirical facts. On the other hand, if we take our most fundamental theory of matter seriously, an entirely different view emerges. If we consider quantum mechanics not only as a bunch of highly successful pragmatic working rules, but as our best candidate for a really fundamental theory of the material world, then we are compelled to admit that individual objects cannot exist in an absolute sense (with the possible but uninteresting exception of the whole universe). Quantum mechanics predicts that nature is nonseparable. This prediction has been verified in recent years in a series of beautiful experiments beyond any reasonable doubt. The empirists' claim that objects are given in an absolute sense has proved to be inadequate – *objects do not exist in perfect isolation.*

To be sure, even in classical science the environment of an object system cannot be left out of consideration but in classical physics it is tacitly assumed that one can describe the influence of the environment by appropriate forces or interactions. From our everyday experience we believe to know that certain things are quite independent

---

[1]   A discussion of such phenomena from the viewpoint of algebraic quantum mechanics can be found in the monographs by Dubin [10], Strocchi [53] and Sewell [48].

of others, notably those distant in time or space. We are so accustomed to accept this separability of nature as something self-evident that we easily forget how artificial this doctrine really is.

It is a mathematical property of classical mechanics that the individual states of subsystems determine the state of the whole system. This property is called *separability*. A system is called holistic if it does not possess the property of separability. Quantum mechanics is the first – and up to day the only – logically consistent and empirically well-confirmed mathematically formalized holistic theory. The nonseparability of nature is described by the so-called Einstein-Podolsky-Rosen correlations which exist even in absence of any interactions. According to quantum mechanics, all subsystems of the world are inextricably entangled so that it is impossible to describe them by pure states. Quantum mechanics describes the material world as a whole, in fact as *a whole which is not made out of parts*.

The world view of classical physics was in the main reductionistic and mechanistic. The fundamental theory was about basic objects and their interactions, its aim was to explain complex phenomena in terms of a few elementary objects. Even today we often say that nuclei can be understood as composed of protons and neutrons, and that molecules can be understood to be composed of nuclei and electrons. What we actually mean is that we can break molecules into protons, neutrons and electrons, and that free protons, free neutrons and free electrons behave *elementary* under the actions of the appropriate kinematical group. However, if we take quantum mechanics seriously, *atomism is dead*. In the terminology of quantum theory atomism assumed that there is a God-given tensor-product decomposition of material reality. That seems not to be so. At least quantum mechanics portrays the material world as a whole. The fact that we can *disintegrate* nuclei, atoms and molecules into elementary systems does not imply that nuclei, atoms and molecules are *made* out of elementary systems. According to quantum mechanics, a molecule can be described as an *entangled system* of neutrons, protons and electrons. Yet these three kinds of elementary quantum systems are not only interacting but Einstein-Podolsky-Rosen-correlated so that it is impossible to attribute to them any kind of individual existence. Matter, as described by the first principles of quantum mechanics, resembles matter in the Aristotelian sense: it is not a substance, but the capacity to receive patterns.

THE CONCEPT OF AN OBJECT IN QUANTUM THEORY

Do physical objects have a real, objective existence? This discussion has been plagued by some deeply ingrained but invalid modes of thought. In order to have a precise language, we will distinguish between the concepts "system" and "object". By a *system* we just mean the referent of a theoretical discussion, without any ontological commitment whatsoever. On the other hand, we choose the term *object* as the most general ontological expression for something that *persists when not perceived*, for something having *individuality* and *properties*. We do not restrict the notion of an "object" to concrete things or to entities which are localized in space. Objects may be quite abstract individuals which not necessarily can be isolated from the rest of the world. Objects are entities which retain their identity in the course of time. They may change their actualized properties but they keep their identity. They stand on their own, we have names for them and talk about them. That is, objects are the referents of genuine names.

If we single out an object system, we have divided the world into two parts. All we have not singled out will be called the *environment* of the system. We adopt the following characterization [36, 37, 38, 39]:

Definition

An object is defined to be an open quantum system, interacting with its environment, but which is not Einstein-Podolsky-Rosen-correlated with the environment.

It follows that objects are exactly those quantum systems for which at every instant a maximal description in terms of pure states is possible. This fact is of conceptual importance since one can interprete a quantum state as an ontic state only if it is pure. Here the notion of an "ontic state" refers to a *mode of being*, describing intrinsic characteristics existing independently of any observation, while the notion of a "pure state" refers to a merely mathematical concept, meaning an extremal positive linear functional on the algebra of observables.

If a quantum system is not an object it does in general not possess a maximal description in terms of pure states, so that no ontic or individual description is possible. However, an epistemic statistical description may be feasible since the reduced statistical state (e.g. given by reduced density operators) can be a system-theoretical state allowing statistical predictions. *The root of many troubles in the interpretation of quantum mechanics lies in the failure to distinguish clearly between objects and more general systems.*

A realistic interpretation of individual quantum systems is possible if and only if there are no Einstein-Podolsky-Rosen correlations between the object system and its environment. Adopting the primacy of quantum mechanics, we have not only the burden but also the chance to derive the conditions for the existence of objects, the classical behavior of their environments, and of patterns and phenomena in our world. We say a mathematical model of an object system is *robust* if the inclusion of the effects of the environment does not change the description in a qualitative way. Unfortunately, most Hamiltonian models with finitely many degrees of freedom are not robust. Therefore the environment of an object system should never be left out of consideration. Yet, the environment of every material system always includes the electromagnetic radiation field, a system having infinitely many degrees of freedom. Accordingly, a proper discussion of the existence of quantum objects is not possible in terms of traditional Hilbert-space quantum mechanics.

QUANTUM OBJECTS IN ALGEBRAIC QUANTUM MECHANICS

It is a theorem of algebraic quantum mechanics that interaction does not necessarily imply entanglement, and that objects are either classical or must have classical environments:

Theorem (Raggio [42])

Every classical system is an object. A nonclassical open system is an object if and only if its environment is classical.

That is, quantum objects exist if and only if classical environments exist. This theorem explains why the problem of reality did not arise in classical physics.

Often it has been claimed that the adherence to classical concepts in the sense of Bohr is not required by the formalism of quantum mechanics. In algebraic quantum mechanics Raggio's theorem proves the contrary: In order to be able to speak about quantum objects, it is necessary to abstract from the EPR-correlations between the object and its environment (which may contain a measuring instrument). Hence *it follows that the environment has to admit a classical description* (in the sense of Boolean logical structure).

All objects we discuss in natural science are contextual objects, their existence depends both on the environment, and on the abstractions we are forced to make in every scientific discussion. In algebraic quantum mechanics this is reflected by the fact that the environment of a quantum object has to be classical. The meaning of the notion «classical» depends, however, on our abstractions and is therefore context-dependent. That is, all objects in a quantum world are *contextual objects*. Contextual objects are abstraction-dependent but they are not free inventions. They represent *patterns of reality,* yet they are not building stones of reality. For example, photons, electrons, atoms, molecules, crystals, objects of common experience are contextual objects, not absolute entities. In algebraic quantum mechanics, a context can be specified by an appropriate topology on the state space of the C*-algebra of intrinsic observables. Usually this will be realized in a rather indirect way, say by selecting a reference state for the GNS-construction of the algebra of contextual observables.

6. INTERPRETATIONS

An interpretation of a physical theory refers to a logically consistent and empirically well-confirmed theoretical formalism. That is, we assume that we have a mathematically rigorous codification of the theory and a minimal interpretation which allows an operationalization of the theoretical propositions. Without such a minimal instrumentalist interpretation the mathematical formalism alone cannot be considered as a physical theory but would be just a physically meaningless game with symbols. On the other hand, working only with the minimal interpretation, we cannot understand "how the world ticks". Interpretations which are richer than the minimal interpretation have nothing to do with experimental facts but are *ways of thinking*. They can neither be extracted from the formalism nor from experience, they have to be freely created. Therefore we adopt the following characterization:

Definition

An interpretation of a physical theory is characterized by a set of normative regulative principles which can neither be deduced nor be refused on the basis of the mathematical codification and the minimal interpretation.

Epistemic interpretations refer to our knowledge of the properties or modes of reactions of systems. *Ontic interpretations* refer to the properties of the object system itself, regardless of whether we know them or not, and without regard to perturbations by observing acts. The operationalistic view uses always an epistemic and statistical interpretation, while a realistic world view refers to an ontic and individual interpretation. To sum up it can be said that epistemic interpretations *describe* und *predict*, while ontic interpretations *explain*.

EXAMPLE FOR A MINIMAL INTERPRETATION

As a pertinent codification of traditional quantum mechanics we can take the mathematical formalism summarized in von Neumann's [28] book but without the projection postulate. In this framework, the minimal instrumentalist interpretation is given by the expectation-value postulate:

Von Neumann's expectation-value postulate

Let A be a selfadjoint observable with spectrum Λ, let Σ be the σ-algebra of Borel sets of Λ and E be the spectral measure of A. The probability that a predictive measurement of the first kind of the observable A gives a value lying in the Borel set $B \in \Sigma$ is given by the Kolmogorov probability measure $\mu : \Sigma \to \mathbb{R}$, defined by

$$\mu(B) = \langle \Psi | E(B) | \Psi \rangle \ ,$$

where Ψ is the normalized state vector immediately before the measurement.

Conditioned by a measurement of the first kind, this postulate implies that the observable A can be considered as a random observable on the Kolmogorov probability space (Λ, Σ, μ). The expectation value $\mathcal{E}(A)$ of this random variable is given by

$$\mathcal{E}(A) = \int_\Lambda \lambda \, \mu(d\lambda) = \langle \Psi | A | \Psi \rangle \ .$$

The expectation-value postulate assigns an empirical meaning to the abstract mathematical formalism. An understanding of this postulate relies on our previous intuitive notions about physical measurements. From a conceptual point of view much more needs to be said, for example about the nature of time (e.g. the earlier-later relationship), about the status of facts (does quantum mechanics admit factual descriptions?), or about the feasibility of making experiments (does a theory with an automorphic time evolution allow observers having free will?). If the world external to the quantum object is taken to be given phenomenologically, this vague intuitive formulation of the expectation-value postulate is sufficient to compare quantum-mechanical predictions with experimental results. In this sense, the expectation-value postulate allows a minimal operational assessment of quantum theory which is of crucial importance for all engineering applications of quantum mechanics. In spite of the associated unrealistic idealizations[1], the expectation-value postulate is a well-confirmed working rule whose success has to be explained by a *fullfledged* interpretation of quantum mechanics.

AN EXAMPLE FOR AN ONTIC INTERPRETATION: CLASSICAL POINT MECHANICS

The Hamiltonian formalism of classical point mechanics alone is not yet a physical theory since every scientific theory also needs a referent and an ontology. The historical development of classical mechanics illustrates the fact that *every* interpretation is of metaphysical nature. In the early days of the development of classical mechanics, a guideline given to the readers of Copernicus' *De Revolutionibus* pointed out that the heliocentric theory was merely a mathematical representation of the observed facts of planetary motion without claiming it to be true. Later Kepler discovered that this preface was not written by Copernicus but by the editor Osiander, and that Copernicus himself advocated just the opposite view and had convinced himself of the *reality* of the Earth's motion around the Sun. Since that time the Copernican revolution was the paradigm for a *realistic interpretation* of a physical theory, strongly endorsed by Kepler, Galilei and Newton. *Scientific realism* is the thesis that the objects of scientific inquiry exist and act independently of the knowledge of them. This thesis cannot be decided empirically and has the status of a *metaphysical regulative principle.*

In a realistic interpretation of classical mechanics the referents of the theory are material objects in the external reality which are posited to exist and to have properties not depending on being observed. In this interpretation, the individual state $\omega_t \in \Omega$ characterizes the *mode of being* of the object at time t. Whether we know this state or not, is considered as entirely irrelevant for the realistic interpretation of the theory. In this usual objectivistic and realistic interpretation of classical Hamiltonian point mechanics we say the *ontic state* of the physical system at time t is represented by a point ω_t of the phase space Ω. Note that such ontic states are not accessible experimentally, since in classical statistical mechanics there are no normal pure states. Moreover, from the Hamiltonian formalism it does not follow that the physical object described by the theory always is in a definite ontic state. Such a predication is of purely metaphysical nature but it is *compatible* with experience and the formalism. Since it can neither be proved nor disproved by experiments or by the formalism, we are free to accept (or to reject) it as a regulative principle.

AN ONTIC INTERPRETATION OF ALGEBRAIC QUANTUM MECHANICS

The main problem of a realistic individual interpretation of quantum mechanics is that according to this theory there are no noncontextual individual objects at all. If we assume that quantum mechanics is universally valid, then every subsystem of the material world is Einstein-Podolsky-Rosen-correlated with everything else. Open quantum systems which are entangled with their environments cannot be in pure states. No progress can be made unless we carefully distinguish between arbitrary quantum systems and quantum *objects. A necessary and sufficient condition for the feasibility of an individual interpretation of algebraic quantum mechanics is that its referents are objects.* Quantum systems which are not objects have no individuality and allow only an incomplete description in terms of statistical states.

---

[1] It is doubtful whether measurements of the first kind can be realized experimentally. But this idealization is not necessarily pointless since one can develop a theory of realistic measurements on the basis of the expectation-value postulate which is not restricted to repeatable measurement. The engineering theory of quantum measurements and quantum communication channels (compare e.g. the work of Holevo [17]) is highly developed and in a much better shape than the rather naive discussion by operationally oriented philosophers.

The referents of a theory in an ontic interpretation are *individual objects*. In algebraic quantum mechanics the potential properties of an object are represented by the selfadjoint elements of an appropriate C*-algebra $\mathcal{A}$ of intrinsic observables. Intrinsic observables describe independently of any observation what is physically real. The *ontic state* of an object at time t is characterized by the set of all potential properties which are actualized at time t. That is: *the potential properties characterize the object while the actualized properties characterize the state of the object*. This delineation fixes the *ontology* of algebraic quantum theory. It is not radically different from the ontology traditionally accepted for classical physical theories. The restriction of this interpretation to the classical part of the system corresponds to the generally adopted realistic individual interpretation of the traditional classical physical theories while for the quantal part we just have to distinguish carefully between *potential* and *actualized* properties. Quantum mechanics does not force us to give up realism but it forces us to change the classical conception of reality. The preconception that all properties have to be actualized properties may come from classical physical theories where all potential properties are always actualized. The distinction between potential and actualized properties is, however, not at all mysterious but well known from everyday life. Bohr's example of goodness and justice as complementary properties shows that both goodness and justice may be potential ways of behavior of a father which cannot be actualized at the same time.

Using this characterization and the general framework of algebraic quantum mechanics one can prove the following theorems relevant for the ontic interpretation of quantum theory:

Theorem 1

There is a one-to-one correspondence between the intrinsic ontic states and the extremal, normalized positive linear functionals on the C-algebra $\mathcal{A}$ of intrinsic observables.*

Theorem 2

A property represented by an intrinsic observable $A \in \mathcal{A}$ is actualized at time t if and only if $\rho_t(A^2) = \{\rho_t(A)\}^2$ where $\rho_t \in \mathcal{A}^*$ represents the intrinsic ontic state at time t.

Theorem 3

There is a one-to-one correspondence between the contextual ontic states and the normal extremal, normalized positive linear functionals on the W-algebra $\mathcal{M}$ of contextual observables.*

Theorem 4

A property represented by a contextual observable $M \in \mathcal{M}$ is actualized at time t if and only if $\varphi_t(M^2) = \{\varphi_t(M)\}^2$ where $\varphi_t \in \mathcal{M}_*$ represents the contextual ontic state at time t.

Theorem 5

Classical observables are defined to be selfadjoint elements of the centre $\mathcal{Z}(\mathcal{M})$ of the algebra $\mathcal{M}$ of contextual observables. At every instant, every classical observable $Z \in \mathcal{Z}$ is actualized and has the dispersionfree value $\varphi_t(Z)$.

Of course, we do not claim that quantum mechanics will not be superseded sometime, so our reference to an *independent reality* makes only sense as a *theoretical construct*. The *intrinsic ontic interpretation* is based on the C*-algebra of intrinsic observables. It is a *strongly objective* theory in the sense of d'Espagnat [12, 13] since in the first place it makes no reference to observers or probabilities. It may describe reality in itself *but not the phenomena we observe*. A theory which describes observable phenomena cannot keep the human means of data processing out of consideration but these means are not described by the C*-algebra of intrinsic observables. The observables which describe the outcomes of measurements are context-dependent, they are represented by the selfadjoint elements − or more generally by the positive operator-valued measures − of the W*-algebra $\mathcal{M}$ of contextual observables. This algebra is not intrinsically given but it can be constructed by the GNS-construction from the context-independent C*-algebra $\mathcal{A}$ by specifying an appropriate reference state in $\mathcal{A}$*. The W*-algebraic formalism describes the *empirical reality,* it is context-dependent hence only *weakly objective* in the sense that given a context, there is intersubjective agreement. In the words of Wolfgang Pauli [30]: "Once the observer has chosen his experimental setup, he has no influence anymore on the result of the measurement, which, objectively registered, is available to everybody". In all science we have to make abstractions, yet typically we have no scruples to restrict scientific concepts to contextual ones without explicitly mentioning the associated abstractions. So we may be allowed to speak in a manner «as if» the contextual objects (like electrons, atoms or molecules) would really exist. In this sense we call the individual states of contextual objects *contextual ontic states.*

7. ENDO- AND EXOPHYSICAL DESCRIPTIONS

All the so-called first principles of fundamental physical theories refer to *strictly closed systems* whose dynamics are given by one-parameter groups of automorphisms[1]. Since strictly closed systems cannot be observed from outside, such descriptions are not operational. Abner Shimony [49] distinguishes between intrinsic and operational theories. An *intrinsic theory* characterizes the properties and the states of a physical system without explicit reference to other physical systems. By contrast, an *operational theory* makes explicit reference to other physical systems than the one singled out for special study. It typically characterizes the properties of the object system in terms of test procedures which must be described in terms of external systems. According to Shimony, "the choice between an intrinsic and an operational formulation ... is not *ipso facto* a philosophical commitment, concerning either epistemology or the ontological status of physical entities. Each choice, to be sure, is accompanied by proclivities. An intrinsic theory is attractive to a physical realist, who maintains that there are physical entities with existence independent of human knowledge and who wishes not to conflate statements about these entities themselves with statements about knowledge of them. An operational theory is attractive to someone who is skeptical of the meaningfulness of a thesis like physical realism or who doubts that it can be endowed with meaning unless it is explicitly linked with procedures for obtaining knowledge."

A related distinction has been made by David Finkelstein [14] and Otto Rössler [45] by introducing the notions of endophysics and exophysics .

Definition

A physical system without an external observer is called an endosystem. If the observer is external, we speak of an exophysical description of the system under investigation. The world of the observer with his communication tools is called the exosystem.

Both endo- and exophysical descriptions have a proper and important place in science but they must not be confused. The first principles of a universally valid theory refer to endophysics but they are not sufficient for a characterization of an exosystem. Therefore, as stressed by Rössler, *endophysics is different from exophysics.* It is an imperative and nontrivial task to exhibit their interdependence. In quantum theory, even the mathematical formalism for quantum endophysics is different from the formalism for quantum exophysics. Many of the conceptual difficulties and alleged paradoxes of quantum mechanics are due to the failure to distinguish properly between endophysical

---

[1] We consider thermostatics and irreversible thermodynamics as phenomenological and not as theoretically fundamental theories. All the well-known difficulties to specify a dynamics for such theories are due to the fact that they are not given by automorphisms. If a dynamics is not given by a one-parameter group, we have no first principles to specify it. All we can hope for is that we are able to derive it from a more basic theory using some additional assumptions.

and exophysical descriptions. The first principles of a universally valid theory refer to *strictly closed* systems, they describe *endophysics* but they are not sufficient for a characterization of exosystems.

Endophysics is the domain where one can hope to find universally valid first principles. «Universally valid» does not mean that in its domain of validity the theory can describe everything at the same instant, but only that it can describe any selected partial object. This state of affairs has nothing to do with quantum mechanics but is a logical necessity.

Every logically consistent closed physical system which is rich enough to admit internal observers is incomplete similar to the incompleteness which Gödel [16] demonstrated for all sufficiently rich axiomatic systems in mathematics[1]. Using the precise language of symbolic logic, Gödel proved the existence of meaningful but undecidable mathematical propositions in every rich logical system. Trying to fix up this logical system by choosing the undecidable propositions as an axiom and thereby declaring them to be true, new undecidable propositions will crop up. If a whole infinite sequence of sentences were to be obtained by successive applications of Gödel's method, and added to the original logical system, the same process could still be applied to find another true sentence still unprovable. Hence all consistent axiomatic formulations of number theory include meaningful but undecidable propositions. That is, Gödel's incompleteness theorem implies that mathematics is open-ended.

The key feature of both Gödelian and endophysical systems is self-reference. Endophysical propositions are by observers within the system but which are also about the system. Any system of internal representation shares the self-referential properties of Gödelian systems. Gödel's first theorem can be rephrased as a semantic incompleteness theorem [23, chapt.1.7] so that it makes sense for every formalized endosystem. In every sufficiently rich and logically consistent endophysical system there are meaningful and conceptually correct propositions which cannot be verified by an observer belonging to this system. A verification of such propositions has to use methods which are not available within the endosystem. That is, we can assign an operational interpretation only to systems which do not include the observer [46].

A comprehensive description would include also the observer and his experimental set-up. Assuming the universal validity of the basic theory, the dynamics of an observer would be governed by the strictly bidirectional deterministic time evolution of fundamental endophysics, so that the observer had no freedom of choosing the experimental set-up. Hence an endophysical observer cannot make experiments. On account of self-reference problems, he also cannot have a complete knowledge of the endophysical ontic state (which is theoretically characterized by those potential properties which are actualized at some instant), since it is logically impossible for an observer to know his own ontic state. This is true in classical mechanics as well as in quantum mechanics.

An operationally acceptable definition of truth (in the sense of Tarski) for an endosystem must be formulated in another language, the so-called metalanguage. This metalanguage has to be essentially richer than the language of the endosystem. If the two languages would be identical (or translatable into each other) we would have a semantically closed language with self-referential sentences[2] [54, 55]. Semantically closed systems engender antinomies like the paradox of the liar : "Enimenides, the Cretan, says that Cretans always lie". Such semantical paradoxes arise because one identifies the concept of the truth of the endo-language with that of an external metalanguage.

---

[1] Gödel [16] has explicitly constructed a proposition about natural numbers which mathematicians could recognize as being true under the intended interpretation of the symbolism of *Principia Mathematica* by Russel and Whitehead, but which is undecidable (i.e. which can neither be proved or disproved) from the axioms by the rules of inference of this system. Gödel's proof applies as well to any other not too limited finite axiom system of mathematics. In a somewhat vague formulation : in *any* formal system adequate for number theory there exists an undecidable formula.

[2] A sentence S is called self-referential if S asserts the fact that S itself is true or that it is false.

8. HOW CAN ONE MOVE BEYOND A SYSTEM FROM WITHIN THAT SYSTEM?

The crucial problem with quantum endophysics is the necessity that endosystems have to be *strictly closed*. But systems we can investigate experimentally cannot be closed systems. In order to get an endophysical description of an experimental set-up, the world has to be divided into "a part which sees, and a part which is seen" [51]. However, the endoworld does not present itself already divided. *We have to divide it.* Such a division of the holistic endoworld inevitably implies that every exophysical description can be at most a *partial description*. In a holistic theory like quantum mechanics, the notions "patterns" and "phenomena" have no a priori meaning. Every exophysical pattern recognition is a *projection* of the non-Boolean endoworld onto a Boolean exosystem. This projection is neither arbitrary nor unique. It is not arbitrary since we have to respect normative restrictions. It is not unique since every description of reality is just a projection of the non-Boolean structure of reality on a Boolean context. To encompass the whole material reality we have to adopt many complementary viewpoints [35]. None of them is more authentic than the others, none can replace the others, all are necessary, none is sufficient.

We may presume that the quantum endoworld represents the material aspects of reality, we may further presume that the first principles of quantum endophysics have the maximal symmetry and that there are no God-given classical observables. The world we experience is, however, full of broken symmetries and classically describable phenomena. How can we reconcile the first principles of endophysics with the observable phenomena in the exoworld?

The usual but ad hoc postulates for the evaluation of expectation values and for an operational description of the measuring process (the so-called "collapse of the wave packet", or the projection postulate) are pragmatic working rules belonging to exophysics and not to endophysics. It will be an important problem to discuss the interplay of endophysics and exophysics, and to develop procedures which enable us to relate experimentally accessible epistemic states with internal ontic states. Experimentally inaccessible ontic states are not meaningless but play a particularly interesting role in classical mechanics [36, sect. 3.6], they lead to the phenomenon of the so-called deterministic chaos[1], that is a dynamical process which ontically is bidirectionally deterministic and epistemically completely nondeterministic.

In order to make quantum endophysics operational, Johann von Neumann [28, chapt. V.1] added to his endophysical codification the notorious projection postulate which formalized the so-called reduction of the wave packet. The projection postulate describes a stochastic time evolution, an instantaneously acting, discontinuous and non-causal change generated by an experimental intervention. This recipe works, but to use two fundamentally different types of time evolution is theoretically unacceptable. The idea that the stochastic time evolution might be derived from the bidirectional deterministic automorphic dynamics of endophysics is the starting point of von Neumann's [28, chapt.VI] theory of measurements. However, von Neumann's attempt to discuss measurements from the inside in terms of the deterministic endodynamics is self-referential and leads to an infinite regress of observing observers. Compared with classical endophysics, in quantum endophysics the situation is aggravated since an endophysical observer is entangled with the observed system by Einstein-Podolsky-Rosen correlations. That is, the reduced state of an endophysical observer is not pure, hence he can have no individuality, and we should actually not use the concept of an «observer».

---

[1] The analogy with endomathematical systems with undecidable true propositions is striking and seems to imply a conceptually deep relationship between chaos and semantically undecidable propositions (compare Agnes and Rasetti, [1]).

It has become popular to conjecture that the impossibility to solve von Neumann's measurement problem is due to the limited validity of von Neumann's codification of quantum mechanics. It is certainly true that Hilbert-space quantum mechanics is inappropriate for the description of macroscopic measurement instruments but this remark does not hit the crux of the problem. The endo-physical self-reference problem exists owing to purely logical reasons and remains also in codifications which can deal with systems having infinitely many degrees of freedom and classical observables. Arthur Komar [21] has shown that there is in general no effective procedure for determining whether or not two arbitrary given states of a quantum endosystem having an infinite number of degrees of freedom are disjoint. That is, there is in general no effective endophysical procedure for determining whether two arbitrarily given states show quantum-mechanical interference effects or not. The corresponding purely mathematical decision problem is known to be recursively unsolvable. Hence there are true physical statements which are endophysically inaccessible.

The impossibility of a complete endophysical description of the measuring process is not a flaw of the theory but a logical necessity in any theory which is self-referential, as it attempts to describe its own means of verification [32]. Every apparatus which realizes the reduction of the wave packet is necessarily a metatheoretical object and has to be discussed exophysically. That is, for a detailed physical description of the measuring process we need an at least two-leveled theory where the second level represents the metatheory [6, 15, 56].

How can one move beyond a system from within that system? Using only arguments from endophysics, this is impossible. But we may assume that the endosystem is very large and includes everything which can be described with the tools of physics. Then we can rephrase the initial question: How can one divide the endoworld into an object system and an observing system? Thereby we have to give thought to the fact that the division of an endoworld into an object system and an observing system is neither arbitrary nor unique. While it is at least logically consistent to posit that quantum endophysics (in an appropriate codification) is universally valid, an exophysical operationalization never can exhaustively describe the whole world. Something must remain unanalyzed. The need for contingent elements is the price we have to pay for making the world visible.

In a quantum endosystem a division into a part "which sees" and a part "which is seen" is possible if and only if between the two parts there are at every instant no Einstein-Podolsky-Rosen correlations [38, 39]. This is possible if and only if one of the parts is a purely classical system, characterized by a commutative algebra of observables. Such a situation can arise only if the basic endophysical system is very rich, that is, if the C^*-algebra of the endophysical intrinsic observables is essentially larger than the C^*-algebra of compact operators (e.g. an anti-liminary C^*-algebra). In general, classical observables are not intrinsic but do not emerge till a GNS-W^*-representation of the algebra of intrinsic observables has been constructed. The important point is that – with the only exception of the C^*-algebra of compact operators – there are infinitely many physically inequivalent W^*-representations of the basic C^*-algebra of intrinsic observables of endophysics. These inequivalent W^*-representations correspond to different exophysical descriptions of one and the same endosystem.

It is a pleasant surprise that we have in fact the tools in order to step out of the endoworld, namely the GNS-construction. Perhaps less surprising is the fact that such a construction of quantum exophysics from quantum endophysics is highly nonunique. Since the external observing system has to be classical, we will select those GNS-representations whose W^*-algebras have a nontrivial center but this does not yet make the problem unique. Every GNS-representation, hence every exophysical description, is related to a particular abstraction and idealization. An exophysical representation is possible only by a deliberate lack of interest, a decision of what we consider as relevant and what as irrelevant. Such a choice is not written down in the endophysical first principles of the basic theory but has to be introduced by some appropriate and well-chosen abstractions. Abstractions from not directly relevant aspects break the holistic unity of nature and are prerequisite to create a *perceptible* reality.

To summarize, if we have a sufficiently rich quantum endosystem, we can construct many partial exophysical descriptions of it, where the respective observers live in classical worlds so that they are quantum-mechanically disentangled from the object systems. While quantum endophysics is governed by first principles and has no (or very few) classical observables and broken symmetries, every exophysical description is context-dependent, has no univeral first principles and displays many classical observables and broken symmetries. A priori all exophysical descriptions are equally legitimate but they are as a rule mutually exclusive. In order to select one of these complementary descriptions we need in addition to the first principles of endophysics some regulative principles for the description of the classical exoworld in which the observer lives. These additional principles may for example reflect the cognitive viewpoints of the external observer, observer or thinker, that is, they represent a particular set of presuppositions and attitudes associated with some notions of ability, interest and value. If it turns out that these regulative principles can be implemented by physical descriptions, they in general involve breakings of symmetries of the basic endophysical description.

9. BACONIAN NORMATIVE PRINCIPLES

In any experiment there is a division of the universe of discourse into an object system and an external world (including the experimenter). It is a tacit assumption of all engineering sciences that nature can be manipulated and that the initial conditions required by experiments can be brought about by interventions of the world external to the object under investigation. That is, we assume that the experimenter has a certain freedom of action which is not accounted for by the first principles of physics. Without this freedom of choice experiments would be impossible. Man's free will implies the ability to carry out actions, it constitutes his essence as an actor. We act under the idea of freedom but the present topic under discussion is neither man's sense of personal freedom as a subjective experience, nor the question whether this idea could be an illusion or not, nor any questions of moral philosophy, but that *the framework of experimental natural science requires the freedom of action as a constitutive though tacit presupposition.* If we would like to derive a theoretical framework for experimental science from an endophysical basic theory, we have to make the very same preassumptions and abstractions as the experimenter makes in his work in the laboratory.

The methodology of traditional scientific research is essentially determined by Francis Bacons's motto *dissecare naturam* (to dissect nature), and has developed a preferred way of dividing the world into objects and observing systems. The *regulative principles* of Baconian science stress the facticity of the past, the probabilistic predictability of the future, and the idea that we can learn from the past for the future. These regulative principles can be realized by choosing a particular GNS-representation, generating a spontaneous breakdown of the time-reflection symmetry of basic endophysics and realizing backward deterministic and forward nondeterministic processes in the classical part of the exosystem. Till today, all well-confirmed applications of quantum mechanics belong to such a Baconian science. That is, the human actions are performed via an intermediate *engineering level*. All the tools of this engineering level have a valid description in terms of *classical* physics and the associated engineering sciences. Since human free will is prerequisite for experimental research, it cannot be directly a subject matter of Baconian science. In the experimental sciences we have therefore to distinguish between at least three hierarchical levels: (i) a first level, governed by the first principles of the basic theory, (ii) a second level, governed by classical physics and engineering science, (iii) a third level which includes the phenomena of man's free will and consciousness.

From our present point of view, the main regulative principles of Baconian exophysics are the following:

(i)   It is possible to dissect nature in such a way that the second level of its exophysical description is not Einstein-Podolsky-Rosen-correlated with the first level of quantum objects. That is, the second level represents a domain with a Boolean logical structure.

(ii) On the second level there are mechanistic but no teleological processes. That is, all processes turning up on the second level are backward deterministic, and there are some processes which are forward purely nondeterministic.

(iii) On the second level, all hardware is realizable by non-anticipatory input-output systems, i.e. by systems whose present states are determined by the past history (*the cause must be prior to the effect*, a property named "causality" by the engineers).

(iv) There exists a third level which can influence the lower levels without being influenced by them, and which is not subject matter of the basic physical theory.

The first condition refers to the (approximate) separability of the external world from the quantum object, it is the prerequisite for the existence of all exophysical descriptions. The second requirement pertains to time-inversion symmetry. It is not enough to break the time-inversion invariance, one must also select a particular solution. If the Baconian regulative principles are justifiable at all for an exophysical description of a bidirectional deterministic endophysical reality, then with the very same arguments the corresponding teleological regulative principles are also compatible with the first principles of endophysics. From this perspective, the admissability of final descriptions depends in a crucial way on the logical coherence of the Baconian scientific world view. The tacit assumptions about the third level are highly problematic. Nevertheless, there is no direct logical contradiction since determinism refers to endophysics while free will and consciousness refer to a higher level necessary for exophysics only. Considering the present state of the art, we may accept the assumptions for the third level at least as a tentative working hypothesis.

If we adopt quantum theory as basic, the first level can be described by a phenomenological quantum theory of open systems. This theory is different from fundamental quantum endophysics, but if we have enough information about the presupposed structure of the higher levels, we can hope to be able to derive the appropriate exophysical theory of open quantum system from the endophysical first principles. In Baconian science the intermediate level depends on the existence of classical hardware, described mechanistically (i.e. non-teleologically) by classical physics and engineering science, both being based on a Boolean logical structure[1]. This second classical level can be generated by an appropriate GNS-representation of the endophysical reality, but it cannot be considered as "the macrophysical limiting case of quantum physics". The dynamical behavior of the tools of this engineering level is affected by influences determined by human actions in the past which are described at the second level by time-dependent external parameters. The temporal asymmetry of the behavior of the open quantum system rests on the asymmetric time evolution of the second level. The Baconian endophysical quantum-theoretical description at the first level is then no longer deterministic but probabilistic, where the probabilities are primary and refer to possibilities in the future. *Probabilities appear in quantum mechanics exclusively in connection with state changes induced by external classical influences*. In the phenomenological description of the measuring process in historical quantum mechanics this breakdown of the time-reflection symmetry of basic endophysics is usually summarized by the phrase "collapse of the wave packet". In a full theoretical description, the coupling between the object system and the classical measuring tools leads in the exophysical description in general to a non-automorphic time evolution of the combined W*-system. The corresponding reduced description of the open object system is then given by a nonlinear semigroup (e.g. by a nonlinear Schrödinger equation for the state vector [40]).

---

[1] This requirement corresponds to Bohr's [3] postulate that the external conditions of a quantum-mechanical experiment have to be described in the language of classical physics.

The predictions of quantum mechanics are context-dependent but in no way more observer-dependent than in classical physics. It is a simple fact of experimental physics that all physical experiments which have been performed so far are in their last stages describable in terms of classical physics and classical engineering science. In principle, all physical experiments can be automated to the extent that the role of the human observer is reduced to simple acts of cognition of the numeric displays of classical measuring instruments. Hence the free will and the awareness of the observing scientist play exactly the same role they have in classical physics and engineering science.

The Baconian world view is compatible with but not implied by the first principles of quantum theory. It is supported by its scientific fertility and the achievements of engineering science, and is widely though not universally accepted. But its support is not so strong that we could say that another choice of the regulative principles of endophysics is inconceivable. If the basic theory is time-inversion invariant and if the algebra of observables contains noncompact intrinsic observables, then there exist symmetry-broken GNS-representations representing purely anticipating or purely non-anticipating processes. But starting with first principles, there is no logical route to select a particular symmetry-broken realization. There are good reasons for believing that both realizations are legitimate in appropriate contexts, *yet contexts are not given by first principles*. That is, «the arrow of time» is part of an epistemic description and should neither be fixed universally by decree nor be transferred to strongly objective ontic interpretations. Francis Bacon restricted science to the investigation of material and efficient causes and attacked in particular the use of final causes. Since almost any effect could be regarded as useful for some purpose, he ridiculed any teleologically oriented research method : "like a virgin consecrated to God it produces nothing" [2]. Still, the dissipativity and the backward determinism of Baconian science are *posited, not derived*. They are derivable from the basic endophysical laws only if we add appropriate boundary conditions. The Baconian method not always works but this mode of explication sets its own well established normative standards of judgement we often tacitly adopt for our research [43]. However, we have to note that final causes are not forbidden by the first principles of quantum physics. A quantum system can manifest itself in a variety of classical modes of appearance. It would be contrary to reason to eliminate one of these in favor of another. It is quite conceivable that backward causal and final descriptions, the structure-function duality, and the paradox of determinism and free will are just examples of complementary classical descriptions of a non-Boolean reality [41].

In summary, let us draw some conclusions. I hope I have made clear that von Neumann's codification is simply not appropriate for discussing philosophical inquiries into the foudation of quantum theory. Yet, a straightforward generalization – algebraic quantum mechanics – can explain rigorously and without any ad hoc assumptions the existence of spontaneous symmetry breakings and the emergence of superselection rules and classical observables. It must be emphasized, however, that the first principles of endophysics cannot explain exophysical normative principles like the Baconian rejection of the existence of final processes, our presupposed freedom to create initial conditions, or the feasibility of «detached observers». The proper choice of normative principles is context-dependent. Accepting Baconian normative principles one can prove the necessity of a *Boolean* description of the exosystem, and explain how probability arises in quantum theory. Whether someone will succeed in extending this substantiation to non-Baconian exophysical descriptions, only the future will tell.

REFERENCES

[1]   C. Agnes and M. Rasetti: *Word problem, undecidability and chaos*. In: *Advances in Nonlinear Dynamics and Stochastic Processes II*. Ed. by G. Paladin and A. Vulpiani. Singapore. World Scientific. 1987. Pp. 1–24.

[2]   F. Bacon: *De Dignitate et Augumentis Scientarum*. 1623. English translation: *Of the Advancement and Proficience of Learning*. Oxford. 1640.

[3]   N. Bohr: *Quantum physics and philosophy, causality and complementarity.* In: *Philosophy in the Mid-Century. A Survey.* Ed. by R. Klibansy. Florence. La Nuova Italia Editrice. 1958. Pp. 308–314.

[4]   O. Bratteli and D. W. Robinson: *Operator Algebras and Quantum Statistical Mechanics. C*- and W*-Algebras, Symmetry Groups, Decomposition of States.* Vol. 1. New York. Springer. 1979.

[5]   O. Bratteli and D. W. Robinson: *Operator Algebras and Quantum Statistical Mechanics. Equilibrium States. Models in Quantum Statistical Mechanics.* Vol. 2. New York. Springer. 1981.

[6]   M. L. d. Chiara: *Logical self reference, set theoretical paradoxes and the measurement problem in quantum mechanics.* J. Philosophical Logic 6, 331–347 (1977).

[7]   P. Curie: *Sur la symétrie dans les phénomènes physiques.* Journal de Physique 3, 393 (1894). Reprinted in: *Oeuvres de Pierre Curie.* Paris. Gauthier-Villars. 1908. Pp. 118–141.

[8]   P. A. M. Dirac: *The Principles of Quantum Mechanics.* London. Oxford University Press. 1930.

[9]   J. Dixmier: *C*-Algebras.* Amsterdam. North-Holland. 1977.

[10] D. A. Dubin: *Solvable Models in Algebraic Statistical Mechanics.* Oxford. Oxford University Press. 1974.

[11] A. Einstein: *Einleitende Bemerkung über Grundbegriffe.* In: *Louis de Broglie. Physicien et Penseur.* Paris. Editions Albin Michel. 1953.

[12] B. d'Espagnat: *Conceptual Foundations of Quantum Mechanics.* Second ed. London. Benjamin. 1976.

[13] B. d'Espagnat: *Physics and reality.* In: *Atti del Congresso Logica e Filosofia della Scienza, oggi. Vol. II Epistemologia e logica induttiva.* Bologna. CLUEB. 1986. Pp. 73–78.

[14] D. Finkelstein: *Finite physics.* In: *The Universal Turing Machine. A Half-Century Survey.* Ed. by R. Herken. Hamburg. Kammerer & Unverzagt. 1988. Pp. 349–376.

[15] D. Finkelstein and S. R. Finkelstein: *Computational complementarity.* Int. J. Theor. Phys. 22, 753–779 (1983).

[16] K. Gödel: *Über formal unentscheidbare Sätze der Principia Mathematica und verwandter Systeme I.* Monatshefte für Mathematik und Physik 38, 173–198 (1931).

[17] A. S. Holevo: *Probabilistic and Statistical Aspects of Quantum Theory.* Amsterdam. North-Holland. 1982.

[18] R. V. Kadison and J. R. Ringrose: *Fundamentals of the Theory of Operator Algebras. I. Elementary Theory.* New York. Academic Press. 1983.

[19] R. V. Kadison and J. R. Ringrose: *Fundamentals of the Theory of Operator Algebras. II. Advanced Theory.* Orlando. Academic Press. 1986.

[20] A. N. Kolmogorov: *Grundbegriffe der Wahrscheinlichkeitsrechnung.* Berlin. Springer. 1933.

[21] A. Komar: *Undecidability of macroscopically distinguishable states in quantum field theory.* Phys. Rev. 133, B542-B544 (1964).

[22] B. O. Koopman: *Hamiltonian systems and transformations in Hilbert space.* Proc. Nat. Acad. Sci. U.S. 17, 315–318 (1931).

[23] W. Kuyk: *Complementarity in Mathematics.* Dordrecht. Reidel. 1977.

[24] W. E. Lamb: *Schrödinger's cat.* In: *Reminiscences About a Great Physicist: Paul Adrien Maurice Dirac.* Ed. by B. N. Kursunoglu and E. P. Wigner. Cambridge. Cambridge University Press. 1987. Pp. 249–261.

[25] L. H. Loomis: *Note on a theorem by Mackey.* Duke Math. J. 19, 641–645 (1952).

[26] G. W. Mackey: *A theorem of Stone and von Neumann.* Duke Math. J. 16, 313–326 (1949).

[27] J. v. Neumann: *Die Eindeutigkeit der Schrödingerschen Operatoren.* Mathematische Annalen 104, 570–578 (1931).

[28] J. v. Neumann: *Mathematische Grundlagen der Quantenmechanik.* Berlin. Springer. 1932.

[29] J. v. Neumann: *Zur Operatorenmethode in der klassischen Mechanik.* Annals of Mathematics 33, 587–642, 789-791 (1932).

[30] W. Pauli: *Naturwissenschaftliche und erkenntnistheoretische Aspekte der Ideen vom Unbewussten.* Dialectica 8, 283–301 (1954).

[31] G. K. Pedersen: *C*-Algebras and their Automorphism Groups.* London. Academic Press. 1979.

[32] A. Peres and W. H. Zurek: *Is quantum theory universally valid?* Amer. J. Phys. 50, 807–810 (1982).

[33] C. Piron: *Quantum mechanics: fifty years later.* In: *Symposium on the Foundations of Modern Physics. 50 Years of the Einstein-Podolsky-Rosen Gedankenexperiment.* Ed. by P. Lahti and P. Mittelstaedt. Singapore. World Scientific. 1985. Pp. 207–211.

[34] M. Polanyi: *Personal Knowledge*. London. Routledge and Kegan Paul. 1958.

[35] H. Primas: *Theory reduction and non-Boolean theories*. J. Math. Biology **4**, 281–301 (1977).

[36] H. Primas: *Foundations of theoretical chemistry*. In: *The New Experimental Challenge to Theorists*. Ed. by R. G. Woolley. New York. Plenum Press. 1980.

[37] H. Primas: *Chemistry, Quantum Mechanics and Reductionism*. Berlin. Springer, second edition 1983. 1981.

[38] H. Primas: *Objekte in der Quantenmechanik*. In: *Grazer Gespräche 1986. Ganzheitsphysik*. Ed. by M. Heindler and F. Moser. Graz. Technische Universität Graz. 1987. Pp. S. 163–201.

[39] H. Primas: *Contextual quantum objects and their ontic interpretation*. In: *Symposium on the Foundations of Modern Physics, 1987. The Copenhagen Interpretation 60 Years after the Como Lecture*. Ed. by P. Lahti and P. Mittelstaedt. Singapore. World Scientific. 1987. Pp. 251–275.

[40] H. Primas: *Induced nonlinear time evolution of open quantum objects*. In: *Sixty-two Years of Uncertainty: Historical, Philosophical and Physics Inquiries into the Foundations of Quantum Mechanics*. Ed. by A. I. Miller. New York. Plenum. 1990.

[41] H. Primas: *Time-asymmetric phenomena in biology. Complementary exophysical descriptions arising from deterministic quantum endophysics*. In: *Proceedings of the International Workshop "Information Biothermodynamics"*. Torun. In press.

[42] G. A. Raggio: *States and Composite Systems in W*-algebraic Quantum Mechanics*. Thesis ETH Zürich, No. 6824, ADAG Administration & Druck AG, Zürich. 1981.

[43] F. Rapp: *Kausale und teleologische Erklärungen*. In: *Formen teleologischen Denkens*. Ed. by H. Poser. Berlin. TUB-Dokumentation, Heft 11, Technische Universität Berlin. 1981. Pp. 1–15.

[44] A. Rosenberg: *The number of irreducible representations of simple rings with no minimal ideals*. Amer. J. Math. **75**, 523–530 (1953).

[45] O. E. Rössler: *Endophysics*. In: *Real Brains, Artificial Minds*. Ed. by J. L. Casti and A. Karlqvist. New York. North-Holland. 1987. Pp. 25–46.

[46] J. Rothstein: *Thermodynamics and some undecidable physical questions*. Philosophy of Science **31**, 40-48 (1964).

[47] S. Sakai: *C*-Algebras and W*-Algebras*. Berlin. Springer. 1971.

[48] G. L. Sewell: *Quantum Theory of Collective Phenomena*. Oxford. Oxford University Press. 1986.

[49] A. Shimony: *The Quantum World*. Manuscript for a yet unpublished book, distributed at a seminar at the ETH Zürich, Nov. 1984.

[50] A. Shimony: *Events and processes in the quantum world*. In: *Quantum Concepts in Space and Time*. Ed. by R. Penrose and C. J. Isham. Oxford. Clarendon Press. 1986. Pp. 182-203.

[51] G. Spencer Brown: *Laws of Form*. London. Allen and Unwin. 1969.

[52] M. H. Stone: *Linear transformations in Hilbert space. III. Operational methods and group theory*. Proc. Nat. Acad. U.S. **16**, 172–175 (1930).

[53] F. Strocchi: *Elements of Quantum Mechanics of Infinite Systems*. Singapore. World Scientific. 1985.

[54] A. Tarski: *Logic, Semantics, Metamathematics*. Oxford. Clarendon Press. 1956.

[55] A. Tarski: *Truth and proof*. Scientific American **220**, 63-77 (1969).

[56] M. Zwick: *Quantum measurement and Gödel's proof*. Speculations in Science and Technology **1**, 135-145, 422-423 (1978).

INDUCED NONLINEAR TIME EVOLUTION OF OPEN QUANTUM OBJECTS

Hans Primas

Laboratory of Physical Chemistry
ETH-Zentrum
CH-8092 Zürich
Switzerland

1. A THEORY OF OPEN SYSTEMS SHOULD BE BASED ON AN INDIVIDUAL DESCRIPTION

In the first part [42] I have emphasized the fact that there is a logically consistent formulation of quantum mechanics of *individual* systems which is compatible with all empirical data. A necessary and sufficient condition for the feasibility of such an individual interpretation is that the referents of the theory are *objects*. We recall that we distinguish between the concepts «system» and «object». By a system we mean nothing but the referent of a theoretical discussion (specified, for example, by a Hamiltonian), without any ontological commitment. An object is defined to be an *open* quantum system which is *interacting with its environment* but which is not Einstein-Podolsky-Rosen-correlated with the environment. Quantum systems which are not objects are entangled with their environments, they have no individuality and allow only an incomplete description in terms of statistical states. Since in quantum theories the set of all statistical states is not a simplex, statistical states have no unique decomposition into extremal states. This fact leads to grave problems for a purely statistical interpretation of quantum mechanics.

Sometimes it has been claimed that macroscopic objects cannot be assumed *to be* in definite pure states [60,61]. It is certainly true that macroscopic systems never can be considered as isolated and that we cannot get an exhaustive knowledge of their states, but the conclusion that this implies the necessity of a description in terms of density operators is due to a confusion of ontic and epistemic descriptions. Ontic interpretations refer to the properties of the object *itself,* a feature of the world. Epistemic interpretations refer to our *knowledge* of the properties or the modes of reactions of the object; an epistemic state provides merely a "state of knowledge" and not an objective description of some aspects of independent physical reality. Nevertheless, every object, microscopic or macroscopic, strongly or weakly interacting with its environment, has always a description in terms of *pure states*. The distinction between epistemic and ontic states is important since the *ontic* states – and not our knowledge of them – determine the effects of the interactions with the environment. It would be a bold undertaking to use quantum mechanics in its epistemic statistical ensemble interpretation to develop a fundamental theory of open systems.

Sixty-Two Years of Uncertainty
Edited by A. I. Miller
Plenum Press, New York, 1990

It is dangerous to switch thoughtlessly between an ontological and an operationalistic way of speaking. As a first example for difficulties which may arise from such an confusion, we may ponder about the meaning of the usual wording «the environment *is* in a thermal state». A thermal state may indeed represent the best information *we* can get. But according to the individual interpretation, the combined system of the object system and its environment always is in a pure state. We know from mathematically exactly soluble models that these two ways of description may give qualitatively different results. Another example refers to the time-inversion symmetry. In engineering physics, only retarded solutions which fulfill the causality requirement «no response before the stimulus» are relevant. However, this does not imply that the actual interactions are of this kind. Without an object-environment separation the physically relevant interactions are time-inversion-invariant.

Besides the greater explanatory power, the individual formulation of quantum mechanics has important *technical* advantages. For the present problems it concerns the question how to characterize states of composed quantum systems which do not show Einstein-Podolsky-Rosen-correlations (EPR-correlations, for short). For pure states the solution is trivial: *a pure state is free from EPR-correlations if and only if it is a product state.* A nonpure state is defined to be free from EPR-correlations if there exists a decomposition into a sum (or integral) of pure states without EPR-correlations. However, till today no applicable criteria are known for the distinction between EPR-correlations and classical statistical correlations also present in an ensemble description.

2.   NONLINEAR SCHRÖDINGER EQUATIONS

STATISTICAL TIME EVOLUTIONS ARE LINEAR

In the literature there is some confusion about the meaning of the concept of *linearity* of the equations of motion[1]. In the traditional irreducible Hilbert-space formulation of quantum mechanics this confusion is due to the fact that the individual states and the statistical states are elements of the very same Banach space[2], nevertheless their equations of motion may be different. If we accept Mackey's axiom IX [32, p.81], which postulates the commutativity of the operation of mixing with the time evolution semigroup for a statistical ensemble, then it can be shown that under weak technical conditions the dynamics can be uniquely extended to a one-parameter semigroup of *linear* operators acting in the predual of the algebra of global observables [16]. That is, *the equations of motion for statistical states are necessarily linear.* This result has nothing to do with the quantum-mechanical superposition principle, it is also true for classical statistical theories. Well-known examples are Koopman's [28] Hilbert-space formalism for classical statistical mechanics, and the equivalence of *nonlinear* stochastic differential equations in the sense of Itô (which provide an individual description) with the *linear* Fokker-Planck equations (which give a statistical ensemble description)[1]. The rash argument "that linear equations are usually an approximation to a more adequate theory" [56] certainly does not apply to the linear

---

[1] For a recent example, compare the discussion between Peres and Weinberg [Peres, 40].

[2] Given the Hilbert space $\mathcal{H}$, the individual states are elements of the dual $(\mathcal{B}_\infty)^*$ of the C*-algebra $\mathcal{B}_\infty$ of compact operators acting in $\mathcal{H}$. The statistical states are elements of the predual $(\mathcal{B})_*$ of the algebra $\mathcal{B}$ of all bounded operators acting in $\mathcal{H}$. A possibility for confusion arises from the generically exceptional relation $(\mathcal{B}_\infty)^* = (\mathcal{B})_*$.

Fokker-Plack equations, for the same reason it does not apply to the time-dependent Schrödinger equation in the statistical ensemble interpretation[1].

THE TIME EVOLUTION OF AN INDIVIDUAL OBJECT MAY BE NONLINEAR

An object is at every instant in a pure state, so that *the time evolution for an object necessarily transforms pure states into pure states.* If the object is a system with only finitely many degrees of freedom (the environment of the system may be an infinite system), then there exists an irreducible Hilbert-space representation for the object, so that the state of the object can be represented in the familiar way by a state vector. The time evolution then transforms state vectors into state vectors. If the time evolution is a semigroup, then there exists an analogue to the time-dependent Schrödinger equation but this evolution equation is neither necessarily linear nor deterministic. Quite generally, the reduced dynamics for the state vector of an object is given by a *nonlinear non-autonomous Schrödinger–Langevin functional–differential equation.* Thereby the dissipative environmental effects are represented by classical fluctuating forces and damping terms. Furthermore, polarizations and reaction fields can lead to feedback effects which give rise to *nonlinear* dynamical equations. The usefulness of such nonlinear Schrödinger equations is well documented in the literature, but a clear appreciation of the nature and the validity of the description by nonlinear Schrödinger equations seem to be lacking. Nevertheless, it may be useful to review some of the physically well-motivated phenomenological approaches[2].

NONLINEAR FEEDBACK EFFECTS

The first conceptually sound discussion of feedback effects via the environment of molecular systems is due to Onsager [36]. He observed that the severe disagreement of Debye's dipole theory with the experiment is due to the neglect of the *reaction field* in Debye's theory. Rephrasing Onsager's physical ideas in a phenomenological quantum-mechanical language [59], one may write the effective molecular Hamiltonian H in the form $H = H_0 + V$, where H_0 represents the molecular Hamiltonian in the vacuum, and V denotes the interaction energy between the molecular system and its environment. This effective interaction V can be approximated by the interaction energy between the molecular dipole moment and the reaction field due to the polarization of the surrounding medium,

$$V = -\mu E_R \quad ,$$

where μ denotes the electric dipole moment operator of the molecular system, and E_R is the electric field strength of the reaction field which may be represented by Onsager's formula

$$E_R = \lambda \langle \Psi | \mu | \Psi \rangle \quad ,$$

where λ can be regarded as a parameter representing the degree of the polarizability of the environment. In this way one gets the following phenomenological *nonlinear* Schrödinger equation

---

1   As pointed out first by Carleman [6], any nonlinear differential equation can be converted (in numerous ways) into a linear differential equation of infinite order. Quite generally, with any system of nonlinear ordinary differential equations there is associated a linear first-order Liouville equation [46].

2   In the literature there are numerous papers on ad hoc modifications of the Schrödinger equations with the sole purpose of achieving reductions of wave packets, spontaneous localizations, or damping phenomena. These approaches may be heuristically interesting but we will not discuss them here, just as little as purely mathematically inspired investigations.

$$i\hbar \frac{\partial \Psi_t}{\partial t} = \{H_0 - \lambda \mu \langle \Psi_t | \mu | \Psi_t \rangle\} \Psi_t \ ,$$

whereby the nonlinearity is due to the feedback term $\langle \Psi_t | \mu | \Psi_t \rangle$ which itself depends on the state of the molecular system.

THE SCHRÖDINGER–MAXWELL EQUATIONS

A similar problem has been discussed by Ulmer and Hartmann [55]. They start with the usual time-dependent Schrödinger equation of a particle of mass m, charge e, and spin zero,

$$i\hbar \frac{\partial \Psi}{\partial t} = \frac{1}{2m} \{P - eA(Q)\}^2 \Psi + eV(Q) \Psi \ ,$$

where the potentials V and A obey Maxwell's equations

$$\varepsilon \Box V = -\rho \ , \quad \Box A = -\mu J \ .$$

If the potentials are considered as *classical* fields, their sources are given by the expectation values of the charge and the current density operators,

$$\rho = e \Psi \Psi^* \ , \quad J = \frac{\hbar e}{2im} \{\Psi^* \nabla \Psi - \Psi \nabla \Psi^*\} \ .$$

Solving Maxwell's equations and eliminating the potentials, one gets a non-local integro-differential equation for the state vector Ψ. In a nonrelativistic approximation one expects that the dominant contribution comes from the electrostatic field,

$$\varepsilon \triangle V = -\rho \ .$$

In this case the corresponding nonlinear Schrödinger equation is given by

$$i\hbar \frac{\partial \Psi(q,t)}{\partial t} = \frac{1}{2m} P^2 \Psi(q,t) + \frac{e^2}{4\pi\varepsilon} \int_{\mathbb{R}^3} \frac{\Psi(r,t) \Psi^*(r,t) \Psi(q,t)}{|q-r|} dr \ .$$

A related approach – using the coupled Dirac–Maxwell equations – has been applied by Barut et al. [2,3] to discuss the radiative processes in quantum electrodynamics, in particular to compute the Lamb shift, spontaneous emission, and the Casimir–Polder van der Waals forces without field quantization. I do not necessarily agree with the basic philosophy of these authors, but their results are remarkable: from our present viewpoint they show that *nonlinear feedback effects from a classical environment can account for phenomena usually considered as purely quantal.*

3. *THE DYNAMICS OF OBJECTS*

A FACTORIZATION THEOREM FOR INDIVIDUAL STATES

The concept of an endophysical description by a C*-algebra of intrinsic observables and contextual exophysical descriptions by W*-algebras of global observables leads to important questions regarding the practical application of the formalism of algebraic quantum mechanics. If we are able to specify an appropriate topology on the endophysical state space, then we can rigorously derive the dynamics relevant for the corresponding exophysical description. As discussed in the first part[1], an explicit construction of an algebra of contextual observables via the GNS-representation with respect to an appropriate reference state is in general unworkable. *Nevertheless, we are in the position to write down the general structure for the equations of motion of a quantum object interacting with its classical environment.* Thus we can get important structural information about the nonlinear Schrödinger equations which govern the motion of objects.

Raggio's Theorem[2] says that the environment of a quantum object necessarily is represented by a *commutative* *-algebra. In an exophysical description, the algebra $\mathcal{M}$ of contextual observables of the object and environment is therefore given by a W*-tensor product,

$$\mathcal{M} = \mathcal{M}_{obj} \otimes \mathcal{M}_{env} \ ,$$

where $\mathcal{M}_{obj}$ is the W*-algebra of contextual observables of the object, and $\mathcal{M}_{env}$ the W*-algebra of contextual observables of the environment. While these W*-algebras characterize the *statistical* description, they do not yet contain all information necessary for an *individual* description. In the following, we will assume that the object is a finite and purely quantal system so that $\mathcal{M}_{obj}$ is a factor of type I, which can be represented by the algebra $\mathcal{B}(\mathcal{H})$ of all bounded operators acting in some separable Hilbert space $\mathcal{H}$,

$$\mathcal{M}_{obj} \cong \mathcal{B}(\mathcal{H}) \ .$$

The contextual statistical states of the object are elements of the predual $\mathcal{B}(\mathcal{H})_*$ of $\mathcal{B}(\mathcal{H})$, which is isomorphic to the Banach space $\mathcal{B}_1(\mathcal{H})$ of all nuclear operators acting on $\mathcal{H}$, which in turn is isomorphic to the dual of the C*-algebra $\mathcal{B}_\infty(\mathcal{H})$ of all compact operators,

$$\mathcal{B}(\mathcal{H})_* \cong \mathcal{B}_1(\mathcal{H}) \cong \mathcal{B}_\infty(\mathcal{H})^* \ .$$

There is a one-to-one correspondence between the pure statistical states (i.e. extremal elements of $\mathcal{B}_1$) and extremal states in the dual of the C*-algebra $\mathcal{B}_\infty$. That is, *the corresponding algebra of contextual intrinsic object-observables is uniquely given by the C*-algebra $\mathcal{B}_\infty(\mathcal{H})$ of compact operators acting in $\mathcal{H}$.*

However, for an individual description we need additional information about the structure of the environment. Here we *assume* that the commutative W*-algebra is given by the Lebesgue space $L^\infty(\Omega, \Sigma, \mu)$, where Ω is a separable symplectic phase space, Σ is the σ-algebra of Borel sets of Ω, and μ is the Lebesgue measure. In this case, the algebra of intrinsic observables is given by the separable C*-algebra $C_\infty(\Omega)$ of all complex-valued continuous functions on Ω which vanish at infinity. Under these premises, the algebra of contextual observables appropriate for an individual description is given by the separable C*-algebra $\mathcal{A}$,

$$\mathcal{A} = \mathcal{B}_\infty(\mathcal{H}) \otimes C_\infty(\Omega) \ ,$$

where the C*-tensor product is uniquely given[3] and can be understood as the injective tensor product. The individual states of the combined system «object and environment» are given by the extremal normalized positive linear functionals on $\mathcal{A}$.

---

[1] Compare [42, sect. 4].

[2] Raggio [44], compare also [42, sect. 5].

[3] The reason is that one of the C*-algebras is commutative, compare e.g. Sakai [50], prop. 1.22.5 on pg. 62.

The crucial concept for the description of the individual behavior of quantal objects is based on the following proposition:

Theorem (Takesaki [54] p.211)

Let $\mathcal{B}$ be a C*-algebra with unity, and let C be a commutative C*-algebra. Every pure state φ on the C*-tensor product $\mathcal{A} = \mathcal{B} \otimes C$ is of product form,

$$\varphi = \rho \otimes \xi$$

for some pure state ρ on $\mathcal{B}$, and some pure state ξ on C.

The time evolution of the combined system can be characterized by the map $\varphi \rightarrow \varphi_t$, $t \in \mathbb{R}$, where the pure state φ_t characterizes the individual state at time t. Takesaki's theorem implies that at every instant t the individual state is a *product state*, i.e

$$\varphi_t = \rho_t \otimes \xi_t \ .$$

This result has the important consequence that in an *individual* description of an object interacting with its environment all *expectation values of product observables factorize:*

$$\varphi_t (B \otimes C) = \varphi_t(B \otimes 1) \, \varphi_t(1 \otimes C) = \rho_t(B) \, \xi_t(C) \ ,$$

for all $B \in \mathcal{B}$ and all $C \in C$.

STATISTICAL STATES DO NOT FACTORIZE

In a *statistical* description of the same system *a corresponding factorization property does not hold.* There are two reasons why a statistical description of an object may be desirable. First of all, for a classical system with an uncountable phase space the individual states are experimentally not accessible, so that every operational description has to be in terms of statistical states. Secondly, in order to integrate the equations of motion for the individual description one has to know the initial state (say at time $t = 0$). For a realistic environment, detailed information about the individual initial state χ of the environment certainly is not available. These facts force us to turn to a statistical description in the classical sense. That is, we specify the statistical state of the environment at time $t = 0$ by a probability distribution function f on the probability space (Ω, Σ, μ), where

$$\int_B f(\omega) \, \mu(d\omega) \ , \quad B \in \Sigma \ ,$$

gives the probability that the individual state characterized by $\omega \in \Omega$ lies in the Borel set B. The pure states of the C*-algebra $C_\infty(\Omega)$ are in a one-to-one correspondence with the points ω of Ω, i.e. for every pure state ξ on $C_\infty(\Omega)$ there is a point $\omega_0 \in \Omega$ such that

$$\xi(C) = C(\omega_0) \quad \text{for every } C \in C_\infty(\Omega) \ .$$

In this statistical ensemble the initial state of the environment is characterized epistemically by a statistical state $\Xi_0 \in C_\infty(\Omega)^*$, defined by the relation

$$\Xi_0(C) = \int_\Omega f(\omega) \, C(\omega) \, \mu(d\omega) \quad \text{for all } C \in C_\infty(\Omega) \ .$$

For classical systems there are no pure statistical states, but for finite and purely quantal systems there is a one-to-one correspondence between individual states and pure statistical states, so there is no compelling reason to average the initial conditions of the object system. If the initial state of the object system is given by $\rho \in \mathcal{B}_\infty(\mathcal{H})^*$, the very same state can also be considered as a pure statistical state, $\rho \in \mathcal{B}(\mathcal{H})_*$, so that the statistical initial state of the combined system is the product state $\phi \in \mathcal{M}_*$,

$$\phi = \rho \otimes \Xi \quad , \quad \rho \in (\mathcal{M}_{obj})_* \, , \quad \Xi \in (\mathcal{M}_{env})_* \quad ,$$

where Ξ denotes the extension of Ξ_0 to $L^1(\Omega, \Sigma, \mu)$. The time evolution $\phi \rightarrow \phi_t$ for the statistical states is defined in terms of the time evolution of the individual states, $\varphi \rightarrow \varphi_t$. Since this time evolution (in general) develops *classical* correlations between the object and its environment, the statistical state φ_t for $t > 0$ will no longer be a product state. Since at present we do not have the technical means to distinguish between such classical statistical correlations and EPR-correlations, a *direct* evaluation of the dynamics of an object interacting with its environment will be very difficult. However, a later statistical averaging over the individual classical trajectories is always possible, and is in fact not only technically but also conceptually preferable since it leads automatically to a well-defined statistical ensemble.

FACTORIZATION OF HEISENBERG'S EQUATION OF MOTION

Let $\mathcal{A} = \mathcal{B} \otimes C$ be the separable C*-algebra of the just discussed contextual individual description of an object interacting with its environment, where $\mathcal{B} = \mathcal{B}_\infty(\mathcal{H})$ and $C = C_\infty(\Omega)$. Assume we know Heisenberg's equation of motion for a suitable set $\{B_1, B_2, \dots\}$ of observables $B_j \in \mathcal{B}$ of the object system, and a suitable set $\{C_1, C_2, \dots\}$ of observables $C_m \in C$ of the environment. In the Heisenberg picture the time evolution can be written as

$$B_j(0) \quad \rightarrow \quad B_j(t)$$
$$C_m(0) \quad \rightarrow \quad C_m(t)$$

where $B_j(0) = B_j \otimes 1$ and $C_m(0) = 1 \otimes C_m$. In many important examples, Heisenberg's equations of motion are of the following general form:

$$dB_j(t)/dt = \sum_k \alpha_{jk} B_k(t) + \sum_n \beta_{jn} C_n(t) + \sum_{kn} \chi_{jkn} B_k(t) C_n(t) \quad ,$$

$$dC_m(t)/dt = \sum_k \zeta_{mk} B_k(t) + \sum_n \eta_{mn} C_n(t) + \sum_{kn} \lambda_{mkn} B_k(t) C_n(t) \quad ,$$

$$\alpha_{jk}, \beta_{jn}, \chi_{jkn}, \zeta_{mk}, \eta_{mn}, \lambda_{mkn} \in \mathbb{C} \quad .$$

For an arbitrary initial state $\phi \in \mathcal{A}^*$ we define the expectation values

$$b_j(t) = \phi\{B_j(t)\} \quad ,$$

$$c_m(t) = \phi\{C_m(t)\} \quad ,$$

and the correlation functions

$$f_{jm}(t) = \phi\{B_j(t) C_m(t)\} - \phi\{B_j(t)\} \phi\{C_m(t)\} \quad .$$

Heisenberg's equations of motion are then transformed into a system of ordinary differential equations:

$$db_j(t)/dt = \sum_k \alpha_{jk} b_k(t) + \sum_n \beta_{jn} c_n(t) + \sum_{kn} \chi_{jkn} b_k(t) c_n(t) + \sum_{kn} \chi_{jkn} f_{kn}(t) \; ,$$

$$dc_m(t)/dt = \sum_k \zeta_{mk} b_k(t) + \sum_n \eta_{mn} c_n(t) + \sum_{kn} \lambda_{mkn} b_k(t) c_n(t) + \sum_{kn} \lambda_{mkn} f_{kn}(t) \; .$$

These equations are *exact* but *not closed* since we have not written down the equations of motion for the correlation functions $t \rightarrow f_{kn}(t)$. In principle, there is no difficulty to set up differential equations for f_{kn} in terms of higher correlation functions. Iterating this procedure, one gets in general an *infinite* system of first order differential equations. However, if the initial state ϕ is a *pure* state, then the factorization theorem implies that the correlation functions f_{kn} vanish, so that we get the following *closed* system of first order differential equations for the real-valued expectation values b_j and c_m:

$$db_j(t)/dt = \sum_k \alpha_{jk} b_k(t) + \sum_n \beta_{jn} c_n(t) + \sum_{kn} \chi_{jkn} b_k(t) c_n(t) \; ,$$

$$dc_m(t)/dt = \sum_k \zeta_{mk} b_k(t) + \sum_n \eta_{mn} c_n(t) + \sum_{kn} \lambda_{mkn} b_k(t) c_n(t) \; .$$

If the environment can be modeled by an infinite linear system (e.g. by infinitely many harmonic oscillators coupled linearly to the object system), then the equations for the variables $t \rightarrow c_m(t)$ can be explicitly solved, so that the equations of motion for the object variables $t \rightarrow b_j(t)$ become a system of non-autonomous *nonlinear* integro-differential equations.

For a large class of simple but important models[1] the reduced dynamics of quantum objects can be represented by *nonlinear stochastic differential equations in the sense of Itô*. The corresponding statistical ensemble description is then given by the associated, mathematically equivalent *linear* Fokker-Planck equation. Such reduced description can be rigorously derived by choosing an appropriate GNS-representation of the C*-algebra of intrinsic observables for an object system interacting linearly with an infinite harmonic environment.

4. CLASSICAL DESCRIPTIONS OF QUANTUM SYSTEMS IN THE FRAMEWORK OF THE TRADITIONAL HILBERT-SPACE FORMALISM

HOW CAN WE GET THE EQUATIONS OF MOTION FOR AN OBJECT?

In the preceding section, we derived rigorously the general structure for the reduced dynamics of an object under the assumption that we know the relevant Heisenberg equations of motion. How can we get these equations? It would be naive to assume that these equations of motion are given by a one-parameter group of automorphisms of the C*-algebra

$$\mathcal{A} = \mathcal{B}_\infty(\mathcal{H}) \otimes C_\infty(\Omega)$$

of contextual observables of the individual description. In such a case the environment could influence the object, but the object could not influence the environment[1]. That is,

---

[1] Compare also section 5.

there could be no polarization effects of the environment, hence no feedback effects giving rise to a nonlinear evolution equation for the object. The time evolution can be expected to have a simple structure only on the level of an *endo*physical description. The exophysical dynamics has to be *derived* from the endophysical one by an appropriate GNS-construction[2] which accounts for the abstractions we are willing to accept. As I have already stressed, as a rule we cannot expect to be able to follow this route. However, we may have a good idea how the exosystem should look like.

If the object system is "small", i.e. if for the object system without environment the uniqueness theorem of Stone and von Neumann ist valid[3], then the appropriate equivalence class of reference states for a GNS-construction is determined by the characteristics of the environment only. Since quantum objects exist if and only if classical environments exist [45], the challenge to guess a suitable reference state is related to the *problem of how to describe classical phenomena in the framework of traditional quantum theory.*

HISTORICAL REMARKS

The idea that the formalism of quantum theory should contain classical mechanics as a limiting case was expressed by Max Planck as early as 1906: *"The classical theory can simply be characterized by the fact that the quantum of action becomes infinitesimally small"* [41, p.143]. Planck's correspondence principle and the later amplifications by Niels Bohr have played an outstanding role in the development of quantum mechanics. There is a widely held but mistaken belief that classical Hamiltonian mechanics is in some sense a limiting case of Galilean quantum mechanics. Many conceptually very different limiting procedures have been discussed extensively in the last 60 years, so for example:

(i)   the limit of high quantum numbers,

(ii)   the macroscopic limit of very many degrees of freedom,

(iii)   the high temperature limit,

(iv)   the limit of vanishing Planck's constant.

However, none of these limiting procedures is sufficient to reduce quantum mechanics to classical mechanics. All text-book treatments are incomplete, often misleading and not too seldom plainly erroneous. For example, high quantum numbers usually refer to energy eigenstates – but with the only exception of ground states, energy eigenstates *never* show a classical behavior. The limit of infinitely many degrees of freedom is ill-defined – there are uncountably infinitely many physically inequivalent representations. The high-temperature limit is legitimate *provided* there are no phase transitions – but the absence of phase transitions is in general extremely difficult to prove. There are a few modern mathematically rigorous expansions in powers of Planck's constant but from a conceptual point of view such merely mathematical investigations also leave much to be desired – Planck's constant $\hbar$ is not a

---

1   This statement follows from the trivial fact that the restriction of an automorphism of an algebra to its center is again an automorphism.

2   Note that even in the case that the endophysical time evolution is given by a one-parameter group of automorphisms, the time evolution of a W*-GNS-representation is in general *not* given by automorphisms.

3   Compare [42, sect. 2].

dimensionless parameter but a fundamental physical constant and cannot go to zero. What we need is a classical description which can be *embedded* in quantum theory *while ℏ retains its physical value*.

All modern investigations show that only for very special initial states and in general only for limited time intervals quantum systems may behave quasiclassically. This conclusion can be found already in Pauli's «New Testament»: *"Man erhält vielmehr nur dann Übereinstimmung mit den aus den klassisch mechanischen Bahnen abgeleiteten Eigenschaften des Systems, wenn man Pakete bauen kann, innerhalb deren die klassische Kraft wenig variiert und sie nur innerhalb solcher Zeiten zu betrachten braucht, während deren die Dimensionen des Paketes sich nur wenig verändern."* [37, Ziff.12, p.166]. Moreover, if a classical description of an individual quantum system exists, it is not given by a deterministic but by a *stochastic* classical dynamical system.

COHERENT STATES AS EXAMPLES FOR CLASSICAL QUANTUM STATES

Examples of purely quantal systems which behave exactly like classical systems are known since the first days of the inception of quantum mechanics. In 1926, Schrödinger [51] constructed a family of Gaussian wave packets which under the time evolution of the one-dimensional harmonic oscillator do not spread and follow the classical trajectory of the harmonic oscillator. These pure states are now called the *coherent states* associated with the harmonic oscillator. Schrödinger speculated that states with similar properties also should exist for the hydrogen atom but he was not able to construct them[1]. The completeness of Schrödinger's family of coherent states has been shown by von Neumann [35], the associated decomposition of unity by Klauder [19]. Schrödinger's coherent states are group-theoretically related to the Heisenberg-Weyl group, for arbitrary Lie groups coherent states were introduced for the first time by Perelomov [39]. The various families of coherent states[2] are the best-known and most useful examples for *classical quantum states*.

FAMILIES OF CLASSICAL QUANTUM STATES

Let Σ be a quantum system with a Hilbert space $\mathcal{H}$ of state vectors $\Psi \in \mathcal{H}$, a time evolution $\Psi \rightarrow \Psi_t \in \mathcal{H}$ for all $t \geq 0$. Let further $\mathcal{F}$ be a symplectic manifold of state vectors, $\mathcal{F} \subset \mathcal{H}$, such that $\mathcal{F}$ is total with respect to $\mathcal{H}$.

Definition

$\mathcal{F}$ *is called a family of classical states for* Σ,

if $\Psi \in \mathcal{F}$ *implies* $\Psi_t \in \mathcal{F}$ *for all* $t \geq 0$.

The time evolution of a quantum system with a classical family $\mathcal{F}$ of states and an initial state $\Psi \in \mathcal{F}$ is isomorphic to the time evolution of a classical Hamiltonian system with the phase space $\mathcal{F}$. The algebra of observables of this classical Hamiltonian system is given by the commutative W*-algebra $L^\infty(\mathcal{F})$.

---

[1] Compare the letter of H. A. Lorentz to Schrödinger on June 19, 1926, reprinted in [43].

[2] An elementary but decent crash-introduction to the theory of boson-coherent states can be found in chapter 7 of the book by Klauder and Sudarshan [27]. The reprint volumes edited by Mandel and Wolf [33], and by Klauder and Skagerstam [26] collect many important papers on coherent states. A nice summary of the group-theoretical approach to coherent states has been given by Perelomov [38].

As an example for a classical description of a quantum system, we consider a quantum system Σ with a Hilbert space $\mathcal{H}$ of state vectors $\Psi \in \mathcal{H}$, a connected Lie group G, a fixed cyclic vector $\Omega \in \mathcal{H}$, a family

$$\mathcal{F} = \{T(g)\,\Omega \mid g \in G\} \subset \mathcal{H}$$

of Perelomov coherent states, where $T : G \to \mathcal{B}(\mathcal{H})$ is an irreducible representation of G on $\mathcal{H}$. If the Hamiltonian of the time evolution $\Psi \to \Psi_t$ is generated by the elements of the Lie algebra of G, then $\mathcal{F}$ is a family of classical states for Σ. The corresponding classical phase space is given by G/G_Ω, where G_Ω is the stability subgroup of Ω.

QUASICLASSICAL DESCRIPTION OF QUANTUM SYSTEMS

If the time evolution is not generated by some Lie algebra, it may be difficult to find a family of classical quantum states. Sometimes it is, however, quite easy to find families of quantum states which behave in a very good approximation like classical quantum states.

Definition

$\mathcal{F}$ is called a family of (ε,T)-quasiclassical states for Σ, if for every $\Psi \in \mathcal{F}$

there is a vector $\Phi \in \mathcal{F}$, such that $\|\Psi_t - \Phi\| < \varepsilon$ for all $t \in [0,T]$.

The time evolution of a quantum system with a family $\mathcal{F}$ of (ε,T)-quasiclassical states and an initial state $\Psi \in \mathcal{F}$ is approximatively given by the time evolution of a classical Hamiltonian system with the phase space $\mathcal{F}$ and a commutative W*-algebra $L^\infty(\mathcal{F})$ of observables.

Many (but by no means all!) macroscopic systems have at least one family $\mathcal{F}$ of (ε,T)-quasiclassical states such that ε is so small and T so large that the quantum mechanical and the classical descriptions are indistinguishable within the domain validity of both theories.

As an example for a quasiclassical description of a macroscopic quantum system, we consider a purely quantum-mechanical description of the astronomical Kepler problem with the Hamiltonian

$$H = \frac{1}{2m}\,P^2 - \frac{\kappa}{|Q|} \quad , \quad \kappa > 0 \quad ,$$

where (P,Q) is a canonical Schrödinger pair with the Cartesian components P_1,P_2,P_3 and Q_1,Q_2,Q_3. If we choose as familiy $\mathcal{F}$ of quasiclassical states the coherent states of the Lie group $SU(2) \otimes SU(2)$ under the Kustaanheimo-Stiefel constraint[1], then $\mathcal{F}$ is a family of (ε,T)-quasiclassical states for the quantum mechanical Kepler problem. As an illustration, consider the motion of the earth in the gravitational field of the sun. There is a one-to-one correspondence between the classical Kepler orbits and the motions of initial state vectors $\Psi \in \mathcal{F}$ such that both solutions coincide with an accuracy

with respect to position: $\quad \Delta Q \approx 10^{-26}\ \mathrm{m}$,

with respect to velocity: $\quad \Delta(dQ/dt) \approx 10^{-33}\ \mathrm{m/s}$,

for all $t \in [0,T]$ with $\quad\quad\quad\quad T \approx 10^{36}\ years$.

---

1   Compare Boiteux [5], Kibbler and Négadi [17, 18], Gerry [10, 11], Bhaumik et al. [4].

Nobody will claim that either classical point mechanics or quantum mechanics is accurate to this degree, so we may say that for the astronomical Kepler problem and *for appropriate initital states,* quantum mechanics gives the same results as classical mechanics.

5. *A VARIATIONAL PRINCIPLE FOR THE DYNAMICS OF OBJECTS*

A STRATEGY FOR FINDING THE DYNAMICS OF OBJECTS

The environment of an object is never an intrinsically classical system in an absolute sense but a quantum system which allows − under the abstractions necessary for the scientific investigation at hand − a classical description. In spite of the fact that the environment is an infinite system, we can get useful information if we approximate it by a finite system, and try to get an appropriate description in terms of traditional Hilbert-space quantum mechanics. This can suggest ways for a rigorous algebraic approach. A feasible strategy could be as follows: First we have to select a family of classical or quasiclassical pure states, describing the classical features of the environment without interactions with the object. In the next step we choose in the usual (heuristic!) way a Hamiltonian, describing the object system, the finite environment, and their interactions. Then we can use the time-dependent variational principle of traditional quantum mechanics. Since we already know that the exact description is by product states, we restrict the admissible states for the variational principle to *pure product states.* If our approximations are well chosen we may hope that in the limit of infinite systems the exact solution is approached. It seems not to be attractive to give a rigorous proof that this or similar procedures are legitimate. The idea is, to get a good guess of the solution, to use this (or a slightly modified) solution to conjecture the appropriate GNS-reference state, and then to use the tools of algebraic quantum mechanics to construct an exact solution.

THE TIME-DEPENDENT VARIATIONAL PRINCIPLE FOR THE SCHRÖDINGER EQUATION

The variational principle for the time-dependent Schrödinger equation has been known since long and is discussed in the text books by Frenkel [8] and by Morse and Feshbach [34, Vol.1, §3.3]. It stems from the action functional

$$\int_{t_1}^{t_2} L_t \, dt \;\; ,$$

where the Lagrangian L_t is a functional of $\Psi_t \in S$ and $\Psi_t^* \in S$, and S is an appropriate subset of the Hilbert space $\mathcal{H}$ of state vectors,

$$L_t = \langle \Psi_t | i\hbar \, \partial/\partial t - H | \Psi_t \rangle \;\; ,$$

where H is the Hamiltonian of the system. The equations of motion are determined by requiring the stationarity of the action

$$\delta \int_{t_1}^{t_2} L_t \, dt = 0$$

with respect to fixed end points and variations of Ψ_t and Ψ_t^*. They are given by

$$i\hbar \frac{\partial \Psi_t}{\partial t} = \frac{\delta \langle \Psi_t | H | \Psi_t \rangle}{\delta \Psi_t^*} .$$

The choice $S = \mathcal{H}$ leads to the time-dependent Schrödinger equation $i\hbar \partial \Psi_t / \partial t = H \Psi_t$. If $S \subset \mathcal{H}$ is not an invariant subspace, then one gets a well-defined constrained dynamics if and only if $\mathcal{F}$ is a phase space (i.e. a symplectic manifold), in this case the constrained dynamics is Hamiltonian[1]. If this symplectic manifold S is taken to be the manifold of coherent states, one recovers Klauder's variational and correspondence principle [20, 23, 24, 25] which is based on his continuous representation theory [21, 22].

If we have some ideas for choosing an overcomplete continuously parametrized family $\mathcal{F} \subset \mathcal{H}_{\mathrm{env}}$ of state vectors for the environment, say a family of appropriate generalized coherent states over some phase space Ω, then we can constrain the trial vectors to remain in product form, so that the manifold $S \subset \mathcal{H}_{\mathrm{obj}} \otimes \mathcal{H}_{\mathrm{env}}$ is given by

$$S = \{ \psi_t \otimes \chi_t \mid \psi_t \in \mathcal{H}_{\mathrm{obj}}, \chi_t \in \mathcal{F} \}$$

where object state vector ψ_t lies in the unconstrained Hilbert space $\mathcal{H}_{\mathrm{obj}}$.

6. A SIMPLE EXAMPLE

SPECIFICATION OF THE MODEL

It may be helpful to illustrate these general considerations by a transparent example. The simplest model for a quantal object is a two-level quantum system (called «spin» in the following), interacting with a harmonic environment. In the first instance, we restrict our discussion to an environment with only finitely many mechanical degrees of freedom. In a strictly quantum mechanical treatment, we start with the Hamiltonian

$$H = H_{\mathrm{obj}} \otimes 1 + 1 \otimes H_{\mathrm{env}} + V ,$$

where H_{obj} is the Hamiltonian of the object system, H_{env} is the Hamiltonian of the environment, and V represents the interaction. The most general spin-$\frac{1}{2}$ Hamiltonian is given by

$$H_{\mathrm{obj}} = -\tfrac{1}{2}\hbar \sum_{v=1}^{3} \Omega_v \sigma_v ,$$

where $\Omega_v \in \mathbb{R}$, and $\sigma_1, \sigma_2, \sigma_3$ are the Pauli matrices with the commutation relations

$$[\sigma_\mu, \sigma_v] = 2i \sum_{\alpha=1}^{3} \varepsilon_{\mu v \alpha} \sigma_\alpha .$$

---

[1]   More precisely, $\mathcal{F}$ has to be a submanifold of the *projective* Hilbert space, this manifold is Kähler, i.e. it has both Riemannian and symplectic metrics. For further details compare [29, 47, 48, 49].

Without restricting the generality, we may assume that the harmonic oscillators of the free environment are in normal form, so that

$$H_{\text{env}} = \sum_v \sum_k \omega_k a_{kv}^* a_{kv} \ ,$$

where $\omega_k > 0$, and a_{kv} are boson operators fulfilling the commutation relations

$$[a_{j\mu}, a_{kv}] = 0 \ , \quad [a_{j\mu}, a_{kv}^*] = \delta_{jk}\, \delta_{\mu v} \ .$$

If the interaction operator V is *linear* in the boson operators, we get a soluble model, therefore we put

$$V = \sum_{v=1}^{3} \sigma_v \otimes A_v \ ,$$

$$A_v = \sum_k \{\lambda_{kv}^* a_{kv} + \lambda_{kv} a_{kv}^*\} \ , \quad \lambda_{kv} \in \mathbb{C} \ .$$

EQUATIONS OF MOTION

In the spirit of the time-dependent variational principle we evaluate the constrained dynamics with respect to product state vectors

$$\Psi_t = \psi_t \otimes \chi_t \ , \quad \psi_t \in \mathbb{C}^2 \ , \quad \chi_t \in \mathcal{F} \ .$$

The assumed linear coupling implies that the structure of the family $\mathcal{F}$ is uniquely determined by the Hamiltonian as the manifold of the usual boson-coherent states $|z\rangle$, defined by

$$a_{j\mu}|z\rangle = z_{j\mu}|z\rangle \ , \quad z = (z_{11}, z_{12}, z_{13}, z_{21}, \ldots) \ , \quad z_{j\mu} \in \mathbb{C} \ ,$$

$$\langle z|z\rangle = 1 \ .$$

Using the manifold

$$S = \{\psi_t \otimes \chi_t \mid \psi_t \in \mathbb{C}^2, \chi_t \in \mathcal{F}\}$$

for the time-dependent variational principle, we get for the expectation values

$$M(t) := \langle \psi_t | \sigma | \psi_t \rangle \ ,$$

$$z_{j\mu}(t) := \langle \chi_t | a_{j\mu} | \chi_t \rangle \ ,$$

$$\alpha(t) := \langle \chi_t | A | \chi_t \rangle \ ,$$

the following Hamiltonian equations of motion:

$$\dot{M}(t) = M(t) \times \Omega + M(t) \times \alpha(t) \ ,$$

$$\dot{\alpha}_{j\mu}(t) = -i\omega_j \alpha_{j\mu}(t) + \tfrac{1}{2}i\,\lambda_{j\mu} M_\mu(t) \ .$$

The last equation represents the motion of the environment under the polarizing influence of the object, and can be integrated easily. Choosing the *retarded* solution – thereby selecting a preferred direction of time – we get

$$\alpha_v(t) = \sum_k \{ \lambda_{kv}^* z_{kv}(t) + \lambda_{kv} z_{kv}^*(t) \}$$

$$= g_v(t) + K_v(0) M_v(t) - K_v(t) M_v(0) - \int_0^t K_v(s) \dot{M}_v(t-s) \, ds \quad , \quad t \geq 0 \quad ,$$

where we introduced an external force function $t \rightarrow g(t)$ depending only on the initial state of the environment

$$g_v(t) = \sum_k \{ \lambda_{kv}^* e^{-i\omega_v t} z_{kv}(0) + \lambda_{kv} e^{+i\omega_v t} z_{kv}^*(0) \} \quad ,$$

and a memory kernel K_v,

$$K_v(t) := \sum_k \frac{|\lambda_{kv}|^2}{\omega_k} \cos(\omega_k t) \quad .$$

The function $t \rightarrow K(t)$ depends only on the parameters occurring in the Hamiltonian and represents the relaxation properties of the combined system. For finitely many degrees of freedom $t \rightarrow K(t)$ is an almost periodic function, so that

$$\limsup_{t \rightarrow \infty} |K_v(t)| = 1 \quad .$$

This example shows that no genuine relaxation can be expected for only finitely many degrees of freedom. If the condition $K_v(0) < \infty$ is fulfilled, then our derivation can be rigorously justified also in the case of infinitely many degrees of freedom. In this case the function $t \rightarrow K_v(t)/K_v(0)$ is the characteristic function of a probability distribution function. According to Lebesgue's decomposition theorem, there exists a unique decomposition of characteristic functions into a part coming from a purely discrete distribution, a part coming from an absolutely continuous distribution, and a part coming from a singular distribution (compare e.g. [31]). A physically reasonable relaxation behavior can be expected only in the case of an absolutely continuous distribution, in this case the Riemann-Lebesgue lemma implies that

$$\lim_{t \rightarrow \infty} K_v(t) = 0 \quad .$$

The simplest physically reasonable example (which can be justified as the first Padé-approximation to the memory kernel) ist the Cauchy distribution, whose characteristic function is an exponential, so that

$$K_v(t) = \gamma_v \exp(-|t|/\tau_v) \quad , \quad 0 < \gamma_v < \infty \quad , \quad 0 < \tau_v < \infty \quad .$$

With this choice of a Cauchy-type environment we get a closed system of integro-diffential equations for the motion $t \rightarrow M(t)$. It is always possible – and indeed very advantageous – to dilate such equations to a system-theoretical form. The Cauchy environment is the simplest one in the sense that a system-theoretical dilation requires only *one* dummy state variable, which we call R. It is defined by

$$R_\nu(t) = \frac{1}{\gamma_\nu} K_\nu(t) M_\nu(0) - \frac{1}{\gamma_\nu} \int_0^t K_\nu(s) \dot{M}_\nu(t-s) \, ds \quad,$$

and fulfills the differential equation

$$\dot{R}_\nu(t) = -\frac{1}{\tau_\nu} R_\nu(t) + M_\nu(t) \quad, \quad R_\nu(0) = M_\nu(0) \quad.$$

With this we get a non-autonomous system of six ordinary first-order differential equations:

$$\dot{M}(t) = M(t) \times \{ \mathbf{\Omega} + \Gamma M(t) - \Gamma R(t) + g(t) \} \quad,$$

$$\dot{R}(t) = -T^{-1} R(t) + M(t) \quad.$$

where Γ and T are diagonal tensors with the eigenvalues $\gamma_1, \gamma_2, \gamma_3$ and τ_1, τ_2, τ_3, respectively.

The external force $t \to g(t)$ depends on the initial state of the environment. Our requirement that the relaxation function is related to a distribution function with an absolutely continous spectrum implies that every mechanical description of the environment necessarily requires infinitely many degrees of freedom. Certainly we never can know the initial values for every single mechanical degree of freedom. It is therefore appropriate to use a statistical approach and to specify only the statistical invariants of the initial state of the environment. If in the mean the environment does not exert a force, then we can assume that the ensemble expectation values for all initial states of every degree of freedom vanishes

$$\mathcal{E}\{z_{k\nu}(0)\} = 0 \quad.$$

If the initial state of the free environment is stationary, then the covariance tensor Σ is given by

$$\Sigma_{j\mu,k\nu} := \mathcal{E}\{z_{j\mu}(0) \, z_{k\nu}^*(0)\} = \delta_{jk} \delta_{\nu\mu} \Sigma_\nu^2 \quad.$$

Usually it is practically impossible to obtain any information about the higher order moment, so the information-theoretical optimal choice is to assume that the initial states have a *Gaussian* probability distribution. With this it follows that

$$\mathcal{E}\{g_\nu(t)\} = 0 \quad,$$
$$\mathcal{E}\{g_\mu(t) \, g_\nu(s)\} = \delta_{\mu\nu} \Sigma_\nu^2 \exp(-|t-s|/\tau_\nu) \quad.$$

That is, the components of the external force $t \to g(t)$ are three stochastically independent Ornstein-Uhlenbeck processes,

$$g_\nu(t) = \frac{1}{\tau_\nu} U_\nu(t) \quad,$$

where the normalized dimensionless Ornstein-Uhlenbeck process $t \to U_\nu(t)$ is defined as solution of the Itô differential equation

$$dU_\nu(t) = -\gamma_\nu U_\nu(t) dt + \sqrt{2D_\nu} \, dW_\nu(t) \quad.$$

274

The diffusion constant D_v is given by

$$D_v = \tau_v \, \Sigma_v^2 \; \geq 0 \; .$$

The one-dimensional standard Wiener process $\{W(t) \,|\, t > 0 \,\}$ fulfills the relations

$$\mathcal{E}\{W_v(t)\} = 0 \; , \quad \mathcal{E}\{[W_v(t) \, W_v(s)]^2\} = |t-s| \quad .$$

Collecting all results, denoting the diagonal tensor of the diffusion constants by D, we get the following system of nonlinear stochastic differential equations in the sense of Itô:

$$
\begin{aligned}
\dot{M}(t) &= M(t) \times \{\boldsymbol{\Omega} + \Gamma M(t) - \Gamma R(t) + T^{-1}U(t)\} \\
\dot{R}(t) &= -T^{-1}R(t) + M(t) \\
dU(t) &= -\Gamma \, dU(t)dt + \sqrt{2D} \; dW(t)
\end{aligned}
$$

The three diagonal tensors Γ, T and D represent nine independent non-negative parameters. While Γ and T are determined by the Hamiltonian alone, the diffusion tensor D also depends on the initial conditions of the free environment. This system of *nonlinear* stochastic differential equations describes the behavior of a two-level quantum object in a Cauchy-environment with Gaussian initial conditions[1]. They can be rewritten as a *nonlinear* stochastic Schrödinger equation describing the individual state, or as *linear* Fokker-Planck equations, describing the time evolution of the statistical state.

SIMPLIFICATIONS FOR THE WHITE-NOISE LIMIT

In many applications the correlation times τ_1, τ_2, τ_3 are very short in comparison to the other dynamical times, so that one may approximate the Ornstein-Uhlenbeck processes by the Wiener processes. With the approximations

$$R(t)dt \approx T dM(t) \; ,$$

$$dU(t)dt \approx T\sqrt{2D} \; dW(t) \quad .$$

the evolution equation for M reduces to the following simple nonlinear stochastic differential equation in the sense of Stratonovich[2]

$$dM(t) = M(t) \overset{\circ}{\times} \{\boldsymbol{\Omega} \, dt + \Gamma M(t) \, dt - H \, dM(t) + \sqrt{2D} dW(t)\} \quad ,$$

where H is the diagonal tensor with the eigenvalues η_1, η_2, η_3 with

$$\eta_v := \gamma_v \tau_v \geq 0 \; .$$

The symbol $\circ$ denotes the Stratonovich multiplication which satisfies all formal rules of ordinary calculus, including integration by parts, the chain rule of differentiation and the

---

[1] These equations can also be derived with the help of the factorization theorem, discussed in section 2.

[2] A theorem by Wong and Zakai [57, 58] implies that the white-noise limit of a smooth (i.e. essentially band-limited) noise problem is described by stochastic differential equations in the sense of Stratonovich.

ordinary rules of variable substitution. Of course, one can easily transform any Stratonovich equation into a mathematically more convenient Itô-equation, but we prefer to keep the equation in its Stratonovich form because only this formulation has a direct physical meaning. Evidently this Stratonovich equation is equivalent to the following nonlinear Schrödinger equation

$$i\hbar\, d\psi_t = -\tfrac{1}{2}\hbar\,\sigma\,\psi_t \circ \{\boldsymbol{\Omega}\, dt + \Gamma M(t)\, dt - H\, dM(t) + \sqrt{2D}dW(t)\}\quad,$$

$$M(t) = \langle\psi_t\,|\,\sigma\,|\,\psi_t\rangle\quad.$$

This nonlinear Schrödinger–Langevin equation has a simple intuitive physical meaning. If we imagine that the two-level object is an elementary system with spin $\tfrac{1}{2}$ and gyromagnetic ratio γ, then we can write

$$i\hbar\,\frac{\partial\psi_t}{\partial t} = -\frac{\gamma\hbar}{2}B(t)\sigma\,\psi_t\quad,$$

and interpret $B(t)$ as the magnetic field which acts at time t on the object system. In the spirit of the Stratonovich calculus we assume that $dW(t)/dt$ exists in the ordinary sense as "almost white noise", so that we can write (at least heuristically)

$$\gamma B(t) = \boldsymbol{\Omega} + \Gamma M(t) - H\dot{M}(t) + \sqrt{2D}\,\dot{W}(t)\quad.$$

This decomposition of the effective field $B(t)$ can be interpreted as follows:

(i) $\gamma^{-1}\boldsymbol{\Omega}$ is an external static magnet field, say applied by the experimenter,

(ii) $\gamma^{-1}\Gamma M(t)$ is a magnetic Onsager-type reaction field, due to the instantaneous polarization of the environment and an instantaneous feedback mechanism,

(iii) $\gamma^{-1}H\dot{M}(t)$ is a magnetic Onsager-type reaction field due to a dynamical polarization and a dynamical feedback mechanism (e.g. the Faraday induction law),

(iv) $\gamma^{-1}\sqrt{2D}\,\dot{W}(t)$ is a classical fluctuating magnetic field, a direct influence of the environment on the object system.

This nonlinear Schrödinger–Langevin equation exhibits a fantastically rich range of phenomena. We mention here only two rather special simplifications.

NONLINEAR LANDAU-LIFSHITZ RELAXATION

As first example we discuss the Landau–Lifshitz relaxation. For the isotropic case, $\gamma_1 = \gamma_2 = \gamma_3$; $\eta_1 = \eta_2 = \eta_3 = \eta$ the instantaneous reaction field vanishes, and one gets the following equations[1]

$$dM(t) = \xi\, M(t) \overset{\circ}{\times} \{\boldsymbol{\Omega}\, dt - \eta\, M(t)\times\boldsymbol{\Omega}\, dt + \sqrt{2D}\, dW(t) - \eta\,\sqrt{2D}\, M(t)\overset{\circ}{\times} dW(t)\}\quad,$$

$$\xi := \frac{1}{1+\eta^2}\quad,$$

which correspond to the Langevin extension of the classical Landau–Lifshitz [30] equations for ferromagnetic spin relaxation. The equivalent *nonlinear* stochastic

---

[1]   For more details and further discussions, compare Funck [9].

Schrödinger equation for ψ_t is remarkable insofar, as it is linearizable in the sense of Gisin [12,13,14,15], that is, ψ_t can be obtained from the solution ϕ_t of the following *linear* stochastic differential equation

$$d\phi_t = -\tfrac{1}{2}\xi(\eta+i)\,\boldsymbol{\sigma}\,\phi_t \circ \{\boldsymbol{\Omega}\,dt + \sqrt{2D}\,dW(t)\} \quad,$$

by the simple nonlinear procedure of normalization

$$\psi_t = \frac{\phi_t}{\|\phi_t\|} \quad.$$

THE DYNAMICS OF MEASUREMENT-TYPE PROCESSES

In the special case of a pure Landau–Lifshitz relaxation, the static polarization term is ineffective. We now consider an example where just this term is dominant. We choose the following set of parameters:

$$\begin{aligned}
\boldsymbol{\Omega} &= (\Omega,0,0) \quad, \\
\boldsymbol{H} &= (0,0,\eta) \quad, \\
\boldsymbol{\Gamma} &= (0,0,\gamma) \quad, \\
\boldsymbol{D} &= (0,0,D) \quad,
\end{aligned}$$

The resulting equations of motion for $\boldsymbol{M}$ are then given by

$$\begin{aligned}
dM_1(t) &= -\eta\,\Omega M_2^2(t)dt + \gamma M_3(t)M_2(t)dt + \sqrt{2D}\,M_2(t)\circ dW(t) \ , \\
dM_2(t) &= -\Omega M_3(t)dt + \eta\,\Omega M_1(t)M_2(t)dt - \gamma M_3(t)M_1(t)dt - \sqrt{2D}\,M_1(t)\circ dW(t) \ , \\
dM_3(t) &= +\Omega M_2(t)dt \ .
\end{aligned}$$

Clearly, $M^2 = \boldsymbol{M}(t)\boldsymbol{M}(t)$ is a constant of motion, so that these equations in fact transform pure states into pure states. The corresponding nonlinear Schrödinger-Langevin equation

$$d\psi_t = -\tfrac{1}{2}\{\Omega\sigma_1 + \eta\,\Omega M_2(t)\sigma_3 - \gamma M_3\sigma_3\}\,\psi_t\,dt - \sqrt{2D}\,\sigma_3\,\psi_t\circ dW(t) \quad,$$

cannot be linerarized in the sense of Gisin, it describes genuinely nonlinear processes.

For a proper choice of models of this kind (our simple example is restricted by the fact that there are only four disponible parameters, Ω,η,γ and D), such genuinely nonlinear stochastic Schrödinger equations can describe the main features of the quantum-mechanical measurement process. If the initial state vector for $t=0$ is given by $\psi = c_\alpha|\alpha\rangle + c_\beta|\beta\rangle$, then the time evolution $\psi \to \psi_t$ transforms state vectors into state vectors, but the long-time behavior depends in an extremely sensitive way on the initial conditions of the environment. After a short initial time, the trajectory $\psi \to \psi_t$ is attracted by either $|\alpha\rangle$ or $|\beta\rangle$. Later the probability that the trajectory switches from a neighborhood of $|\alpha\rangle$ to a neighborhood of $|\beta\rangle$ becomes increasingly small. For a "good" measuring instrument (i.e. for a proper choice of the parameters Ω, γ, η and D), the probability that after a long time the state is in a neighborhood of $|\alpha\rangle$ becomes approximately $|c_\alpha|^2$. By an appropriate choice of the parameters, one can approximate the result predicted by the projection postulate to any degree of accuracy, but one cannot reach it precisely. This seems to be a general feature since the projection postulate implies the existence of a constant of motion which freezes the measuring dynamics.

Where are the irreducible probabilities of exophysical descriptions of quantum systems coming from? Even if the initial data are perfectly known, the motion $\psi \to \psi_t$ is unpredictable for times much longer than the correlation time. In this respect, the trajectory $\{ \psi_t \mid t \in \mathbb{R} \}$ of an individual state shows a chaotic character. The chaotic nature is due to the fact that under the model assumptions, the environment behaves like a classical *K-flow-type dynamical system*. The probabilities of quantum mechanics do not appear in quantum endophysics, they belong to exophysics and emerge by strong enough interaction of a quantum object with the K-flow type classical exosystem, they are contextual but irreducible.

7. SOLVED AND UNSOLVED PROBLEMS

If objects exist all all, and provided quantum mechanics is the proper theory for the description of the properties of individual objects, then it follows *without any ad hoc modifications of the first principles of quantum mechanics* that the individual description of finite objects is given by a *nonlinear, non-autonomous functional–differential Schrödinger equation*. In important special cases, this equation reduces to a *nonlinear stochastic Schrödinger equation*. A nonlinear stochastic Schrödinger–Langevin equation in the sense of Itô has a dilation to a one-parameter group dynamics which is *deterministic, but not empirically determinable*. The Kolmogorov structure of this stochastic Schrödinger equation implies a deterministic chaotic instability, so that the associated stochastic process is neither predictable nor controllable. In this frame, the measurement problem of quantum mechanics, including the enigma of the «reduction of the wave packet», does not meet with unsurmountable obstacles. There is no "problem of the actualization of potentialities", at every instant a maximal set of potential properties *is* actualized, but the time evolution may change actualized to potential, and potential to actualized properties. In a measurement situation this change appears to the observer chaotic, but from an ontic viewpoint the so-called «reduction of the wave packet» is a continuous and strictly deterministic process whose epistemic description inevitably is probabilistic (in the sense of the classical "deterministic chaos"). In every concrete situation there remain difficult mathematical problems, but I claim that the problem of the nature of the measurement process can be considered as solved *in principle*. This is, however, not the case with the problem of *localization*. Mathematically this may be a similar problem, but I do not expect that the problem of localization is conceptually of the same nature as the measurement problem.

There are many open questions. The first concerns the existence of *intrinsically classical* domains. There are indications that the infrared part of the quantized electromagnetic radiation field could be intrinsically classical [52,53]. This question certainly deserves a more penetrating analysis. If it should turn out that indeed the electromagnetic field has an *intrinsic* classical part, then already the C*-algebra of intrinsic observables has a large center, so that objects could exist not only in a contextual but even in an absolute sense. The same type of problems applies also to the gravitational field.

Questions regarding the role of gravitation in noncosmological quantum mechanics I must leave unanswered. If one could formulate a reasonable dynamical Galilei-relativistic quantum theory of gravitation, then one could include the gravitational field in the environment of objects. This should not pose grave technical difficulties since in the molecular and the mesoscopic domain, gravitation manifests itself in a purely classical manner. The nonlinear stochastic Schrödinger equation we expect looks like a parameter-free Itô-type Schrödinger equation recently proposed by Diósi [7], but is not identical with it. Nevertheless, one may speculate that the inclusion of gravitation in the environments of quantum objects could be responsible for spontaneous position localizations.

REFERENCES

[1] L. Arnold: *Stochastische Differentialgleichungen. Theorie und Anwendung.* München. Oldenbourg. 1973. English translation: *Stochastic Differential Equations: Theory and Application.* New York. Wiley. 1974.

[2] A. O. Barut and J. P. Dowling: *Quantum electrodynamics based on self-energy, without second quantization: The Lamb shift and long-range Casimir-Polder van der Waals forces near boundaries.* Phys. Rev. A **36**, 2550–2556 (1987).

[3] A. O. Barut and J. F. Van Huele: *Quantum electrodynamics based on self energy: Lamb shift and spontaneous emission without field quantization.* Phys. Rev. A **32**, 3187–3195 (1985).

[4] D. Bhaumik, B. Dutta-Roy and G. Ghosh: *Classical limit of the hydrogen atom.* J. Phys. A: Math. Gen. **19**, 1355–1364 (1986).

[5] M. Boiteux: *The three-dimensional hydrogen atom as a restricted four-dimensional harmonic oscillator.* Physica **65**, 381–395 (1973).

[6] T. Carleman: *Application de la théorie des équations intégrales linéaires aux systèmes d'équations différentielles non linéaires.* Acta Math. **59**, 63–87 (1932).

[7] L. Diósi: *Models for universal reduction of macroscopic quantum fluctuations.* Phys. Rev. A **40**, 1165–1174 (1989).

[8] J. Frenkel: *Wave Mechanics. Advanced General Theory.* Oxford. Clarendon. 1934.

[9] P. Funck: *Die Landau-Lifshitz-Gleichung als nichtlineare Schrödingergleichung.* Diplomarbeit, Laboratorium für Physikalische Chemie der ETH Zürich. 1989.

[10] C. C. Gerry: *On coherent states for the hydrogen atom.* J. Phys. A: Math. Gen. **17**, L737–L740 (1984).

[11] C. C. Gerry: *Coherent states and the Kepler-Coulomb problem.* Phys. Rev. A **33**, 6–11 (1986).

[12] N. Gisin: *Brownian motion of a quantum spin with friction.* Progr. Theor. Phys. **66**, 2274–2275 (1981).

[13] N. Gisin: *A simple nonlinear dissipative quantum evolution.* J. Phys. A: Math. Gen. **14**, 2259–2267 (1981).

[14] N. Gisin: *Spin relaxation and dissipative Schrödinger like evolution equations.* Helv. Phys. Acta **54**, 457–470 (1981).

[15] N. Gisin: *A model of irreversible deterministic quantum dynamics.* In: *Quantum Probability and Applications to the Quantum Theory of Irreversible Processes. Lecture Notes in Mathematics, vol. 1055.* Berlin. Springer. 1984. Pp. 126–133.

[16] W. Guz: *On quantum dynamical semigroups.* Reports on Mathematical Physics **6**, 455–464 (1974).

[17] M. Kibler and T. Négadi: *On the connection between the hydrogen atom and the harmonic oscillator.* Lett. Nuovo Cimento **37**, 225–228 (1983).

[18] M. Kibler and T. Négadi: *On the connection between the hydrogen atom and the harmonic oscillator: the continuum case.* J. Phys. A: Math. Gen. **16**, 4265–4268 (1983).

[19] J. R. Klauder: *The action option and a Feynman quantization of spinor fields in terms of ordinary c-numbers.* Annals of Physics **11**, 123–168 (1960).

[20] J. R. Klauder: *Restricted variations of the quantum mechanical action functional and their relation to classical dynamics.* Helv. Phys. Acta **35**, 333–335 (1962).

[21] J. R. Klauder: *Continuous-representation theory. I. Postulates of continuous-representation theory.* J. Math. Phys. **4**, 1055–1058 (1963).

[22] J. R. Klauder: *Continuous-representation theory. II. Generalized relation between quantum and classical dynamics.* J. Math. Phys. **4**, 1058–1073 (1963).

[23] J. R. Klauder: *Weak correspondence principle.* J. Math. Phys. **8**, 2392–2399 (1967).

[24] J. R. Klauder: *The fusion of classical and quantum theory.* Preprint of a Lecture presented at the "Conference on Quantum Theory and the Structures of Time and Space", Tutzing, West Germany, July, 1980. (1980).

[25] J. R. Klauder: *Coherent-state path integrals for unitary group representations.* In: *Proceedings of 14th ICGTMP.* Ed. by Y. M. Cho. Singapore. World Scientific. 1986. Pp. 15–23.

[26] J. R. Klauder and B.-S. Skagerstam: *Coherent States. Applications in Physics and Mathematical Physics.* Singapore. World Scientific. 1985.

[27] J. R. Klauder and E. C. G. Sudarshan: *Fundamentals of Quantum Optics.* New York. Benjamin. 1968.

[28] B. O. Koopman: *Hamiltonian systems and transformations in Hilbert space.* Proc. Nat. Acad. Sci. U.S. **17**, 315–318 (1931).

[29] P. Kramer and M. Saraceno: *Geometry of the Time-Dependent Variational Principle in Quantum Mechanics.* Berlin. Springer. 1981.

[30] L. D. Landau and E. Lifshitz: *On the theory of the dispersion of magnetic permeability in ferromagnetic bodies.* Phys. Z. Sowjet. **8**, 153 (1935).

[31] E. L. Lukacs: *Characteristic Functions.* London. Griffin. 1970.

[32] G. W. Mackey: *The Mathematical Foundations of Quantum Mechanics.* New York. Benjamin. 1963.

[33] L. Mandel and E. Wolf: *Selected Papers on Coherence and Fluctuations of Light, Vol. 1, 1850–1960, Vol. 2, 1961–1966*. New York. Dover. 1970.

[34] P. M. Morse and H. Feshbach: *Methods of Theoretical Physics*. New York. McGraw-Hill. 1953.

[35] J. v. Neumann: *Mathematische Grundlagen der Quantenmechanik*. Berlin. Springer. 1932.

[36] L. Onsager: *Electric moments of molecules in liquids*. J. Amer. Chem. Soc. **58**, 1486–1493 (1936).

[37] W. Pauli: *Die allgemeinen Prinzipien der Wellenmechanik*. In: *Handbuch der Physik*. Ed. by H. Geiger and K. Scheel. Second ed.. Berlin. Springer. 1933.

[38] A. Perelomov: *Generalized Coherent States and Their Applications*. Berlin. Springer. 1986.

[39] A. M. Perelomov: *Coherent states for arbitrary Lie groups*. Commun. Math. Phys. **26**, 222–236 (1972).

[40] A. Peres: *Nonlinear variants of Schrödinger's equation violate the second law of thermodynamics (with a reply by S. Weinberg)*. Phys. Rev. Lett. **63**, 1114–1115 (1989).

[41] M. Planck: *Vorlesungen über die Theorie der Wärmestrahlung*. Leipzig. Barth. 1906. English translation: *Theory of Heat Radiation*. New York. Dover. 1959.

[42] H. Primas: *Mathematical and philosophical questions in the theory of open and macroscopic quantum systems*. In: *Sixty-two Years of Uncertainty: Historical, Philosophical and Physics Inquiries into the Foundations of Quantum Mechanics*. Ed. by A. I. Miller. New York. Plenum. 1990.

[43] K. Przibram: *Briefe zur Wellenmechanik. Schrödinger, Planck, Einstein, Lorentz*. Wien. Springer. 1963.

[44] G. A. Raggio: *States and Composite Systems in W*-algebraic Quantum Mechanics*. Thesis ETH Zürich, No. 6824, ADAG Administration & Druck AG, Zürich. 1981.

[45] G. A. Raggio and H. Primas: *Remarks on "On completely positive maps in generalized quantum dynamics"*. Foundations of Physics **12**, 433–435 (1982).

[46] G. Rosen: *Linear correspondents of nonlinear equations*. Lett. Nuovo Cimento **20**, 617–618 (1977).

[47] D. J. Rowe: *Constrained quantum mechanics and coordinate independent theory of the collective path*. Nucl. Phys. **A 391**, 307–326 (1982).

[48] D. J. Rowe, A. Ryman and G. Rosensteel: *Many-body quantum mechanics as a symplectic dynamical system*. Phys. Rev. **A 22**, 2362–2373 (1980).

[49] D. J. Rowe, M. Vassanji and G. Rosensteel: *Density dynamics: A generalization of Hartree-Fock theory*. Phys. Rev. **A 28**, 1951–1956 (1983).

[50] S. Sakai: *C*-Algebras and W*-Algebras*. Berlin. Springer. 1971.

[51] E. Schrödinger: *Der stetige Übergang von der Mikro– zur Makromechanik*. Naturwissenschaften **14**, 644-666 (1926).

[52] H. P. Stapp: *Exact solution of the infrared problem*. Phys. Rev. **D 28**, 1386–1418 (1983).

[53] H. P. Stapp: *On the unification of quantum theory and classical physics*. In: *Symposium on the Foundations of Modern Physics*. Ed. by P. Lathi and P. Mittelstaedt. Singapore. World Scientific. 1985.

[54] M. Takesaki: *Theory of Operator Algebras I*. New York. Springer. 1979.

[55] W. Ulmer and H. Hartmann: *On the application of a Gauss transformation in nonlinear quantum mechanics*. Nuovo Cimento **47A**, 359–376 (1978).

[56] J. Waniewski: *Mobility and measurements in nonlinear wave mechanics*. J. Math. Phys. **27**, 1796–1799 (1986).

[57] E. Wong and M. Zakai: *On the convergence of ordinary integrals to stochastic integrals*. Ann. Math. Statist. **36**, 1560–1564 (1965).

[58] E. Wong and M. Zakai: *On the relation between ordinary and stochastic differential equations*. Internat. J. Engrg. Sci. **3**, 213-229 (1965).

[59] S. Yomosa: *Nonlinear Schrödinger equation on the molecular complex in solution*. J. Phys. Soc. Japan **35**, 1738–1746 (1973).

[60] H. D. Zeh: *On the interpretation of measurement in quantum theory*. Foundations of Physics **1**, 69–76 (1970).

[61] H. D. Zeh: *On the irreversibility of time and observation in quantum theory*. In: *Proceedings of the International School of Physics "Enrico Fermi", Course 49. Foundations of Quantum Mechanics*. Ed. by B. d'Espagnat. New York. Academic Press. 1971. Pp. 263–273.

QUANTUM LOGIC: A SUMMARY OF SOME ISSUES

E.G. Beltrametti

Dipartimento di Fisica dell'Università di Genova
and Istituto Nazionale di Fisica Nucleare
Genova, Italy

1. INTRODUCTION

The use of the name "quantum logic" is rather broad in the literature: it points at a variety of different mathematical objects, of different approaches to the foundations of quantum mechanics, and of different versions of a nonclassical logic.

Whatever exact meaning we give to "quantum logic", its first origin can be drawn back to the 1936 paper of G.Birkhoff and J.von Neumann[1]. There, the peculiar ordered structure of the projectors of a Hilbert space was examined, along with its departures from the classical counterpart formed by the Boolean algebra of the subsets of a phase space, and the question was raised of what logical calculus has that ordered structure as an algebraic model.

Quantum mechanics is in any case on the background of quantum logic; more specifically it is the nonrelativistic quantum mechanics that is called into play, with its limitations in the description of natural phenomena and with its reference to the usual notions of physical system, of its states, of its physical quantities (or observables).

2. REMARKS ON HILBERT-SPACE QUANTUM MECHANICS

In the usual Hilbert-space formulation of quantum mechanics, to each physical system is attached a separable Hilbert-space H (generally infinite dimensional) over the complex field. The states of the system are represented by density operators, i.e., linear, bounded, self-adjoint, positive, trace-class operators on H of trace one. The physical quantities are represented by linear, self-adjoint (not necessarily bounded)

Sixty-Two Years of Uncertainty
Edited by A. I. Miller
Plenum Press, New York, 1990

operators on H. If we deal with a strictly quantum system then the correspondence between states and density operators as well as the correspondence between physical quantities and self-adjoint operators is one-to-one. If however the behaviour of the physical system retains some classical features, so that superselection rules are operating, then it is no longer true that every density operator represents a state nor that every self-adjoint operator represents a physical quantity. In the sequel we shall however refer to the strictly quantum case.

The so-called spectral theorem ensures that a self-adjoint operator A determines, for every Borel set E of the real line $\mathbb{R}$, a projection operator $P_A(E)$ such that $P_A(\phi) = 0$, $P_A(\mathbb{R}) = I$ (the identity operator), and $P_A(\cup_i E_i) = \sum_i P_A(E_i)$ for every disjoint sequence $< E_i >$ in the family $B(\mathbb{R})$ of Borel subsets of $\mathbb{R}$. Then, if the physical system is in the state (represented by the density operator) D then the probability that the value of the physical quantity (represented by the self-adjoint operator) A lies in the Borel set E is given by $tr(DP_A(E))$. Notice that the function $E \to tr(DP_A(E))$ has indeed the properties of a probability measure on $\mathbb{R}$.

The set of density operators in a Hilbert-space is a σ-convex set. This means that if $< D_i >$ is a countable sequence of density operators and $< w_i >$ is a corresponding sequence of positive numbers such that $\sum_i w_i = 1$ then the convex combination $\sum_i w_i D_i$ (or its uniform limit when the sum is infinite) is still a density operator. Notice that if $\sum_i w_i D_i$ is the state of the system then the probability that the physical quantity A takes a value in $E \in B(\mathbb{R})$ becomes $\sum_i w_i tr(D_i P_A(E))$.

It may happen that a state cannot be written as a convex combination of other states: in this case it is called a pure state, or an extreme element of the convex set of states. In the other case it is called nonpure or a mixture. We have that[2]

(i) a state is pure if and only if it is (represented by) a projector onto a one-dimensional subspace;

(ii) a nonpure state can always be written as a convex combination of pure states.

Item (i) says that the pure states correspond to the vectors of H up to a factor. A projector onto a one-dimensional subspace will be written as $P^{[\psi]}$ where ψ is any vector in that subspace. Notice that for pure states the probability distribution $tr(P^{[\psi]} P_A(E))$ of a physical quantity A takes the form $(\psi, P_A(E)\psi)$ provided ψ is a unit vector (we denote $(.\ ,\ .)$ the inner product of H).

The linearity of the Hilbert space generates infinitely many new pure states out of any family of them: if ψ_1, ψ_2 are unit vectors then $\lambda_1 \psi_1 + \lambda_2 \psi_2$, $\lambda_1, \lambda_2 \in \mathbb{C}, |\lambda_1|^2 + |\lambda_2|^2 = 1$, is another pure state, a superposition of the states ψ_1, ψ_2.

With reference to nonpure states it is worth remarking that their convex decomposition is never unique. This fact is intertwined with the superposition principle of quantum theory. Indeed, if the sequence $< \psi_i >$ of pure states carries a convex decomposition of a nonpure state

$$D = \sum_i w_i P^{[\psi_i]} \quad ,$$

then any other sequence $< \varphi_i >$ that spans the same subspace as $< \psi_i >$ carries another decomposition

$$D = \sum_i w_i' P^{[\varphi_i]} \quad .$$

A mixture has therefore infinitely many decompositions into pure states (as well as into nonpure ones). This fact represents a peculiar feature of the quantum behaviour, and a basic departure from the classical case.

3. MATHEMATICAL STRUCTURES EMERGING FROM THE HILBERT SPACE DESCRIPTION OF QUANTUM MECHANICS

We can isolate several mathematical substructures in the edifice of quantum theory. It is typical of the literature on the foundations of quantum theory - also of the literature on quantum logic - to focus attention on them, on their mutual relations, on the extent each of them, assumed as a primitive object, contribute to shape the quantum theory.

A significant structure was alluded to just at the end of the previous section: the convex set of states. The nonunique decomposition of mixtures implies that it cannot be a simplex (contrary to the classical case); the shape of its boundary - which inherits a high degree of "smoothness" from the superposition principle - embodies many features of the physical system, as one can visualize, e.g., from the Poincaré sphere representing the states of polarization of a photon or of a spin-$\frac{1}{2}$ particle.

Other significant mathematical structures might be the closure space formed by the set of pure states under the superposition operation, and the transition probability space generated by the set of pure states under the modulus squared of the scalar product that maps the pairs of them into the real segment $[0, 1]$.

Another relevant structure is the involutive algebra generated by the bounded physical quantities, with the states behaving as complex-valued, normalized, positive linear functions on that algebra. This structure has played an important role in the so called algebraic approach to quantum mechanics.

We shall however restrict, here on, to the ordered structure of projectors of a Hilbert space, for this structure has played an emblematic role in the studies referring to quantum logic. Actually that ordered structure is often by its own referred to as quantum logic.

The projectors of a Hilbert space represent a special class of physical quantities: those that can take just two values, 0 and 1. The notion of dichotomic physical quantities is of special importance. Several terms are commonly used to refer to them: propositions, properties, events, tests, questions, filters, yes-no experiments, all alluding to particular ways of picturing the same idea of dichotomy (a proposition is either true or false, a property either possessed or not, an event either occurring or not, etc.). The set $\mathcal{P}(H)$ of projectors on H has a natural ordered structure. If $P^M, P^N \in \mathcal{P}(H), M, N$ being the closed subspaces onto which P^M, P^N project, we have that the $\leq$ relation defined by

$$P^M \leq P^N \quad \text{iff} \quad P^M P^N = P^N P^M = P^M \quad ,$$

(or, equivalently, iff $M \subseteq N$), is transitive, reflexive and antisymmetric thus being on order relation.

Actually, $\mathcal{P}(H)$ is a partially ordered set (poset) of a highly structured kind.

i) Given the elements P^M, P^N there exists in $\mathcal{P}(H)$ their meet (or greatest lower bound) $P^M \wedge P^N = P^{M \cap N}$ and the join (or least upper bound) $P^M \vee P^N = P^{\overline{M \cup N}}$, where $\overline{M \cup N}$ denotes the closure of $M \cup N$, i.e., the smallest closed subspace of H containing M and N. The same is true for every family of elements so that $\mathcal{P}(H)$ qualifies as a complete lattice. The least element of $\mathcal{P}(H)$ is the null projector 0 while the greatest element is the identity I.

ii) $\mathcal{P}(H)$ contains "atoms", i.e. elements that majorize no other element than 0: they are the projectors onto one-dimensional subspaces. Thus the existence of atoms of $\mathcal{P}(H)$ merely reflects the existence of pure states. Moreover, $\mathcal{P}(H)$ is atomic, i.e., every non-zero element majorizes at least one atom, and even atomistic, i.e., every nonzero element is the join of (all) the atoms it contains.

iii) $\mathcal{P}(H)$ satisfies the Birkhoff covering property, namely the join of an element P^M with an atom not contained in it "covers" P^M, i.e., is a minimal majorization of it (there is no element in between).

iv) The unary relation $P^M \longmapsto (P^M)^\perp$ defined by $(P^M)^\perp = P^{M^\perp}$ with $M^\perp = \{\varphi \in H : (\psi, \varphi) = 0 \quad \forall \psi \in M\}$ satisfies the properties

$$(P^M)^{\perp\perp} = P^M \quad ,$$

$$P^M \leq P^N \quad \text{implies} \quad (P^N)^\perp \leq (P^M)^\perp \quad ,$$

$$P^M \wedge (P^M)^\perp = 0 \quad , \quad P^M \vee (P^M)^\perp = I \quad ,$$

and it is thus an "orthocomplementation". $\mathcal{P}(H)$ is therefore an orthocomplemented lattice.

v) $\mathcal{P}(H)$ satisfies the law of orthomodularity expressed by

$$if \quad P^M \leq P^N \quad then \quad P^N = P^M \vee (P^N \wedge (P^M)^{\perp}) \quad .$$

This law can be viewed as a weakening of the law of distributivity of the join (meet) with respect to the meet (join). Indeed it amounts to say that the distributivity is ensured only for the triples of the form $(P^M, P^N, (P^M)^{\perp})$ when $P^M < P^N$. Let us recall that the distributivity for all triples of elements holds for the lattice (Boolean algebra) of all subsets of a set: such a lattice is the counterpart of $\mathcal{P}(H)$ for a classical physical system whose "propositions" are indeed in one-to-one correspondence with the subsets of the phase-space of the physical system.

If D is a density operator on H, then the function from $\mathcal{P}(H)$ into the real interval [0,1] defined by

$$P \longmapsto tr(DP)$$

satisfies the properties of a probability measure on $\mathcal{P}(H)$. In fact it maps the least and greatest elements of $\mathcal{P}(H)$ into the numbers 0 and 1, and is additive on orthogonal sequences (if $\{P_i\}$ is such that $P_i \leq P_j^{\perp}$, or equivalently $P_i P_j = 0$, when $i \neq j$, then $tr((\sum_i P_i)D) = \sum_i tr P_i D$). Moreover, due to Gleason's theorem, we have that every probability measure on $\mathcal{P}(H)$ comes from a density operator (if dim $H \geq 3$), so that the set of all states can be identified with the set of all probability measures abstractely defined on $\mathcal{P}(H)$.

$\mathcal{P}(H)$ carries also all informations on the physical quantities of the physical system. In fact, due to the spectral theorem, every self-adjoint operator on H is completely determined by its spectral measure $E \longmapsto P_A(E), (E \in B(\mathbb{R}), P_A(E) \in \mathcal{P}(H))$: thus the physical quantities can be viewed as measures on the real line $\mathbb{R}$ taking values in $\mathcal{P}(H)$.

4. SOME ISSUES OF QUANTUM LOGIC

In this section we shall briefly mention some factual issues that have been considered within the research area referable to quantum logic[2]. They are intended as mere examples, without any claim of exhaustivity.

4.1 The Coordinatization Problem

Here the problem is the one of reconstructing the Hilbert-space formulation of quantum mechanics out of a much less structured structure formed by the proposi-

tions (or yes-no experiments) of the physical system, thought of as primitive, unde-
fined entities. Take the set L of "propositions" and endow it with an order relation
and with some of the properties of $\mathcal{P}(H)$, but without any notion of Hilbert space
on the ground. Consider the set S of all probability measures on L looking at them
as the states of the physical system. Consider the set O of all functions from $B(\mathbb{R})$
into L that have the formal properties of spectral measures, and look at them as the
physical quantities of the system. Then the question: is it possible to determine a
Hilbert space H such that L is identified with $\mathcal{P}(H)$, S with the set of all density
operators on H, and O with the set of all self-adjoint operators on H? In other words:
to what extent is the Hilbert-space description of quantum mechanics, with its highly
structured mathematical edifice, coded into the ordered structure of propositions?

More specifically the starting point is the assumption that the propositions form
a complete lattice L that is orthomodular, atomic and has the covering property.
A structure like this is often called, by its own, a "quantum logic". It is useful
to add the condition that L is irreducible* though, when this is not the case, one
could refer to its irreducible addends. The first step is whether L can be viewed as
(is isomorphic to) some lattice of subspaces of a vector space, or, in other words,
whether L admits a "vector-space coordinatization". Before stating the answer let
us recall some terminology.

An involution of a field K is a mapping $\lambda \longmapsto \lambda^*$ of K into itself such that, for
every $\lambda, \mu \in K$,

$$(\lambda + \mu)^* = \lambda^* + \mu^* \quad , \quad (\lambda\mu)^* = \mu^*\lambda^* \quad , \quad \lambda^{**} = \lambda \quad .$$

If V is a vector space over K we say that a mapping f from $V \times V$ into K is an
hermitean form on V if, for every $v, w, v_i, w_i \in V$ and $\lambda, \mu \in K$,

$$f(\lambda v_1 + \mu v_2, w) = \lambda f(v_1, w) + \mu f(v_2, w) \quad ,$$

$$f(v, \lambda w_1 + \mu w_2) = f(v, w_1)\lambda^* + f(v, w_2)\mu^* \quad ,$$

$$f(v, w) = f^*(w, v) \quad ,$$

$$f(v, v) = 0 \quad \text{iff} \quad v = 0 \quad .$$

---

* In an orthomodular lattice (or poset) we say that $a, b, \in L$ commute if there are
pairwise orthogonal elements $a_1, b_1, c \in L$ such that $a = a_1 \vee c, b = b_1 \vee c$. The center
of L is the set of elements which commute with every element of L. We say that L
is irreducible if its center contains only the least and greatest elements 0, I of L.

exhaustive to avoid dispersion and to recover determinism. The states so conjectured are often called "completed states".

Hidden-variable theories have been widely explored in the context of the Hilbert-space structure but a number of studies and results have been carried out at the more general level of the (L, S) structures quoted in the previous sections.

A way of formalizig the problems goes as follows[2]. To the pair (L, S) we add a space Ω whose elements are understood as the hidden-variables. Since we need averagings over these variables, we request for Ω the structure of a classical probability space, i.e., a triple (Ω, Σ, μ) where Σ is a family (a Boolean σ-algebra) of measurable subsets of Ω and μ is a probability measure on Σ. A completion of a quantum pure state $\alpha \in S^P$ is then a pair (α, ω), with $\omega \in \Omega$, assumed to be dispersion free on L, i.e., a function from L into the pair of numbers 0, 1. To recover the consistency with the prediction of quantum theory we need that the probability distributions assigned to propositions (hence to physical quantities) by the state α be the average over Ω of the values assigned by the completed state (α, ω):

$$\alpha(a) = \int_\Omega (\alpha, \omega)(a)\mu(d\omega)) \qquad a \in L \ , \ \alpha \in S^P \ .$$

If this pattern occurs, we say that the pair (L, S) admits an underlying non-contextual hidden-variable theory, where the word "noncontextual" outlines that a single space Ω is used to generate dispersion-free states on the whole L.

This is not however the only idea of hidden-variable theory that has been considered in the literature. One might imagine that the choice of the hidden-variable space depends upon, or is contextual to, the physical quantity to be dealt with. Since it is natural to associate to each physical quantity a Boolean sub-σ-algebra of L, we say that the pair (L, S) admits an underlying contextual hidden-variable theory if there is a family $\{\Omega_B, \Sigma_B, \mu_B\}$ of probability spaces labelled by the (maximal) sub-σ-algebras of L such that, for every B and every quantum state α, the pair (α, ω_B), with $\omega_B \in \Omega_B$, is a dispersion-free probability measure on B (not necessarily on the whole L), and the function $\omega_B \longmapsto (\alpha, \omega_B)(a)$ from Ω_B into $\{0, 1\}$ is measurable and satisfies the consistency condition

$$\alpha(a) = \int_{\Omega_B} (\alpha, \omega_B)(a) \ \mu_B(d\omega_B) \ .$$

Noncontextual hidden-variable theories are ruled out under standard assumptions for the pair (L, S) implemented by some regularity conditions for the completed states: this is the result of no-go theorems proved under slightly different hypotheses[16,17]. When the premises are weakened enough to escape these theorems it becomes a serious problem to decide whether the noncontextual hidden variable theories that become available[18-20] have any physical interest. The no-go theorems alluded to do not apply to contextual hidden variable theories and several explicit examples of such theories have been built up. These models, while restoring some classical deterministic aspects, have to pay, on other sides, in quite radical departures from properties of classical states. In the last decades it became clear that certain contextual hidden variable theories (like the "local" ones) lead to discrepancies with the prediction of quantum theory, and that there is place for experimental testing. We refer to Bell's inequalities that had the great merit of bringing the discussion on hidden variables theories from the controversies of go and no-go theorems onto the ground of experiments.

4.4. Quantum Logic as a Logical Calculus

We just mention in this section a whole branch of studies, typically referred to as quantum logic, that falls in the territory of logicians and philosophers. Roughly, the starting question is whether an ordered structure like the lattice $\mathcal{P}(H)$ of projections of a Hilbert space can be associated with a propositional calculus and which rules this propositional calculus inherits from the structure of the lattice $\mathcal{P}(H)$.

We have already called "propositions" the two-valued physical quantities - renting a word of logic - because we can associate to them elements of a language (the language pertaining to the description of the physical system under discussion).

For the sake of definiteness we label by "yes" and "no" the two possible outcomes of a two-valued physical quantity. A possible (though not unique) way of associating to each dycotomic physical quantity a logical proposition can then be the one of constructing the sentence "the measurement of a, when the physical system is the (pure) state α, gives the yes outcome". We denote a_α such a sentence.

A crucial difference between the classical and the quantum case already emerges when we consider the truth values of these logical propositions.

In the classical case there is no ambiguity about the semantical values of a_α: it is either true or false. In the quantum case, however, the factual occurrence of the yes outcome of a (when the system is in the state α) is not sufficient to state that a_α is true, because in a replica of the same measurement (with the system in the same state) the yes outcome is not ensured. Indeed, it is peculiar of the quantum phenomena that every (two-valued) physical quantity admits (pure) states that assign a probability for the yes outcome which is neither 0 nor 1, but a value in between 0 and 1.

We might define the truth of a_α as that situation in which the yes outcome of a is certain but, in this case, how to define the falsehood of a_α? We could for instance say that a_α is false whenever it is not true, namely when in the state α the yes outcome of a is not certain. In this case we would still have just two semantical values, true and false, but they would no longer be in symmetric relation, as they are in the classical case. Indeed, by assuming that a_α is false, we cannot ensure that the no outcome of a is certain: we can just say that the yes outcome is not certain but nevertheless possible.

Another attitude could be the one of saying that a_α is false when the no outcome of a is certain. In this case truth and falsehood would recover a symmetric position, but departures from the classical case would emerge on other sides. Indeed we would no longer be allowed to say that a_α is either true or false, for a quantum systems always admits states such that neither the yes nor the no outcome of a are certain. Thus, truth and falsehood would not exhaust all the possibilities. We would be faced with a many-valued logic. Should we collapse all cases in which a_α is neither true nor false into a unique semantical value of "indeterminacy" we would deal with a three-valued logic, as in the well known Reichenbach suggestion[21].

The ordered and orthocomplemented structure generated by the dycothomic physical quantities (propositions) - should it be the Boolean algebra of the classical case of the orthomodular lattice $\mathcal{P}(H)$ of the quantum case - contains algebraic operations such as the orthocomplementation, the meet, the join that are the natural candidates to represent, in the associated language, the elementary logical connectives of the negation, of the conjunction "and", of the disjunction "or", respectively. The propositional calculus or, in loose sense, the logic determined in this way reflects in a natural way the properties of the corresponding ordered orthocomplemented structure. By taking into account that Boolean algebras are the algebraic models of the classical logic, the question arises of which logic has the orthomodular lattice $\mathcal{P}(H)$ as its algebraic model. "Quantum logic" is the name that designates the answer, though there are several views about the precise content of this name.

Let us remark that in $\mathcal{P}(H)$ the meet and the join are given by

$$P^M \wedge P^N = P^{M \cap N} \quad , \quad P^M \vee P^N = P^{\overline{M \cup N}}$$

where $\overline{M \cup N}$ denotes the smallest closed subspace of H containing M and N. From this we see that while the truth table of the sentence "a_α and b_α" is the usual one, the truth table of "a_α or b_α" is not as in the classical case, for we can have states that make true "a_α or b_α" without making true neither a_α nor b_α. Another departure from the classical case emerges when we come to the conditional connective "if a_α then b_α". The usual Whitehead-Russel rule that identifies this conditional with the form "not a_α or b_α" is untenable in any quantum logic, because it misses the most basic requirements a conditional must fulfil (for instance we might have a_α true, b_α false and "not a_α or b_α" true).

Without any further insistence on the structure of a propositional calculus generated by an orthomodular lattice like $\mathcal{P}(H)$, let us only remark that not all tautologies of classical logic can survive in a quantum logic. For instance, if we take into account that the distributivity of the meet with respect to the join (and viceversa) does not hold in an orthomodular, non distributive, lattice, we see that in a quantum logic there is no place for the law of distributivity of the disjunction with respect to the conjunction (and viceversa) which holds in classical logic.

As we have seen, quantum mechanics, with its lattice of projections of a Hilbert space, suggests a new propositional calculus, the quantum logic. This outlines first of all the problem of the unicity or plurality of logics. It is factually unavoidable the historical fact of a pluralistic situations, in the sense that various logics have been mathematically formalized: not only classical and quantum logic but even intuitionistic logic, minimal logic, etc. The thesis of unicity of logic, that had and is having advocates on the grounds of philosophical reasons, must thus take the way of asserting that the various logics which have been formalized are mere mathematical structures with just one exception that should be the "true" logic.

Referring to the role of quantum logic other questions naturally arise. In which sense can a physical theory, like quantum mechanics, bring to a new logic? Is logic empirical or a priori?

In the literature there is a line of thought accepting a dependence of the logic from the experience and accepting the idea that quantum mechanics forces to adopt a nonclassical logic, in the sense that quantum phenomena admit a coherent rational explanation only if one abandons classical logic in favour of quantum logic. So, for instance, it is argued that the two-slit interference phenomenon of light is rationally explained in terms of corpuscular nature of light only if we give up the classical law of distributivity of the disjunction with respect to the conjunction.

On the basis of the fact that classical logic has its algebraic model in the lattice of subsets of the phase space of classical mechanics, and on the basis of the fact that quantum mechanics is more fundamental a theory than classical mechanics, it has also been argued that quantum logic should supersede classical logic, as a more fundamental logic. This thesis is faced with non obvious problems such as the one of justifying our way of thinking and deducing, the use of mathematics and the use of quantum mechanics itself wedded as they are to classical logic. How can quantum theory, which is formulated in terms of classical logic, give rise to a new nonclassical logic? What is the relationship between quantum logic and classical logic? Consistency reasons would require that the metalogic of quantum logic be classical logic: can this be proved? It has been argued that quantum logic becomes equivalent to classical logic when we restrict to the sentences and concepts that form the common language and the body of mathematics.

Quantum phenomena, like, e.g., the quoted two-slit interference of light, admit a coherent description even in the framework of classical logic provided one gives up the idea of assigning to the notion of physical object, or particle, all properties possessed in classical mechanics. Thus it has been given to quantum logic even a weaker role: quantum logic should not be necessary to quantum mechanics, nor it should supersede classical logic, but it constitutes an algebraic structure, included into the formalism of quantum mechanics, whose operations can be interpreted as logical operations. In this way, quantum logic would not have the role of "solving" certain conceptual difficulties (often emphatically called paradoxes) of quantum theory, but at most the role of putting them in a peculiar language.

If logic is thought to be a priori, it becomes natural to ask whether quantum logic can be approached without reading it out of quantum mechanics: the answer appears to be affirmative since quantum logic can be generated within various theoretical situations of indeterministic nature.

REFERENCES

1. G.Birkhoff and J.von Neumann, Ann. Math. 37: 823 (1936)
2. E.G.Beltrametti and G.Cassinelli, "The Logic of Quantum Mechanics", Addison Wesley, Reading, Mass. (1981)
3. F.Maeda and S.Maeda, "Theory of Symmetric Lattices", Springer, Berlin, Heidelberg, New York (1970) (chapter 7)
4. V.S.Varadarajan, "Geometry of Quantum Mechanics", Vol.I, Van Nostrand, Princeton, N.J. (1968) (chapters 3, 4, 5, 7)
5. C.Piron, "Foundations of Quantum Physics", Benjamin, Reading, Mass. (1976) (chapter 3)
6. J.P.Eckmann and P.Zakey, Helv. Phys. Acta 42: 420 (1969)
7. P.A.Ivert and T.Sjödin, Helv. Phys. Acta 51: 635 (1978)
8. E.G.Beltrametti and G.Cassinelli, Found. of Phys. 2: 1 (1972)
9. H.A.Keller, Math. Z., 172: 41 (1980)
10. I.Amemiya and H.Araki, Publications Research Inst. Math. Sci., Kyoto Univ., A2: 423 (1966)
11. D.J.Foulis, Proc. Amer. Math. Soc., 11: 648 (1960)
12. J.C.T.Pool, Commun. Math. Phys. 9: 212 (1968)
13. W.Ochs, Commun. Math. Phys. 25: 245 (1972)
14. G.Cassinelli and E.G.Beltrametti, Commun. Math. Phys. 40: 7 (1975)
15. W.Guz, Rep. Math. Phys. 16: 125 (1979)

16. J.M.Jauch and C.Piron, Helv. Phys. Acta 36: 827 (1963)

17. S.P.Gudder, Proc. Amer. Math. Soc., 19: 319 (1968)

18. N.Wiener and A.Siegel, Nuovo Cimento Suppl., 2: 982 (1955)

19. N.Wiener, A.Siegel, B.Ranking and W.T.Martin, "Differential Space, Quantum Systems and Prediction", M.I.T. Press, Cambridge, Mass. 1966

20. S.P.Gudder, J.Math. Phys. 11: 431 (1970)

21. H.Reichenbach, "Three-valued logic and interpretation of quantum mechanics" in "The Logico-Algebraic Approach to Quantum Mechanics", C.A.Hooker ed., Reidel, Dordrecht, 1979

REMARKS ON THE MEASUREMENT OF
DISCRETE PHYSICAL QUANTITIES

E.G. Beltrametti

Dipartimento di Fisica dell'Università di Genova
and Istituto Nazionale di Fisica Nucleare
Genova, Italy

INTRODUCTION

The measurability of a physical quantity implies the existence of a corresponding experimental macroscopic device, or measuring apparatus, M - equipped with a reading scale and with instructions on how to couple it with the physical system S under consideration - which has to fulfil a number of necessary requirements to be considered in the sequel.

Obvious needs of completeness of quantum theory suggest considering the measuring instrument itself as a physical system belonging to the domain of the theory. But here it becomes unavoidable to recognize that a measuring apparatus is not a purely quantum system: if we agree, as it is natural, that different positions of the pointer label different states of M, we have to agree that we cannot give any physical meaning to the superposition of two states that correspond to different positions of the pointer. Thus, the measuring instruments can be included into the domain of quantum theory at the price of allowing for them the existence of superselection rules. As a matter of fact, the inclusion of superselection rules into the formalism of quantum mechanics makes the theory so flexible and general that no physical system is known to irrecoverably escape from it.

In our case we shall thus assume that the states of the measuring apparatus labelled by different pointer positions are separated by superselection rules. Of course, the acknowledgement of these superselection rules leaves open the problem of their

Sixty-Two Years of Uncertainty
Edited by A. I. Miller
Plenum Press, New York, 1990

origin, in particular the problem of whether the description of a macroscopic object as made of purely quantum systems (its atoms and molecules) can generate its superselection rules as a very consequence of aggregating all these subsystems.

In the sequel we shall adhere to the familiar idealization of measuring apparata that are nondestructive: by this we mean that the physical system S and the measuring apparatus M, initially separated, will form a compound system preserving their own identities during the mutual interaction, and then become again separated so that it is meaningful to speak of final state of S and of M after the measurement. We shall write $D_S^{(i)}$ and $D_S^{(f)}$ for the initial and final state of S, and, similarly, $D_M^{(i)}$, $D_M^{(f)}$ for the initial and final state of M. It should be clear that the nondestructiveness of M is not a logical necessity of quantum theory, nor it corresponds to the common factual situation (think e.g. of the measurement of the energy of a photon by a spectrometer that absorbs the photon).

Though the above idealization allows to speak of final state of S, it is not strong enough to make this final state unambiguously defined: indeed a physical quantity can admit many nondestructive measuring instruments that affect the state of S in different ways. A further idealization, able to induce, or even to eliminate, such an ambiguity, is often called into play: it is the so-called von Neumann's or Lüders's projection postulate[1,2].

It says, in the von Neumann's version, that if a physical quantity is measured twice in succession, then the same result is obtained each time. In case the physical quantity A to be measured is discrete, this idealization amounts to say that if λ_i is an eigenvalue of A and if the measurement gives the value λ_i then the final state of S has to belong to the λ_i-eigenspace of A. If λ_i is nondegenerate then the final state of S is uniquely determined as the corresponding eigenvector, but when λ_i is degenerate some undeterminacy survives for that state. It is removed by the more stringent idealization embodied by Lüders projection postulate which adds the requirement that if the initial state of S is the pure state φ then the final state of S is the projection of φ onto the eigenspace of the measured value λ_i. If the initial state of S is not pure then the Lüders's rule can be written as

$$D_S^{(f)} = \frac{P_A(\{\lambda_i\})D_S^{(i)}P_A(\{\lambda_i\})}{tr(D_S^{(i)}P_A(\{\lambda_i\}))}$$

where $E \to P_A(E)$, E being any element of the family $B(\mathbb{R})$ of the Borel sets of $\mathbb{R}$, is the spectral measure of A.

Von Neumann's and Lüders's projection postulates, despite the emphatic (for historical reasons) name of "postulates", do not appear necessary "a priori" for the internal coherence of quantum theory: they rather define a special class of measuring instruments. Instruments of this sort are usually not used in laboratories, but there is no definite example of a physical quantity that cannot admit a measurement of this kind.

Let us remark that a prescription like Lüders's, which uniquely specifies the state of S after the measurement of some physical quantity, is crucial to give meaning to the notion of probability distribution of a physical quantity given the result of a previous measurement of another physical quantity. In other words it is only through such a prescription, corresponding to the idea of ideal first-kind measurements, that we can go to the notion of conditional probabilities within quantum theory.

A PROBABILISTIC REQUIREMENT

Since both the physical system S under discussion, and the measuring apparatus M will be described within a quantum theoretical framework, we associate to S its Hilbert space H_S and to M its Hilbert space H_M. When we think of $S + M$ as a compound system, the corresponding Hilbert space will become the tensor product $H_S \otimes H_M$ according to the prescription of quantum theory for the composition of nonidentical subsystems (and indeed we obviously assume that S and M are not identical). Let A_S be the self-adjoint operator of H_S that represents the physical quantity to be measured (in the previous section we have just written A for this operator). The counterpart in H_M of A_S will be what we may call the pointer observable (since M is planned to measure A_S) that we shall denote, when necessary, by A_M.

The very notion of measuring instrument of A_S demands that, for any initial state $D_S^{(i)}$ of S, the probability that the interaction of M with S will cause the pointer of M to fall into the reading scale interval $E \in B(\mathbf{R})$ is precisely the probability that the value of A_S lies in E. This is expressed by the equality

$$tr(D_S^{(i)} P_{A_S}(E)) = tr(D_M^{(f)} P_{A_M}(E)) \quad , \quad E \in B(\mathbf{R}) \tag{1}$$

whose role and implications will be overviewed in this section.

We shall restrict to discrete physical quantities so as to achieve a number of significant implications that allow some general results about the role of eq.(1). Let λ_i be the eigenvalues of A_S and n_i the degeneracy of λ_i; let $\{\varphi_{ij}\}$, for given i and $j = 1, ..., n_i$, be an orthonormal basis of the λ_i-eigenspace so that $\{\varphi_{ij} \mid i = 1, 2, ...; j = 1, ..., n_i\}$ is an orthonormal basis for H_S, and we have

$$A_S \varphi_{ij} = \lambda_i \varphi_{ij} \quad .$$

Of course, the eigenvalues of A_M have to coincide with the λ_i's and we take for A_M the simple self-adjoint operator of H_M which has the λ_i's as non degenerate eigenvalues. Let Φ_i be the eigenvector of A_M corresponding to λ_i: it represents the state of M in which the pointer is in the λ_i position.

The compound system $S + M$ will undergo, by the measuring procedure, a state transformation, say:

$$D_S^{(i)} \otimes D_M^{(i)} \rightarrow W(D_S^{(i)} \otimes D_M^{(i)}) \ . \tag{2}$$

Assuming that the mapping W preserves the convex structure of the density operators in $H_S \otimes H_M$, and taking into account the linearity and continuity of the theory as well as the fact that every density operator can be written as a convex combinations of vector states (projectors onto one-dimensional subspaces), we can restrict without loss of generality to initial states of S and M that are vector (pure) states, say

$$D_S^{(i)} = P^{[\varphi]} \quad , \quad D_M^{(i)} = P^{[\Phi]}$$

so that the initial state of $S + M$ is

$$P^{[\varphi]} \otimes P^{[\Phi]} = P^{[\varphi \otimes \Phi]} \ .$$

We can now state the following result[3]. The requirement (1) implies and is implied by the fact that the mapping (2) has the form

$$P^{[\varphi \otimes \Phi]} \rightarrow P^{[U(\varphi \otimes \Phi)]}$$

with U a continuous linear extension on $H_S \otimes H_M$ of a map of the form

$$\varphi_{ij} \otimes \Phi \rightarrow \psi_{ij} \otimes \Phi_i \quad i = 1, 2, ...; \ j = 1, ... n_i \tag{3}$$

where $\{\psi_{ij}\}$ is any set of unit vectors (in H_S) that are orthogonal with respect to the second index. Moreover, U can always be chosen as a unitary (or antiunitary) operator.

This result, which characterizes every measurement of a discrete physical quantity that satisfies the requirement (1), is physically natural. If the physical system S is initially in an eigenstate φ_{ij} of λ_i then, with certainty, the pointer of M has to jump into the λ_i-position so that M goes from the initial state Φ into the final state Φ_i, while S moves from φ_{ij} into a new vector state ψ_{ij} whose nature might depend upon the particular measuring instrument M. Notice that these vectors are not requested, at this stage, to be orthogonal with respect to the index i. Should the λ_i's be non degenerate, the index j would drop out, together with the orthogonality requirement of the ψ_{ij}'s with respect to second index. Notice also that a particular choice for $\{\psi_{ij}\}$ might be $\{\varphi_{ij}\}$ in which case we would be considering a measuring device that obeys von Neumann-Lüders projection postulate.

If the initial state of S is represented by some vector φ then, writing

$$\varphi = \sum_{ij} c_{ij} \varphi_{ij} \ ,$$

we have that $S + M$ evolves from the initial state $P^{[\varphi \otimes \Phi]}$ into the final state

$$D^{(f)}_{S+M} = P^{[\sum_{ij} c_{ij} \psi_{ij} \otimes \Phi_i]} \ . \tag{4}$$

Though this state has the form of a projector into a one-dimensional subspace of $H_S \otimes H_M$ we cannot be sure that it represents a pure state of $S + M$ in view of the superselection rules that we can expect to hold in $H_S \otimes H_M$, as we shall see in the sequel.

Anyhow, no matter whether (4) represents a pure or a nonpure state, we can derive from it the final states of the subsystems S and M. Putting

$$\gamma_i = N_i^{-i} \sum_{j=1}^{n_i} c_{ij} \psi_{ij} \ , \quad N_i^2 = \sum_{j=1}^{n_i} |c_{ij}|^2$$

we get

$$D^{(f)}_S = \sum_i N_i^2 P^{[\gamma_i]} \tag{5}$$

$$D^{(f)}_M = \sum_i \sum_{j=1}^{n_i} N_i N_k < \gamma_i \mid \gamma_k > |\Phi_k >< \Phi_i| \tag{6}$$

(where in the last formula we have used Dirac's bras and kets).

From this we see that the final state of S is a mixture which has a natural convex decomposition into the pure states $P^{[\gamma_i]}$ (notice that in absence of degeneracy the γ_i's reduce to the ψ_i). Also the final state of M is a mixture which, however, has not an immediate form of convex combination of pure states, owing to the fact that the γ_i's are not requested to be orthogonal, as a result of the fact that the ψ_{ij} are not requested, at this stage, to be orthogonal with respect to the first index.

In the particular case of a von Neumann-Lüders measurement, the ψ_{ij} would become the φ_{ij}, the γ_i's would become orthogonal, and the final state of M would be the familiar mixture

$$D^{(f)}_M = \sum_i N_i^2 P^{[\Phi_i]} \ . \tag{7}$$

CORRELATIONS

It fits with the pattern of the measurement process the existence of a correlation between a possible final state of S of the form $P^{[\gamma_i]}$ (or $P^{[\psi_i]}$ in absence of degeneracy of the λ_i's) and a final state of M of the form $P^{[\Phi_i]}$.

Let us first recall that for any projection operator P_S in H_S and any projection P_M in H_M we can define their correlation coefficient, given a state D_{S+M} of the compound system $S + M$, as

$$\rho(P_S, P_M; D_{S+M}) = \frac{tr(D_{S+M}P_S \otimes P_M) - tr(D_S P_S)tr(D_M P_M)}{\sqrt{tr(D_S P_S)(1 - tr(D_S P_S))tr(D_M P_M)(1 - tr(D_M P_M))}}$$

where D_S and D_M are the reduced states of D_{S+M}.

The numerical values of ρ range from -1 to $+1$: we speak of strong correlation when $\rho = \pm 1$ while we say that there is no correlation when $\rho = 0$.

We can now go back to the correlation between a final state of S of the form $P^{[\gamma_i]}$ and a final state of M of the form $P^{[\Phi_i]}$ given the final state $D^{(f)}_{S+M}$ of $S + M$ as expressed by eq.(4). We have indeed the following result[3]: if the probabilistic requirement (1) is met then the property

$$\rho(P^{[\gamma_i]}, P^{[\Phi_i]}; D^{(f)}_{S+M}) = 1 \tag{8}$$

is equivalent to the statement that $\{\psi_{ij} \mid i = 1, 2, ...; \; j = 1, ..., n_i\}$ is an orthonormal system.

Recall that the orthogonality of the ψ_{ij}'s with respect to the second index was already achieved by the condition (1): what the correlation condition (8) adds is the orthogonality of the ψ_{ij}'s with respect to the first index.

The fulfilment of (8), which is by its own physically relevant, entails (and is entailed by) the fact that the final state of M takes the physically natural form (7). Indeed, the orthogonality of the ψ_{ij}'s induces the orthogonality of the γ_i's and hence (6) collapses into (7).

FINAL STATES OF THE OBSERVED SYSTEM
AND OF THE MEASURING APPARATUS

Having seen in the previous sections the conditions that make the final states of S and M of the form

$$D^{(f)}_S = \sum_i N_i^2 P^{[\gamma_i]} \quad , \quad D^{(f)}_M = \sum_i N_i^2 P^{[\Phi_i]} \tag{9}$$

we come now to the physical interpretation of them.

The first point we want to stress is that the superselection rules holding in H_M allow for the mixture $D_M^{(f)}$ the usual "ignorance interpretation". To see this, let us remind a few facts.

In the Hilbert-space description of a strictly quantum system every nonpure state can be written as a convex combination of pure states but this decomposition is never unique[4]. If $\{\psi_i\}$ is a family of vectors of H that spans a subspace $K \subseteq H$ and carries a convex decomposition of a density operator D in the sense that

$$D = \sum_i w_i P^{[\psi_i]} \qquad (\sum_i w_i = 1 \ , \ w_i > 0) ,$$

then any other family $\{\varphi_i\}$ of vectors that spans the same subspace K carries another decomposition of D, i.e.,

$$D = \sum_i w'_i P^{[\varphi_i]} \qquad (\sum_i w'_i = 1 \ , \ w'_i > 0) .$$

Thus we see that, due to the linear vector-space structure of H, or in other words, due to the possibility of superposing pure states to get new pure states, a nonpure state has infinitely many decompositions into pure states. This fact marks a crucial difference between quantum and classical systems. In the latter case a nonpure state has always a unique decomposition into pure states and this justifies the "ignorance interpretation" of mixtures. According to this interpretation, when we write, e.g., $D = w P^{[\psi_1]} + (1 - w) P^{[\psi_2]}$, we mean that the physical system is, in reality, either in the pure state ψ_1 or in the pure state ψ_2, but our ignorance prevents our saying in which one of them the system actually is: our knowledge is limited to saying that the sytem is in the state ψ_1 with probability w, and in the state ψ_2 with probability $1 - w$.

When the decomposition of a mixture into pure states is not unique, as it is for quantum systems, the ignorance interpretation becomes untenable. Think, e.g., of the nonpure state describing an unpolarized beam of photons: it admits a decomposition, with equal weights, into two (pure) states of linear polarization along two orthogonal directions but this fact does not justify saying that a photon has actually one of these linear polarizations for the unpolarized state could equally well be thought of, for instance, as a convex combination, with equal weights, of circularly right and circularly left polarization states.

The pattern of nonunique decomposition of mixtures, with the associated failure of the ignorance interpretation, can however be modified by the presence of superselection rules[4].

When a physical system has a limited quantum behaviour, the correspondence between physical quantities and self-adjoint operators, as well as the one between states and density operators, or the one between pure states and one-dimensional projectors, is no longer bijective. The set $\{A_i\}$ of self-adjoint operators associated to the physical quantities generates a von Neumann algebra $\mathcal{A} = \{A_i\}''$ (the double commutant of $\{A_i\}$) that is a proper subset of $B(H)$, the algebra of all bounded operators on H. Therefore there are nontrivial operators, that is, operators that are not multiples of the identity, that commute with every element of $\mathcal{A}$. They are called superselection operators, and they can also be viewed as the nontrivial bounded operators that commute with every physical quantity. If, for simplicity, we assume that the superselection operators have pure point spectrum, then there will exist a contable set of pairwise orthogonal projectors $\{Q_k\}$, with $\sum_k Q_k = I$, such that every superselection operator will take the form $\sum_k \xi_k Q_k$ with eigenvalues ξ_k.

Let K_k be the closed subspace of H onto which Q_k projects. Any two distinct subspaces $K_k, K_{k'}$ are orthogonal and the direct sum $\oplus K_k$ is the whole H. Every element of $\mathcal{A}$ leaves each K_k invariant. The K_k's are called the coherent subspaces of H.

Every density operator in H assigns a probability distribution to each physical quantity, but there are density operators that are different as operators on H but are equivalent with regard to the probability distributions they assign to the physical quantities. In other words, when superselection rules are present, the physical quantities are not numerous enough to separate the set of all density operators on H. Only in case we restrict to those density operators that belong to $\mathcal{A}$ can the bijective correspondence with the states of the physical system be recovered.

In particular, it turns out that a vector of H represents a pure state if and only if it belongs to a coherent subspace. Thus the superposition principle of quantum mechanics applies only within each coherent subspace. When a vector has components in different coherent supspaces it cannot represent a pure state for it is fully equivalent to a density operator (in the von Neumann algebra $\mathcal{A}$) that represents a mixture.

After these remarks we can go back to our measuring instrument M and to its final state. We have already noticed that the only vectors of H_M that are interpreted as pure states of M are the Φ_i's (the states labelled by pointer positions on the reading scale of M that corresponds to the possible values of the physical quantity to be measured). In other words the very nature of M as a measuring instrument requires that the coherent subspaces of H_M are one-dimensional and can be identified with the Φ_i's. But this means that the expression of $D_M^{(f)}$ given in (9) is the only decomposition of the nonpure state $D_M^{(f)}$ into the pure states of M. As such, this mixture admits the classical "ignorance interpretation": the actual state of M is just one of the Φ_i's, with probability N_i^2.

But now, thanks to the correlation (8), also the final state $D_S^{(f)}$ of S as given in (9) inherits from M an ignorance interpretation: we can say that S is actually in one of the states γ_i, with probability N_i^2.

Of course, the expression (9) gives the states of S and M after their mutual interaction but before one records the actual position of the pointer of M. Once the position is recorded, M collapses into one of the indicator states and S into the state determined by the correlation with M. But, in view of what we said before, this collapse has a purely classical nature: it is nothing else than the reduction of ignorance caused by the new information about the pointer position.

A CONSISTENCY CONDITION

The boundary between the physical system S on which the measurement is performed and the measuring instrument M is, to a good extent, a matter of taste. To outline this fact let us remark that, instead of acting by M on S, we might consider a measuring instrument M' acting on $S' \equiv S + M$, or a M'' acting on $S'' \equiv S' + M'$, and so on, always achieving the same information about S. Thus we have a chain, each step of which corresponds to a different positioning of the boundary between what is observed and the measuring instrument. Since the choice of a step in the chain is just a matter of convention, the description of the measurement has to be neutral with respect to this choice. As we shall see, this amounts to a natural prescription on how the superselection rules of the measuring instrument do propagate into the Hilbert space of the compound system formed by the instrument and the observed system[4].

For simplicity, let us compare the first two steps of the chain: M as a measuring instrument acting on S, and M' as a measuring instrument on $S' \equiv S + M$. We shall assume all the requirements discussed in the previous sections: the probability requirement (1), the correlation condition (8), and the fact that the indicator states of a measuring instrument have to be separated by superselection rules.

The step "M acting on S" has been already discussed. Summarizing: the initial state of S is $\varphi = \sum_i c_{ij} \varphi_{ij}$ and that of M is Φ, so that $S + M$ evolves from the pure state

$$D_{S+M}^{(i)} = P^{[\sum_i c_{ij} \varphi_{ij} \otimes \Phi]}$$

into the final state

$$D_{S+M}^{(f)} = P^{[\sum_{ij} c_{ij} \psi_{ij} \otimes \Phi_i]} \tag{4}$$

from which we deduce the nonpure final states of S and M

$$D_S^{(f)} = \sum_i N_i^2 P^{[\gamma_i]} \quad , \quad D_M^{(f)} = \sum_i N_i^2 P^{[\Phi_i]} \ . \tag{9}$$

We go now to the step "M' acting on $S' \equiv S + M$". Since M' has to "read" the reading scale of M, to each position λ_i of the pointer of M there must correspond a position of the pointer of M' which labels an indicator state Φ'_i of M'. Then we have the following scheme: the initial state of S' is $D_{S+M}^{(i)}$ as given before and that of M' is some Φ' so that $S' + M'$ evolves from the pure state

$$D_{S'+M'}^{(i)} = P^{[\sum_{ij} c_{ij}\, \varphi_{ij} \otimes \Phi \otimes \Phi']}$$

into the final state

$$D_{S'+M'}^{(f)} = P^{[\sum_{ij} c_{ij}\, \psi_{ij} \otimes \Phi_i \otimes \Phi'_i]} \quad,$$

from which we deduce the nonpure final state of M'

$$D_{M'}^{(f)} = \sum_i N_i^2 P^{[\Phi'_i]}$$

and the nonpure final state of S'

$$D_{S'}^{(f)} = \sum_i N_i^2 P^{[\gamma_i \otimes \Phi_i]} \quad, \tag{10}$$

which, in turn, determine the same nonpure final states of S and M as given by (9).

Comparing the two schemes, we see that their coherence simply amounts to the equivalence of the density operators given by (4) and (10). This means that the superselection rules acting in H_{S+M} must be such as to keep the one-dimensional projector (4) outside any coherent subspace of H_{S+M}, thus making it representative of a nonpure state. More precisely, it can be seen that the equivalence of (4) and (10) implies and is implied by the fact that the coherent subspaces $M_k^{(S+M)}$ of $H_{S+M} = H_S \otimes H_M$ take the form

$$M_k^{(S+M)} = H_S \otimes \hat{\Phi}_k \tag{11}$$

where $\hat{\Phi}_k$ denotes the one-dimensional subspace of H_M spanned by Φ_k (i.e., the k-th coherent subspace of M). Since we have all along tacitly assumed that S is a purely quantum system (free of superselection rules), we read the condition (11) by saying that the coherent subspaces of the compound system $S + M$ must be products of the coherent subspaces of the subsystems S and M. Therefore we see that, in order to qualify M as a measuring instrument, it is necessary, besides the conditions (1) and (8), and besides the occurrence of its own superselection rules, to assume that, coupling it with S, we get a compound system that inherits superselection rules according to the product rule (11). As a bibliographic remark, let us quote Landau and Lifshitz's texbook on quantum mechanics[5] where this propagation of superselection rules from M to $S + M$ is in some way accounted for.

Of course, the remarks dealt with in this paper do not have the status of a "theory" of the measurement process. Only in case we should be able to deduce the classical behaviour of M, namely its superselection rules, out of its nature of a nonisolated physical system composed of a large number of (quantum) subsystems, would we reach that status. There have been in the literature several models, of various nature, that try to approach explicitly the classical behaviour of a measuring instrument: they seem to agree on the various features examined in this paper.

REFERENCES

1. J.von Neumann, "Mathematical Foundations of Quantum Mechanics", Princeton University Press, Princeton, N.J. (1955).
2. G.Lüders, Ann. Physik 8: 322 (1951).
3. E.G.Beltrametti, G.Cassinelli and P.J.Lahti, "Unitary Measurements of Discrete Quantities in Quantum Mechanics", J.Math.Phys. (to be published).
4. E.G.Beltrametti and G.Cassinelli, "The Logic of Quantum Mechanics", Addison Wesley, Reading, Mass (1981).
5. L.D.Landau and E.M.Lifshitz, "Quantum Mechanics - Non Relativistic Theory", Pergamon Press, Oxford (1965).

SOME COMMENTS AND REFLECTIONS

Abner Shimony

Departments of Philosophy and Physics
Boston University
Boston, MA

1. With the failure of the program of local hidden variables theories, it is reasonable to return to the naive view that the pure quantum state of a system gives a complete description of the (non-relational) properties of the system. If so, then this complete description is characterized by objective indefiniteness in certain respects and objective probability of the outcomes in case -- somehow -- the indefinite properties are actualized. Hence, the state of a physical system is not just a catalogue of actualities, as in classical physics, but is also a network of potentialities (to use Heisenberg's word).

2. The concept of potentiality, vague though it may be, may help the intuition. For instance, it helps us to understand why symmetry is more important in quantum than in classical physics. (When a vacation is still potential, there may be a high rotational symmetry of directions in which it may be taken, but when one buys a ticket to a destination the symmmetry is broken!) Quantum nonlocality is also illuminated to some extent by the concept of potentiality. Thus, in the singlet state of a pair of fermions, the spin of particle 1 and the spin of particle 2 are strictly correlated with respect to all axes, but that fact is inseparable from the fact that all components of spin are merely potential.

3. I am not convinced by Michael Redhead's discussion of robustness. He formulated this concept and drew consequences from it for the probabilities of outcomes without attention to (a) the typical physical arrangement of a correlation experiment, and (b) the possibility of forming subensembles of the ensemble of interest. Hence, he seems to overlook the fact that when outcome independence fails -- as it does in the quantum mechanics of entangled systems -- one can indeed capitalize upon this failure to send a message from one analyzer-detector assembly to the other. But this possibility of communication is innocuous from the standpoint of relativity theory, because it is slower than light. I do not see how either the possibility of communication because of the failure of outcome independence or the innocuousness of this communication is illuminated by robustness.

4. The term "passion at a distance" has no explanatory power. I merely used the contrast between the Latin words "actio" and "passio" to underline that we have a kind of causality whereby no <u>control</u> of an event may be achieved by operating upon an event with space-like separation from the first. There must be more to say than this. Deeper understanding is needed of one or more of the following: (i) causality, (ii) event, (iii)space-time struc-

ture. Regarding (iii) I once suggested to Wheeler that his idea of wormholes might be used to explain quantum mechanical nonlocality, for the path between two events through a wormhole might be time-like, whereas the path around the wormhole might be space-like. He was not encouraging. In fact, he said, "I am skeptical as hell."

5. The nonlocality involved in geometric phases (e.g., the interference fringes predicted by Ehrenberg and Siday and by Aharonov and Bohm when an electron wave function is split about an infinite solenoid, outside of which the magnetic field vanishes) is somehow related to nonlocality of entangled spatially separated systems. But there has been no good clarification of the relation between these two kinds of nonlocality. Ne'eman has suggested that both involve parallel transport of vectors, but I don't see this in the case of entangled spatially separated systems. It is a historical accident that polarizations were analyzed in the tests of Bell's Inequality so far. Other multi-outcome variables can be correlated, and these need not have a vectorial character.

6. In many ways this is a terrible time in history, but intellectually it is glorious, and we are fortunate to be invited to a feast of ideas.

INVITED LECTURERS

David Z. Albert
Department of Philosophy
Columbia University
New York, NY 10027

Alain Aspect
Laboratoire de Spectroscopie
Hertzienne de l'E.N.S.
24, rue Lhomond
75231 Paris Cedex 05
FRANCE

John S. Bell
Theory Division
CERN CH-1211 Geneva
SWITZERLAND

Enrico Beltrametti
Department of Physics
University of Genova
Via Dodecaneso, 33
Genova
ITALY

Giancarlo Ghirardi
International Centre for Theoretical Physics
Strada Costiera, 11
34014 Trieste
ITALY

J. Hilgevoord
Department of History and Foundations of Science
Buys Ballotlaboratorium
University of Utrecht
P.O. Box 80.000
3508 TA Utrecht
THE NETHERLANDS

Don Howard
Department of Philosophy
University of Kentucky
Lexington, KY 40506-0027
USA

F. Károlyházy
Institute of Theoretical Physics
Eötvös University
Budapest
H-1088 Budapest VIII
Puskin UTCA 5-7
HUNGARY

Arthur I. Miller
Department of Philosophy
University of Lowell
Lowell, MA 01854
&
Department of Physics
Harvard University
Cambridge, MA 02138
USA

Hans Primas
Department of Chemistry
ETH Zentrum
Universitätstrasse
CH-8092 Zurich
SWITZERLAND

Alberto Rimini
Dipartimento di Fisica Nucleare e Teorica
Università di Pavia
Via Bassi 6
27100 Pavia
Italy

Abner Shimony
Department of Philosophy
Boston University
Boston, MA 02215
USA

INDEX